Software Design for
Engineers and Scientists

Software Design for Engineers and Scientists

John A. Robinson
University of York

Newnes

AMSTERDAM · BOSTON · HEIDELBERG · LONDON · NEW YORK · OXFORD
PARIS · SAN DIEGO · SAN FRANCISCO · SINGAPORE · SYDNEY · TOKYO

Newnes
An imprint of Elsevier
Linacre House, Jordan Hill, Oxford OX2 8DP
30 Corporate Drive, Burlington, MA 01803

First published 2004

British Library Cataloguing in Publication Data
A catalogue record for this book is available from the British Library

ISBN 0 7506 6080 5

For information on all Newnes publications
visit our website www.newnespress.com

Typeset by Charon Tec Pvt. Ltd, Chennai, India
Printed and bound in Great Britain by Biddles Ltd, King's Lynn, Norfolk

Contents

Preface   xi
  Acknowledgements   xiii
  Errors   xiv

1 Introduction

1.1	Theme	1
1.2	Audience	2
1.3	Three definitions and a controversy	2
1.4	Essential software design	3
1.5	Outline of the book	4
	Foundations	4
	Software technology	5
	Applied software design	5
	Case studies	6
1.6	Presentation conventions	6
1.7	Chapter end material	7
	Bibliography	7

2 Fundamentals

2.1	Introduction	8
2.2	The nature of software	8
2.3	Software as mathematics	10
2.4	Software as literature	14
2.5	Organic software	18
2.6	Software design as engineering	23
2.7	Putting the program in its place	27
2.8	User-centred design	33
2.9	The craft of program construction	35
2.10	Programmers' programming	36
2.11	Living with ambiguity	37
2.12	Summary	38
2.13	Chapter end material	40
	Bibliography	40

3 The craft of software design

3.1	Introduction	43
3.2	Collaboration and imitation	43
3.3	Finishing	45
3.4	Tool building	45
3.5	Logbooks	46
3.6	The personal library	48
3.7	Chapter end material	50
	Bibliography	50

4 Beginning programming in C++

4.1	Introduction	51
4.2	The programming environment	52
4.3	Program shape, output, and the basic types	54
4.4	Variables and their types	59
4.5	Conditionals and compound statements	62
4.6	Loops	65
4.7	Random numbers, timing and an arithmetic game	67
4.8	Functions	70
4.9	Arrays and C-strings	74
4.10	Program example: A dice-rolling simulation	78
4.11	Bitwise operators	82
4.12	Pointers	84
4.13	Arrays of pointers and program arguments	89
4.14	Static and global variables	92
4.15	File input and output	93
4.16	Structures	97
4.17	Pointers to structures	100
4.18	Making the program more general	102
4.19	Loading structured data	104
4.20	Memory allocation	105
4.21	typedef	108
4.22	enum	108
4.23	Mechanisms that underlie the program	109
4.24	More on the C/C++ standard library	111
4.25	Chapter end material	114
	Bibliography	114

5 Object-oriented programming in C++

5.1	The motivation for object-oriented programming	115
	Objects localize information	115
	In an object-oriented language, existing solutions can be extended powerfully	117
5.2	Glossary of terms in object-oriented programming	119
	Data structure	119
	Abstract Data Type (ADT)	120
	Class	121
	Object	121
	Method	122
	Member function	122
	Message	122
	Base types and derived types	122
	Inheritance	122
	Polymorphism	122
5.3	C++ type definition, instantiation and using objects	123
	Stack ADT example	123
	Location ADT example	126
	Vector ADT example	129
5.4	Overloading	132
	Operator overloading	134
5.5	Building a String class	138
5.6	Derived types, inheritance and polymorphism	145
	Locations and mountains example	145
	Student marks example	151

	5.7	Exceptions	160
	5.8	Templates	163
	5.9	Streams	166
	5.10	C++ and information localization	171
	5.11	Chapter end material	171
		Bibliography	171

6 Program style and structure	6.1	Write fewer bugs!	172
	6.2	Ten programming errors and how to avoid them	173
		The invalid memory access error	174
		The off-by-1 error	175
		Incorrect initialization	176
		Variable type errors	178
		Loop errors	178
		Incorrect code blocking	179
		Returning a pointer or a reference to a local variable	180
		Other problems with **new** and **delete**	181
		Inadequate checking of input data	181
		Different modules interpret shared items differently	183
	6.3	Style for program clarity	184
		File structure: a commentary introduction is essential	185
		Explanatory structure: comment to reveal	185
		Visual structure: make the program pretty	185
		Verbal structure: make it possible to read the code aloud	186
		Logical structure: don't be too clever	186
		Replicated structure: kill the doppelgänger	187
	6.4	Multifile program structure	187
	6.5	A program that automatically generates a multifile structure	188
	6.6	Chapter end material	192
		Bibliography	192

7 Data structures	7.1	Structuring data	194
	7.2	Memory usage and pointers	194
	7.3	Linked lists	196
	7.4	Data structures for text editing	197
		Arrays	197
		Arrays of pointers	197
		Linked lists	198
	7.5	Array/Linked list hybrids	203
		Hash tables	204
	7.6	Trees	205
	7.7	Elementary abstract data types	210
		ADT Ordered List	210
		ADT stack	210
		ADT queue	211
		ADT priority queue	213
	7.8	The ADT table – definition	216
	7.9	Implementing the ADT table with an unordered array	217
	7.10	Alternative implementations	220

	7.11	Chapter end material	221
		Bibliography	221
8	**Algorithms**		
	8.1	Introduction	222
	8.2	Searching algorithms	222
		Unordered linked list – sequential search	222
		Unordered array – sequential search	223
		Ordered array – binary search	223
	8.3	Expressing the efficiency of an algorithm	224
	8.4	Search algorithm analysis	225
		Unordered linked list – sequential search	225
		Unordered array – sequential search	225
		Ordered array – binary search	226
	8.5	Sorting algorithms and their efficiency	226
		Selection sort	227
		Insertion sort	227
		Mergesort	229
		Quicksort	230
		Heapsort	232
	8.6	Exploiting existing solutions	233
9	**Design methodology**		
	9.1	Introduction	235
	9.2	Generic design methodologies	235
	9.3	Reliable steps to a solution? – a sceptical interlude	236
		Brainstorming	237
		Sceptical design for designers	239
	9.4	Design methodology for software	239
	9.5	Design team organization	240
	9.6	Documentation	241
	9.7	Chapter end material	242
		Bibliography	242
10	**Understanding the problem**		
	10.1	Problems	244
	10.2	The problem statement	245
		Examples	246
		Some solutions	248
	10.3	Researching the problem domain	248
		Library research	248
		Getting information from the network	250
	10.4	Understanding users	250
	10.5	Documenting a specification	253
	10.6	Chapter end material	254
		Bibliography	254
		Users and user interfaces	254
11	**Researching possible solutions**		
	11.1	Introduction	255
	11.2	Basic analysis	255
	11.3	Experiment	256
		Example	256
	11.4	Prototyping and simulation	257
	11.5	Notations and languages for developing designs	257
	11.6	Dataflow diagrams	258

11.7	Specifying event-driven systems	259
11.8	Table-driven finite state machines	259
	Example	262
11.9	Statecharts	263
	Basic statechart syntax	264
11.10	Using statecharts – a clock radio example	266
	Problem statement	266
	Understanding the problem domain	266
	Developing and specifying a solution	267
11.11	State sketches – using state diagrams to understand algorithms	268
	Example: binary search	271
11.12	Chapter end material	274
	Bibliography	274

12 Modularization

12.1	Introduction	276
12.2	Top-down design	277
12.3	Information hiding	278
12.4	A modularization example	279
12.5	Modularizing from a statechart	280
12.6	Object-oriented modularization	282
	Example	282
12.7	Documenting the modularization	284
12.8	Chapter end material	287
	Bibliography	287

13 Detailed design and implementation

13.1	Introduction	288
13.2	Implementing from a higher-level representation	288
13.3	Implementing with data structures and algorithms: rules of representation selection	291
13.4	The ADT table (again)	291
	A specific implementation	292
	Other scenarios and their implications	296

14 Testing

14.1	Introduction	298
14.2	Finding faults	298
14.3	Static analysis	300
	Code inspection	301
14.4	Dynamic testing: (deterministic) black box test	302
	Example 1	303
	Example 2	303
14.5	Statistical black box testing	304
14.6	White box testing	305
	Example	306
	Difficulties with white box testing	306
14.7	Final words on testing for finding faults	307
14.8	Assessing performance	307
14.9	Testing to discover	308
14.10	Release	309
14.11	Chapter end material	309
	Bibliography	309

15 Case study: Median filtering

15.1	Introduction to the case studies	310
15.2	Introduction to this chapter	310
15.3	Background	311
15.4	Why use median filtering?	311
15.5	The application	312
15.6	Approaching the problem	312
15.7	Rapid prototyping	313
15.8	Exploit existing solutions	316
15.9	Finishing	323

16 Multidimensional minimization – a case study in numerical methods

16.1	Numerical methods	329
16.2	The problem	331
	Finding minima in 1D	331
	Finding minima in multidimensions	331
16.3	Researching possible solutions	332
16.4	Nelder–Mead Simplex Optimization	334
16.5	Understanding the method with state sketches	335
16.6	Experiment-driven development	337
	Basic working	337
	Learning from experiments	338
	Minimizing noisy functions	338
16.7	Program code	340

17 stable – designing a string table class

17.1	A perennial problem in data analysis	353
	Collating one type of table into another type	353
	Sifting and computing	353
17.2	Design approach	354
17.3	Rapid prototyping a framework	355
17.4	A quick fix	359
17.5	Reading and writing	359
17.6	Finding things	362
17.7	Matching the requirements	364
17.8	Generalizing **stable** to do more	366
17.9	Size flexibility	367
17.10	Yet more generality: using templates to store other types in **stable**	367
17.11	A final program before refactoring	367
17.12	Refactoring	378

Appendix: Comparison of algorithms for standard median filtering	398
Index	405

Preface

In 1990 I persuaded my colleagues in the Department of Systems Design Engineering at the University of Waterloo to approve two brave and exciting ideas. First, we'd upgrade all our boring old MSDOS computers to NeXT machines. NeXT machines were black, beautiful, and soon-to-be discontinued. We all enjoyed programming them, so the cost in terms of obsolete software when we went back to DOS/Windows computers must have been dozens of development-years. My second, rather better, idea was to put a software design course in our undergraduate programme.

Systems Design at Waterloo taught a wide-ranging curriculum in engineering, with emphasis on systems theory and design, and our students emerged as 'superb generalists'. That was our claim, and the careers of former students prove it was – and still is – true. But their exposure to software amounted to an early programming course in Pascal, then Matlab alongside Numerical Methods. In later courses they might pick up FORTRAN or C. I argued for a course distilling the insights of computer science (CS), without the depth of focused courses in a full CS curriculum, but providing key methods and tools for designing reliable, efficient, maintainable software. The new course would be broad – from data structures to software engineering – but selective, buttressed by a challenging, integrative software project.

I wanted to teach software design because I love to program, I've done a lot of it on medium-sized industry projects, and I had something to say about methodology. As an engineer I like systematic methodology, but as a scientist I've been taught to look for evidence, and there seemed to be surprisingly little evidence that the design processes and techniques in software engineering are optimal. To convey a suitable scepticism as well as give worthwhile advice and promote good practice was a challenge I wanted to tackle.

The department approved a new core course in software design and I taught it for the following five years. The framework of my lectures and notes was: introduce with deduction, support with induction and promote with abduction. First, the class would look at software from a particular perspective, highlight the features that are prominent in that way of seeing things, and deduce some rules for design (deduction). Next we'd review the experimental evidence testing those rules, and in most cases supporting them, and demonstrate their application in practical exercises (induction). Finally, I'd turn up the polemic, using anecdote and the voice of experience to promote good practice, and tales of doom of what happens when you neglect the rules. (Abduction means reasoning by analogy from similar cases. It also might mean kidnapping a person for a particular cause – in this case, good software design.) This framework was, I thought, a refreshing change from design approaches that merely tell you what to do and don't say why (maybe because they don't know). But engineers can only take so much scientific scepticism and balance. They want to get on and make things, so the more searching inductive parts of the course gradually gave way to added advice based on the collective experience of programmers.

This book grew out of that course, and still reflects it. Chapter 2, the book's core, deduces principles and rules from nine perspectives on what software is. *Software Design for Engineers and Scientists* is unusual among books on software and programming in laying this foundation. The chapters on applied software design include reflection on the empirical (inductive) evidence for how good methods are. But the majority of the text is advice, supported inductively and abductively by numerous examples. Much of this is straightforward information – on writing C++ or on the details of algorithms, for example. But some advice is open to question, so I hope the book also conveys how to challenge and test received ideas about software design.

After I left Waterloo, I continued to develop the material that's now this book. It has been tried, tested, augmented and enhanced in courses at Memorial University of Newfoundland, Canada, and the University of York, UK. Numerous graduate students have learned software design through it, and provided candid, valuable feedback. But the world of software design has changed significantly since the early drafts. Three important things have happened that affected the shape of the book:

Important thing number 1: The tension between programming as craft and traditional software engineering has been sharpened by the introduction and advocacy of development paradigms like Extreme Programming and Agile Programming. Extreme Programming doesn't just challenge traditional assumptions, it overturns them. As a revolutionary approach, it is promoted with confident, sometimes bellicose rhetoric. But some of the big ideas of Extreme Programming are as lacking in empirical support as those of traditional software engineering. There's no shortage of anecdotes about cutting development time and happy programmers, but controlled experiments on design methodology are so difficult to do that we really don't know the true benefits and costs yet. What's needed is to identify, or at least suggest, the best elements of the competing paradigms. This I try to do here; admittedly not through controlled experiments, but as a practising programmer, I take all the advice I can get and test it on real projects. This book shares what I've learned.

An aside: because it contains so many code examples in the context of the whole software design process, this book could claim to be the first Extreme Programming text that really gives code the status it should have – at the centre. Unfortunately it can't claim that, because it doesn't buy into the whole Extreme Programming package.

See page 2 for more on this controversy.

Important thing number 2: The rise of the web has meant that students are increasingly familiar with getting computers to do what they want via an artificial language. Although their practical exposure may amount only to HTML tweaking, the upshot is they are able to launch into learning a 'difficult' programming language like C++ without a prior, gentle, introduction to programming and computers in general. Consequently *any* science or engineering student ought to be able to use this book, even if they have not had a first programming course. To ensure this is possible, the tutorial introduction to C++ in Chapters 4 and 5 is selective, but at first gentle.

Important thing number 3: The programming language landscape has changed. Java emerged in the mid-1990s and is dominant in many fields. C# could be the language of the future, at least for Windows-based applications. The language used for this book, C++, has been standardized. This environmental change has affected both the text and the program examples, and it has meant that I will be providing code in other languages on the companion website. Chapter 4 includes some comments about the trade-offs between languages and idioms (page 51).

The testimonials I've had from Waterloo students who still use their course notes suggest that the original idea of software design for engineers (and scientists) was a good one – perhaps even good enough to compensate for the NeXT machine debacle. I think it still is.

Acknowledgements

Software Design for Engineers and Scientists explains how many different perspectives there are on what software is, and how we can learn from them all. But there's no doubt that the greatest benefit to a designer comes from working with people who are better designers. I've been lucky to work alongside some great programmers and learn from them. Chronologically, rather than in order of importance (which I couldn't even estimate), these are the people who have left a permanent stamp on my thinking and ability as a software designer: Jürgen Foldenauer, Tim Dennis, Chris Toulson, Guy Vonderweidt, Charles Nahas, Lawrence Croft, Michael Chambers, Jason Fischl, Steffen Lindner, Mehran Farimani, Li-Te Cheng. Thank you all – wherever you are.

My Waterloo students had to provide feedback on the notes that became this book. Among those writing helpful critiques, the most important was Todd Veldhuizen, and I also learned a lot from Don Bowman. Many graduate students read later drafts and provided their corrections and comments. Thanks to my colleagues at Waterloo for approving the software design course, particularly Ed Jernigan, Shekar, and the TAs who were so encouraging about the course when the project was driving the students (and us) mad.

Thanks to my colleagues in the Electrical and Computer Engineering discipline at Memorial for providing the most supportive working environment I've experienced: particularly (in the context of this book) to Theo Norvell for letting me pontificate on software design during his courses, and to John Quaicoe for wise management and inspiring teaching quality.

Thanks to the staff and students of the Department of Electronics at York, especially Stuart Porter and the students in Data Structures and Algorithms who have experienced some of this book as it went through yet more iterations.

Finally, my thanks to Gill, Luke and Stephen for everything else.

Errors   Errors are my responsibility. If you find them I'd like to know, and I'll ensure that you're acknowledged in any future edition. You can contact me by writing to the publisher.

I'm happy to hear about any kind of error, from missing evidence about the merits of a particular method, to program bugs. However, there is one kind of code error that you don't need to tell me about.

```
char buf[80];
cin >> buf;
```

probably is a bug, whereas

```
string buf;
cin >> buf;
```

probably isn't. By the end of Chapter 6, all readers will know that. But I've left things like the first example in Chapters 4 and 5 (suitably flagged as bug-spotting exercises), because they are classic examples of a programming error you *must* grapple with to be forearmed against (see page 174).

1 Introduction

1.1 Theme

This book is about:

- how to design good software
- how to program in C++
- data structures and algorithms
- scientific and engineering programming.

The book is modular but integrated. On its four themes it says both less and more than specialized texts.

- Software design is not only a big subject, it's also fast-changing and controversial. Current debates include: traditional software engineering versus extreme programming, UML versus ad hoc notation, C++ versus Java versus C#. This book outlines the issues, summarizes advice from both sides, then plumps for a particular approach to show practical real examples. A lot gets left out. But by starting with *why* we design software the way we do, the book doesn't just prescribe, it explains too.
- C++ is a big language now. Some people say the way to learn is by full immersion in the standard library. We take a much more modest and traditional approach. The book could, in principle, have featured Java, C#, or some other language (but see page 51). C++ is here, with a tutorial, because the essence of software is the program code, so a book like this has to give a central place to real programs in a real programming language.
- Data structures and algorithms are a fundamental and relatively unchanging part of software design, so we need to talk about them, giving lots of standard examples. But full coverage would demand a full book and there are plenty of good ones already. We focus on how the software designer uses data structures and algorithms to solve practical problems.
- This book includes a few recipes for scientific programming: case studies in Chapters 15 to 17, and examples of programs elsewhere that might be useful for practical science and engineering applications. But I hope you'll leave the book as a designer, not just with more knowledge about particular applications. Indeed, one message of the book is to treat recipes with care. Although Chapter 2 includes the rule 'Exploit existing solutions', it also emphasizes that 'Every program has surprises'. Uncovering and understanding any surprises in existing solutions, including other people's software recipes, is a vital skill.

The book's overriding purpose is to help you to think and act as a good software designer.

1.2 Audience

Computer programs manage society, enable every kind of telecommunication, mediate almost all impersonal business tasks. In engineering and science, they conduct experiments, analyse data, simulate complex phenomena and control complex systems. Every engineer and scientist can understand the way these programs are designed. *Software Design for Engineers and Scientists* is written to explain how. It brings together important ideas about what software is and how it is made, building from fundamentals to principles and practices. It overviews all the important aspects of software technology and the key steps in design. It shows how to program well, design substantial pieces of software correctly and efficiently, and apply the tools of computer science in an effective way. It provides copious examples and exercises to help you learn by doing. And if, having read it, you want to pursue further directed learning in computer science, the book provides a broad and reliable foundation for advanced software studies.

Software design also provides opportunities for creative problem solving, elegant crafting, and the conversion of imaginative ideas into real systems – it is stimulating and enjoyable, as well as important. As author, I have to come clean and admit that I'm excited by the subject: I enjoy writing programs and I want to make them as good as possible. I'm not in the business of trying to convert you, but I hope you'll capture at least some of the delight of software design through this book!

1.3 Three definitions and a controversy

In this book three terms appear many times: *software design*, *software engineering* and *programming*.

Software design means everything to do with creating a computer program for a particular purpose. It includes any needs analysis, specification, high-level and low-level design, modularization, coding (i.e. writing a program in a programming language), integration, debugging, testing, verification and validation, maintenance. It isn't necessary to worry about all those steps right now, just to know that *software design* encompasses them. Now comes a terminological difficulty: the term *software engineering* also includes all the steps, and, depending on who you talk to, *programming* does too! There are differences of meaning, but they aren't always shared by different people. So here is an explanation of how the three terms are used in this book.

Software design itself is a generic term with no implication about scale or standards. It applies to the whole process of creating the smallest 'Hello, world' program and to the whole process of creating a 20 million line operating system.

Software engineering means using systematic methodologies in the design process. It emphasizes that the non-coding steps of design become more important as the project gets larger and it draws on other engineering disciplines (and management) for planning techniques, processes, and standards that organize and control those steps. It also applies to coding techniques that enforce good practice and protection against errors. *Software engineering* takes the big picture and looks inwards, with coding as just one step in the process.

Programming looks outwards from coding and sees planning, specification, testing, etc. all through the lens of the program text. This is not to say that programming is *just* coding, but that is its core, and many perspectives on programming use the coding process as the organizing mechanism for all other parts of the design.

If this all seems a little abstract, let's see how *software engineering* and *programming* contrast. A 'software engineer' might criticize 'programming' by saying that code-centric design does not put enough emphasis on being defensive, allowing safety factors in performance estimation, and considering the implications of specification changes. For example, many books have been written on the analysis and design of data structures and algorithms. But some forget to tell the reader the most certain fact about a specification: it will change. (Chapter 13 of this book demonstrates how completely proper reasoning from a specification to a data structure and algorithm implementation can be blown apart by a minor change to the specification.) To a 'software engineer', this sort of oversight typifies the danger of focusing on the code.

On the other hand, a 'programmer' might protest that some approaches to software engineering neglect coding all together. At best, they emphasize what a small part coding has in the total software development task. At worst, they give the impression that the important jobs in software system design can be done without knowing anything of the code-writing dirty work. And yet the code is the final essential product – the place where the difference between good and bad design really counts.

Since the mid-1990s, *programming* has been challenging the traditional *software engineering* view by providing, through methodologies like extreme programming, a framework for the design of big systems in which the code is central. (This challenge could be seen as redefining software engineering to encompass code-centric views, but I am going to stick with using *software engineering* in the traditional way.) Software engineering promotes systematic organization at every level of a project, but now programming (bearing uncompromising tags like *extreme*) is blatantly advocating design in ways that seem to traditionalists chaotic and unmanageable. This book flags the major tensions, and tries to find insights from both sides wherever possible. Software engineering is made essential to programming by treating modularization, data structure design and program style from a project-wide problem-solving viewpoint. Programming is made essential to software engineering by a high view of the code, summed up in the 'textual principle' that the essence of the software product is in the source code text. Particular practices are compared, selected or rejected on their own merits.

> The Association for Computing Machinery (the professional body that represents programmers in the USA and develops curriculum guidelines for computer science programs) recommends that software engineering concepts be taught even in the earliest programming courses. One valuable consequence is an emphasis on good testing methods (which everyone agrees are important). One questionable consequence is the early introduction of loop invariants, which then tend to be forgotten (see page 14).

> The extreme programming FAQ at www.jera.com/techinfo/xpfaq.html gives a summary of the extreme philosophy and methods.

1.4 Essential software design

Software Design for Engineers and Scientists is pragmatic when it comes to deciding between traditional software engineering and extreme programming. But it does have an underlying view on the fundamentals. The book is unusual in the central role it gives to a way of looking at software pioneered by noted computer scientist, Fred Brooks. In the April 1987 issue of *IEEE Computer* magazine, Brooks wrote an article which has become a classic of software design: 'No silver bullet: essence and accidents of software engineering'. Brooks said that to find good strategies for designing software, we must first understand

Objects, components, or what?

One of Brookes' arguments was that, for all the hype, object-oriented programming (OOP) is not a silver bullet – it doesn't change the essential problems of software design. Many people disagree. OOP does two very important things: it provides a strong model for modularity, and it allows old code to call new code. These two mean that OO code is reusable more effectively than other kinds of code. But Brookes is probably right: even together these don't make a silver bullet to slay the software werewolf.

what software really is. His own understanding of software's essential nature, and the implications for design, are summarized in the organic software section of Chapter 2 (page 18). But the most important insight of Brooks' paper is that 'How should we design software?' only makes sense if we first ask 'What *is* software?' Brooks inspired the underlying philosophy of this book: principles, rules, practices, methods, application should all be built on an understanding of software's essential properties. This paradigm is called *essential software design* – design that engages with the essence of software.

It turns out that there are many perspectives on what software is. We will examine nine in detail. For each, we begin with an appropriate definition for 'program', review the development of insights from that perspective, and derive principles about the nature of software and rules for its design. Although this approach does not identify a single essence for software, it demarcates the subject. Everything that comes after is built on this integrated understanding of the nature of software. And where there are controversies about methods and design choices, we refer back to the principles and rules to inform our decision.

1.5 Outline of the book

The structure of the book is illustrated in Figure 1.1. The earlier chapters are lower in the diagram because they form the foundation for what comes after.

Foundations

Chapter 2 is the book's foundation, where principles about the nature of software and appropriate design rules are derived from nine different perspectives. Some of the rules are about accelerating your development as a software designer, and these are discussed in Chapter 3. The rest of the book is about the construction of programs based on the two legs of technology and application. Methods and practices are developed based on the core set of principles and rules, which themselves rest on the definitions or understandings of what software is.

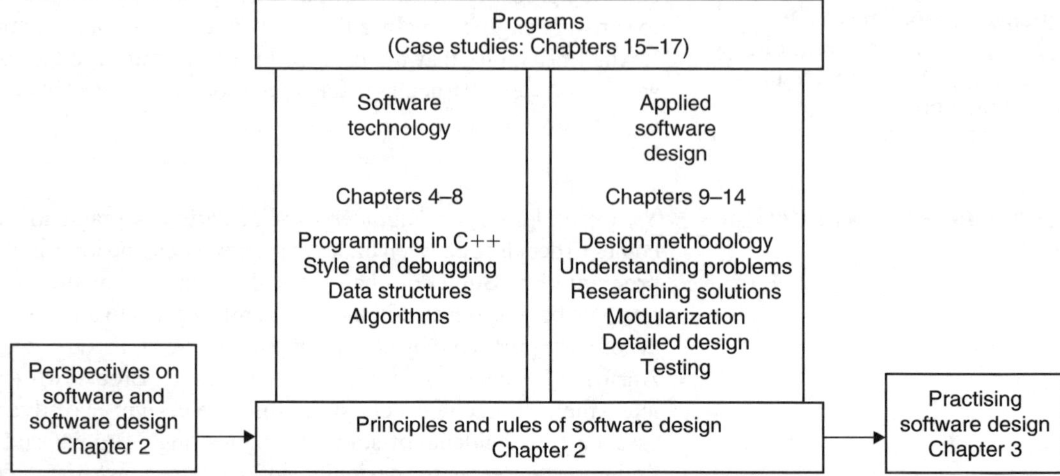

Figure 1.1 *Structure of the book*

Software technology

One of the principles developed in Chapter 2 guides the book's approach to program examples. The choice of programming language is an *accident* rather than an *essential* of the design. But 'The textual principle' says that the program code (in whatever language it happens to be written) is the essence of software. It would be contradictory therefore not to have profuse code examples in a book on software design. And why not give real programs in a real language? C++ is therefore introduced in Chapters 4 and 5 and used heavily in examples thereafter. C++ was chosen because of its wide availability and importance. Indeed, it is arguable that another principle, 'Embrace standards', is served by using C++.

Chapters 4 and 5 provide a tutorial introduction to C++. They can be used in front of a computer with a C++ compiler running, and the compiler documentation at hand. They do not document all the details of C++, and they are not big enough to be a complete survey, but they will equip you to program intelligently in the language. Chapter 6 discusses program structure, style and debugging. In particular it surveys the most common programming errors, and gives advice about program clarity and how to split a program into files.

Chapters 7 and 8 are about low-level program design. Chapter 7 introduces elementary data structures, and by looking at Abstract Data Type (ADT) implementation, raises the issues in choosing data structures and algorithms. To provide tools for analysing implementation alternatives, Chapter 8 discusses algorithm analysis. On the way it gives C++ implementations for popular searching and sorting algorithms.

Applied software design

Chapters 9 to 14 cover the steps in software problem solving. Chapter 9 introduces design methodology with generic design, and develops a minimalist methodology for software. It includes overviews of design team organization and documentation for medium-sized projects. It also includes a sceptical critique of some solution generation methods, to alert you to open questions in problem solving.

Chapter 10 is about defining and understanding problems. It gives a simple recipe for writing a problem statement. The problem domain is understood by background research. This is task specific but there are some generally useful procedures that help in learning as much as possible about the problem domain. The chapter includes material on understanding human users, since they are the common element of most major software systems.

Chapter 11 deals with researching possible solutions to problems. It provides advice on quantitative analysis, experiment and simulation, then introduces diagrammatic notation techniques. Concentrating on state diagrams and statecharts, it shows how these characterize event-driven systems economically, and lead to good event-driven code. It also explains how state sketching can be useful in notating and understanding ordinary algorithms and procedures.

Modularization is increasingly important as program size grows. The treatment in Chapter 12 is founded on the principle of information hiding, and applied to both medium-sized and small programs. It includes discussion of object-oriented design and the use of state diagrams.

Accident is used here in the technical sense of those attributes that are not fundamental to a thing's existence. Some people have felt that software is just one big accident. That is a different meaning of accident ... and probably not true.

Many of C++'s foremost practitioners advocate learning it in a way that emphasizes the full range of its standard capabilities. The kind of C++ taught here is much closer to the machine than that taught, for example by Stroustrup and Deitel. The reasons for this are discussed in Chapter 4.

Chapter 13 on detailed design uses the insights developed in the Technology Chapters (4–8) to show how higher level representations (problem statement, specification, statechart, modularization description) can be turned into code. It includes discussion on ADT implementation and an example design illustrating how a good specification can lend its structure directly to a program.

Chapter 14 discusses testing, both static and dynamic. Code inspection, white box and black box testing are included. The final section of this chapter deals with the final section of the design methodology – release of the product.

Case studies

Three case studies are included at the end of the book.

Chapter 15 considers a signal-processing application – median filtering. Signal processing crops up in many data analysis applications, and median filters are handy for signal conditioning and enhancement. This is, however, a case study, not a recipe, and its purpose is to show how the steps of design fit together in a practical example that is typical of research computing.

The second case study in Chapter 16 tackles another scientific application: multidimensional optimization. Again, we'll see how the program develops as the problem and the method are better understood.

The final case study (Chapter 17) illustrates the development of a text table class that is useful for analysing the data output from experiments or simulations but can also be used more generally for manipulating textual databases. In this example, the growth of the program at many stages is illustrated, showing the interplay of specification, coding and testing. Some of the intermediate code presented contains bugs. But the finished version is heavily tested and is believed to be bug free.

1.6 Presentation conventions

There are five different styles of presentation for textual material in this book.

The running text in Times Roman font is tutorial. It is the backbone that carries the other types of material.

Program examples are in Rockwell font like this and illustrate the concepts being developed in the running text.

Some programs include line numbers like this:

```
1: // Null program to illustrate line numbering
2: int main( ) { return 0; }
```

The line numbers are not part of the text of the program.

Table 1.1 *An example table*

	Presentation type	Purpose
	Table	To supplement the running text
⇒		An arrow symbol, located in the left-hand column of a table shows the most used or most important cases. Not all tables have the arrow symbol column.

An explanatory principle or a design rule

A paragraph in Helvetica like this is a statement of a Principle or a Rule. There are 13 principles and 21 rules, all developed in Chapter 2. Later in the book, they are placed as flags to show which principle is being developed or which rule applies.

> Exercises are shown in boxes like this one. They are placed with the material they exercise. You are encouraged to do the exercises as you go along. Outline solutions are sometimes given at the ends of chapters.

1.7 Chapter end material

Bibliography

All sources alluded to in this chapter, including extreme programming texts and Brooks' *Silver bullet* paper are included in the references list for Chapter 2.

A disaster

With the book deadline only weeks away, a research idea diverted me from its demanding, accusing, unfinished chapters. Immediately I wanted to test a new extension of work I'd done just six months ago. So I pulled up the programs to work out where to add new code. I scanned files, functions, class definitions and comments, written by me, for work that I know intimately. After searching analysis, I expressed my understanding of the old code in the following comprehensive thought: 'What?'

Six months ago, just like today, I had a great idea and the important thing was implementing it and testing it. In a flurry of creative activity, I did good science with a program that grew big, ugly and poorly documented. It was fully comprehensible to me then, because it was growing under my hands, urgent and vital. Trouble is, six months on, it's unreadable.

I had no excuse. First, I've been writing programs for a long time, and some of my best are widely used, read and even (among people who like that sort of thing) admired. But second, and more embarrassingly, I'd already written (and taught) drafts of this book where I'd given lots of advice about designing software. If I'd heeded it, my research code would have been much easier to read and modify.

How could I have avoided going wrong? This book will tell you. But with that kind of record, am I the right person to offer advice? Strangely enough, yes. For although I'm responsible for the software disaster I've just confessed, for ignoring my own best advice, I *had* followed other advice that allowed me to get my old code into comprehensible, usable, shape in ninety minutes. Nothing to be proud of, but not that much of a disaster after all.

The advice I hadn't followed was to do with finishing and documentation. But the advice I *had* followed was systematic construction of test cases and logbook writing. I was able to decipher my old code by going through the tests I developed and reading the contents of test logs against their timestamps. This allowed me to trace the evolution of features and recall my thought processes.

Reasoning backwards from the results of testing is not the recommended way to interpret an old program. But the moral is that even if we're not the great programmers we could be all the time, good habits give us safety nets. This book will not make you a perfect programmer, but will guide you towards habits and practices that, even when you're imperfect, like me, help you to be better.

2 Fundamentals

2.1 Introduction

To help you become proficient and productive in software design, this book informs, advises and gives lots of examples. But where does the advice come from? Experience – certainly. Other people's ideas – definitely. But also from trying to understand what software really is. This chapter examines the nature of software and software design, so that the rest of the book is neither truisms nor anecdotes, but based on fundamental ideas. So important is this foundation that the 13 principles and 21 rules derived here will be cited again and again later in the book.

The principles are statements of the nature of things (e.g. 'A program grows along the framework of its first version'); the rules are advice (e.g. 'Make the first version durable not functional, and get it running early') implied by the principles. Some principles and rules are terse statements of ancient wisdom ('The specification will change'), some revise that wisdom with modern twists ('Hide information in public; reveal it in private'). I have ruthlessly excised principles that are platitudes. Among the *very good things to remember* that this chapter says are: 'Style matters, design systematically, design for your users, see yourself as a craftsperson.' All these were principles in early drafts of this chapter; all have been 'demoted' to commentary on the basis that they are truisms. The principles that remain are either fundamental statements about software and its design or specific practical rules.

2.2 The nature of software

Software is strange stuff. It is invisible, abstract, malleable, and complex. These properties are part of the very nature – the essence – of software. It is also difficult, inelegant, fragile and often wrong. Are these also essential properties, or just accidents of practical programming? Software turns a general-purpose machine to a particular purpose, for as long as a particular program runs. That purpose may be so far removed from the electronic interactions happening inside the computer, that there is no comprehensible way of describing one in terms of the other.

Shaky foundations?

Chapter 1 admits that there's much disagreement about software design, so how sure can we be about so-called 'foundations'? Things can shift, certainly, but I've tried to move from legitimate perspectives to clear principles that will last. For example, one principle is 'test to break'. That means test hard, trying everything you can to break your program. At the time of writing there's debate about whether testing should have a prominent upfront role in coding. Test-driven development is based on the idea that you write a test first, fill in enough of the system to pass the test, then go on to writing the next test. Chapter 14 contrasts this view of unit testing with more conventional methods. But whether or not there is a shift in coding practice towards test-first/code-to-fix, just about everyone is agreed that program testing needs to be organized, extensive, rigorous, and hard. So the principle stands.

Software has independent existence, yet is so reflective of human thinking that Fred Brookes has called it 'pure thought stuff'.

Philosophers have not paid much attention to the nature of software. Perhaps they haven't yet realized how strange it is. But there is a history of the subject, and by reviewing that we can see how software has been understood by its practitioners. In this chapter we will put different ideas about what software is into nine categories, created from definitions of the word 'program', and label the categories as though they described schools of thought or distinct intellectual perspectives. To some extent they do, but the point is to organize and compare ideas, not to invite you to choose your favourite label. Indeed, I hope you'll choose all the labels, because each perspective has some consequences for software design that are worthwhile enough to be called principles.

The nine perspectives fall into three groups. The first three focus on the program in itself and so are perhaps the most abstract. The next three consider the relationship between the program and its context and are therefore engineering oriented. The final three are as much about the psychology of programmers as what software really is. In each case, the purpose is to derive insights that a thinking human (you!) can use to design good programs.

So here are nine ideas about the nature of software, each expressed as a definition of 'program', each from a particular perspective.

The mathematical view	A program is a formal description of the sequence of primitive operations required to reach some result. It is a mathematical statement, formally equivalent to a proof.
The literature view	A program is a document, expressing thoughts verbally. Software is written, read, and indexed; it is a literature.
The organic growth view	A program is a large, complicated, invisible organism, whose development and growth are determined by programming.
The engineering view	A program is an artifact, designed systematically, applying science and mathematics, to solve a problem.
The system view	A program is an organized set of control components in a complex system.
The user-centred view	A program is a machine that makes a computer *usable* for a particular task.
The artisan view	A program is a handmade artifact, fashioned by a craftsperson, who applies creative judgement and experience to achieve elegance and utility simultaneously.
The organizational behaviour view	A program is the consequence of programmers programming.
The multiple analogy view	A program is an X $\{X =$ movie, research paper, communiqué, $\ldots\}$.

In exploring the nine perspectives, we will review some of the history of programming, bringing in ideas from some of the foremost practitioners of computer science and engineering. Discussion of each definition will lead to design principles. On these, the rest of the book rests.

2.3 Software as mathematics

> The mathematical view   A program is a formal description of the sequence of primitive operations required to reach some result. It is a mathematical statement, formally equivalent to a proof.

In the early days of computers, there was no doubt that they belonged to the realm of mathematics. Not only were the original applications of computers mathematical, but the people who theorized about them, designed and built them were mathematicians.

In the early twentieth century, Russell and Whitehead pioneered the formal application of logic to the *process* of mathematical proof. They saw that a 'procedure' (how something is to be done) could be talked about in the same kind of abstract formalism as the objects on which the procedure operates. Gödel's famous theorem about the limitations of mathematical systems was developed by thinking of proof procedures in just this way – as objects that could be manipulated algebraically. The other side of this coin was that the essence of a mechanical process lies not in the machine that does it, but in the abstract control structure for that machine. Church, Post, Turing, and others saw this abstract control structure as a 'program', or sequence of simple operations. In showing that the logical form of a machine can be separated from its physical realization, they postulated that equivalent machines could be created (or 'run') on different hardware, and that some hardware could be general purpose (universal) enough to run many different programs.

The first computers were truly virtual machines, existing only in the imaginations of thinkers like Turing. They were conceived as general-purpose automata that could be 'programmed' to behave as though they were particular specific machines. That is, given a formal specification of a particular machine (in other words, a program), the computer would 'become' that machine. In an exciting and technologically unprecedented way, a single general-purpose computer could become any one of many abstract machines, according to the program it was given. And that program was invisible, an abstraction, 'just' a formalism. This strange stuff we now call software rarely holds for us the same mystique. But at the birth of computer science, the conception of such a universal mechanism, which changes its behaviour according to a mathematical abstraction, was an intellectual wonder.

Most of the early physically realized computers were programmable calculating machines. They were used for numerical calculations, such as those required for the development of war technology in the Second World War. Mathematicians, like Turing, who were directly involved in the war effort, were able to participate in the

Bertrand Russell (1882–1970) was a mathematician, philosopher, author and controversial public figure. Born into a privileged family, married four times, educated and did his early scholarly work at Trinity College, Cambridge, founded and taught at his own progressive school (as well as university teaching at Trinity, in the US, China, and elsewhere), imprisoned in 1918 for pacifism and in 1961 for obstruction while demonstrating with CND, sacked from City College, New York (1940), for being an 'enemy of religion and morality', won the Nobel Prize for Literature (1950). Russell features on several websites including www.mcmaster.ca/russdocs/russell.htm.

Alfred North Whitehead (1861–1947) was Russell's tutor at Trinity College and later moved to Imperial College then Harvard. He made significant contributions to logic before meeting Russell and later developed philosophical ideas combining psychological and physical experience.

Russell and Whitehead published *Principia Mathematica* between 1910 and 1913. It has been called 'the greatest single contribution to logic since Aristotle'.

Kurt Gödel (1906–1978) was born in Brno, Czech Republic, studied and taught in Austria and later worked in the US. His theorem (1931) showed there would be formally undecidable statements in any formal system (including Russell and Whitehead's). An entertaining discussion of the theorem is incorporated in Hofstadter's book *Gödel, Escher, Bach* (1979).

Alan Turing (1912–1954) (www.turing.org.uk/turing) was an English mathematician who studied at Cambridge, then at Princeton. He later worked at the National Physical Laboratory, Bletchley Park, and Manchester University. His idealized computer, the 'Turing machine', was the framework for his fundamental contributions to the early theory of computation. He also invented the idea of the 'Turing test' by which a fair comparison could be made between a computer and a human to decide whether the computer was exhibiting 'intelligence'. Turing's life and genius, his contribution to the UK's codebreaking programme in the Second World War, and the personal and political consequences of his homosexuality, have been dramatized a number of times, notably in the play *Breaking the Code*. Turing committed suicide by eating a poisoned apple.

Von Neumann (1903–1957), educated in Budapest, Berlin, and Zurich, was a child prodigy, wrote his first mathematical paper at 18, and was one of the original six professors to be appointed to Princeton's Institute for Advanced Study. (He was at Princeton during Turing's time there, but they did not collaborate directly.) Von Neumann made important contributions to mathematics, especially in game theory. He promoted the concept of a stored program for computers. See www.gap.dcs.st-and.ac.uk/~history/Mathematics/VonNeumann.html.

building of the first machines that approached their conceptual structures. But it was not until the late 1940s that truly general-purpose computers began to appear. It had been shown that there are many possible structures that could support programming. It was not long before the structure, or architecture, proposed by the Hungarian mathematician John von Neumann, was widely adopted. This structure is familiar in the computers of today. It involves an arithmetic unit for combining data by addition, subtraction, etc., an area of storage for both instructions and data (here again is the idea of the equivalence of primitive operations and data), and a controller that successively fetches and executes instructions. There are facilities for testing conditions and for jumping out of the normal sequence to an instruction elsewhere in the program.

During the 1950s, new types of theoretical computer science grew up. The analysis of algorithms using combinatorial mathematics provided tools for discovering how best to do common jobs like sorting and searching. Numerical analysis developed, as long and complicated procedures became practicable when run on computers. The early emphasis on abstraction motivated the design of the first high-level languages. BASIC, FORTRAN and LISP were all invented in the 1950s, the last of these being particularly interesting. The syntax of LISP was originally obscure and hard to learn. But then John McCarthy,

John McCarthy invented LISP at MIT then spent most of his career at Stanford, where he is now an emeritus professor. His homepage, www-formal.stanford.edu/jmc includes versions of the original paper on LISP.

LISP's inventor, wanted to use it to do some mathematics where the procedures and data would have the same syntactic form. The new syntax he developed (used today) reflects the same mathematical mindset as that of Russell and Whitehead – procedures can be manipulated just like other objects. McCarthy has written at length on the use of LISP as a vehicle for reasoning about mathematics and computers. LISP is still a much-used language in artificial intelligence, though its links to the theory of machines have not been as significant as its flexibility in symbol manipulation.

Right from the start, programmers' thinking diverged from the idea of manipulating mathematical abstractions. Once computing machinery was available, they began to develop programs that grew in size and complexity, such that thinking of them as abstract machines was no longer natural (or even possible). During the 1950s, practical programming became an art, learned chiefly through observation and imitation, almost decoupled from the theory being developed at the same time. The beginning of Computer Science (CS) teaching as a separate discipline, in the early 1960s, was in part a response to this gap between theory and practice.

At the birth of computer science, there was already confusion about what it was. Some university CS departments were grown out of engineering or mathematics departments, some were created fully formed from nowhere. Those who were mathematicians wanted to return programming to its roots. Their main argument was that in seeing the equivalence between a program and a mathematical proof, they could not only teach people to think properly about programming, but they could also write programs that would be guaranteed correct.

Some of the most respected computer scientists of the 1960s were proponents of programming as mathematics, yet they are usually cited for their practical contributions to the art. For example, in several papers Edsger Dijkstra explored methods for proving program correctness. But he is probably better known for seminal work in real-time systems (semaphores, cooperating sequential processes) and for his leadership in promoting structured programming. (Exactly what 'structured programming' means has never been agreed upon, but it certainly includes programming without GOTO instructions – see the sidebar on page 16.) C.A.R. Hoare, whose paper 'An axiomatic basis for computer programming' epitomizes the mathematical view, is better known as a contributor to real-time systems theory (monitors) and particularly as the inventor of Quicksort (see Chapter 8). These researchers would probably insist that their mathematical perspective underlay all their work, and the search for correctness-proving mechanisms inspired their practical discoveries.

Edsger Dijkstra (1930–2002) and Tony Hoare were both contributors to ALGOL, one of the most influential computer languages. Dijkstra (born and educated in the Netherlands, later moving to the US) is also famous for the shortest-path graph algorithm that bears his name, and for his promotion of structured programming. His obituary at www.cs. utexas.edu/users/UTCS/notices/dijkstra/ewdobit.html summarizes many other contributions. Hoare studied at Oxford, then invented Quicksort while a graduate student at Moscow State University. He eventually moved to head the Computing Laboratory at Oxford and is now with Microsoft Research, Cambridge. See research.microsoft.com/users/thoare.

Hoare's 'axiomatic basis' was the first practical system for proving program correctness. He was able to demonstrate it on the non-trivial program 'Find', which solves a subproblem of Quicksort, namely to find the element of an array $A[1 .. N]$ whose value is nth in order of magnitude, and to put it at location $A[n]$. Hoare's success inspired much work throughout the 1970s, culminating in comprehensive correctness-proving schemes, such as that described in David Gries' 1981 book *The Science of Programming*. This is an inspiring book to read, though from a period of over 20 years, it now looks a bit overconfident. The reader would probably smell a rat as early as Dijkstra's foreword, which makes much of the impact that correctness proving will have on programming practice. Dijkstra had said similar things a decade earlier, and the sea-change in programming practice he looked for has still not arrived.

What does the mathematical perspective on programming give to the student of software design? Not the theory of algorithms, nor complexity theory, nor numerical analysis – these are engineering perspectives really: the application of mathematics to programming, rather than interpreting programming as maths. What the mathematical perspective gives is the recognition that programming is about tautology – saying the same thing different ways. Once the requirements for the program are properly understood, the creation of the final product is a sequence of conversions between equivalent representations. This does not belittle programming – mathematicians are all too aware that working out different ways of saying things (like proving theorems) can be fiendishly difficult. Rather, it recognizes that *how* things are represented is more fundamental than *what* is represented. Programming is about making the right choices in representation – how to structure data, how to group items into modules, how to express the complex twists and turns of processing.

The mathematical perspective provides a principle of design:

Principle 1

Software design is representation.

At different stages in the software design process – problem statement, specification, modularization, programming, testing – the same information will be represented in different ways. Within each stage there are a huge number of possible alternatives. The designer has to choose the representations – the frameworks – within which everything (data, events, actions, attributes) is expressed. How those frameworks are then 'filled in' is secondary.

Rule 1

Begin every step in software design by asking: *How* should this information, this activity, this state, this relationship, be represented?

Representation is the first question. But, like practical mathematics, representation is hard. So,

Rule 2

Don't expect to get the representation right first time. Look for opportunities to improve code by *re*-representation.

Representation is *so* important, *so* pervasive, that we always need to be on the lookout for ways to improve it. Re-representation is often

Invariants

The idea of invariants comes from mathematicians such as Floyd and Naur working on program verification. An invariant of an object is something that is true of the object no matter what operations are performed on it. The idea is that initialization makes sure the invariant is true, any operation on the object can then assume the invariant is true, and must make sure it leaves it true. Specifying an invariant is a way of underlining that the object must maintain a well-defined state, which can be checked as a test of program correctness. Loop invariants (which specify something that must remain true on each iteration of a loop) often appear around Chapter 2 of a textbook (i.e. right *here*). Their usual fate, after being explained and illustrated, is to disappear from the rest of the text. I have to admit that *I* was going to advocate loop invariants, but I did not achieve the necessary diligence in my program examples, so to avoid embarrassment, I have relegated them to a sidebar! On the other hand, I am fairly keen about **const** correctness (see Chapters 4 and 5), which could be thought of as a semi-formal technique for improving the chances of program correctness.

called *refactoring*, meaning restructuring or rewriting code with the sole intention of improving it (i.e. not to add functionality).

Arguably, the mathematical perspective could have provided a third rule of design:

**Apply any available formal technique to ensure
program correctness.**

But on pragmatic grounds, this book fails to live up to the demands such a rule would make. In this it is not alone. Frankly, it's hard to find much application of formal correctness-proving techniques to real-life programs. Although software in the small is formally parallel to mathematics in the small, no human mind can extend that parallelism to the size of practical software. What is more, adopting a proof procedure during the creation of programs does not guarantee against mistakes. There are plenty of examples of published but flawed mathematics. These are not the final words on the matter; controversies about program proving still rage. But despite the enthusiasms of many prominent computer scientists, formal techniques for enforcing or proving correctness are often quietly sidelined.

2.4 Software as literature

> The literature view   A program is a document, expressing thoughts verbally. Software is written, read, and indexed; it is a literature.

The idea of 'writing' a program is much older than 'building' a program. It is still the most common description of the process, and it invites consideration of parallels between programming and other forms of writing. While some experts, such as Nicklaus Wirth, the inventor of Pascal and Modula 2, dislike even the term programming *language*, others see programming as true literary expression.

As an illustration of this perspective, Kernighan and Plauger's definitive book *The Elements of Programming Style* opens with these words:

> Good programming cannot be taught by preaching generalities. The way to learn to program well is by seeing, over and over, how real programs can be improved by the application of a few principles of good practice and a little common sense. Practice in critical reading leads to skill in rewriting, which in turn leads to better writing.

This point of view is founded on the literary view of software, and sees well-written code as the essence of a good software product. Kernighan and Plauger reiterate the same point of view in their *Software Tools* book, but there it is promoted from the preface to the front cover!

E. Horowitz, in *Fundamentals of Programming Languages*, speaks of a programming language as 'a systematic notation by which we describe computational processes to others'. This is something beyond a mere recognition that maintainability should be considered when a program is written. It suggests that a program should be addressed to a human audience first, and only second to the machine.

Later writers, like Carolyn Van Dyke, in her article 'Taking "computer literacy" literally', take the parallels between program writing and prose writing further. Van Dyke suggests five ways in which students (and teachers) of programming can learn from students (and teachers) of writing.

- Beware of rigid human standards beyond the explicit constraints of the language. Instead, learn from literate practitioners. Van Dyke cites Donald Knuth's attitude to the GOTO statement. Knuth is an acknowledged master of programming, yet he swims against the tide of wisdom that says GOTOs should be avoided.
- Learn style and syntax from texts that show the language in use. That is, read programs to benefit your own writing.
- Plan, write and rewrite in a recursive (*sic*) way. Don't expect to compose either linearly or in a top-down fashion, rather keep honing the program as you write.
- In teaching programming, emphasize the arts of invention; allow students some latitude in choosing topics, and freedom in their expression.
- Recognize that computer programs are a medium of human communication.

Reading programs to enlarge understanding is good advice, as is revising your program as you go (see Rule 2), and freedom of choice may increase student motivation. But some of Van Dyke's ideas may be overstated, especially if addressed to apprentice programmers who don't have enough experience with rules to be good at deciding when to break them. Her emphasis on literacy though, echoing Kernighan and Plauger and Horowitz, epitomizes the literature view of software.

GOTO: a historic battle

The opening shot in the GOTO wars was probably fired sometime in the early 1960s, but the most famous salvo was Dijkstra's letter to *Communication of the ACM* (*CACM*), 'GOTO statement considered harmful' in 1967 (see www.acm.org/classics/oct95). Dijkstra's argument was that 'GOTO *label*' – a computer-language statement that causes execution to jump to another statement prefixed with *label* – makes for tangled and hard-to-maintain code. The alternative is to structure the program by putting sequences of instructions into blocks, repeating and selecting between blocks through primitive statements, then building bigger blocks out of smaller ones. This is *structured programming*, and it is the universal way in which programming is taught (and mostly done) today. In 1987 GOTO was still provoking religious wars, even in the pages of *Communications of the ACM*, where there would be periodic exchanges about its virtues and vices. Today, although mechanisms for unusual jumps from one part of the program to another are still supported for exceptional situations, GOTO is effectively obsolete. On the other hand, COMEFROM has a charm of its own. See R.L. Clark, 'A linguistic contribution to GOTO-less programming', *CACM*, Vol. 27 (1984), pp. 349–350, reprinted from *Datamation*, December 1973, available at neil.franklin.ch/Jokes_and_Fun/Goto-less_Programming.html.

Donald Knuth (1938–) is perhaps the world's most respected computer scientist. His multi-volume series *The Art of Computer Programming* is as yet unfinished, but recognized as the field's first classic. He designed T$_E$X and Metafont – computer typesetting programs that are models of literate programming, respect for printers' art, culture and history, and the needs of mathematical authors. Knuth is an emeritus professor at Stanford. His webpage is www.cs-faculty.stanford.edu/~knuth.

Donald Knuth, cited by Van Dyke, has been a long-time advocate of 'literate programs'. He and his students developed a system called 'web' that allowed a programmer to compose code and commentary simultaneously, then automatically 'tangled' the program into object code for the computer, and 'weaved' it into a structured, typeset document for the human. Knuth himself wrote big programs using 'web', and two, T$_E$X and METAFONT, have been published as books.

In a 'web' program, the code is introduced in 'natural stages in the order its parts might have been written'. Each section begins with commentary and then presents modified Pascal code that implements something the size of a subroutine. Everything is cross-referenced via the typesetting facilities of T$_E$X.

At the outset, Knuth hoped his 'literate programming' paradigm would be widely adopted and lead programmers to better design, just as it had in his own experience. To give literate programming a boost, Knuth produced two programs for the 'Programming Pearls' column in *Communications of the ACM* (appearing in the May and June 1986 issues). An irregular column, 'Literate Programming', appeared in *CACM* for the next couple of years. Each column included a 'literate program' and a critique.

Knuth's article, followed by the five Literate Programming columns through to March 1990, provide a potted, rather depressing, history. The only people submitting 'literate programs' were those who had designed their own literate programming systems (of which 'web' was but the first). Most of the published programs contained bugs. Some were extremely hard to understand. Knuth himself was guilty in this, in that he chose to introduce a new data structure, the hash trie, within one of his programs. It is all very elegant, and if you struggle hard, you can appreciate the artistry, but many people surely preferred the more conventional solution to the same problem published later by D. Hanson. Hanson's program did have a gigantic bug, however, whereas Knuth's only had a little one. The 'Literate Programming' editor's tone in his final column is resigned: Literate programming is worthwhile, but perhaps its time is not quite come.

What, then, are we to make of literate programming? Granted that good style and comments are necessary, is it worth going further and writing programs as literature?

There are at least two insights from the literature view that are essential. First, it establishes the fundamental point, so obvious that it is easy to miss:

Principle 2 (The textual principle)

The text of the program itself – the code – is the canonical expression of the design.

A plot outline or a scenario does not capture the essence of a book – it is not literature. In the same way, project documentation is not the software – only the program is. The code is the essence, the core, the canonical representation.

The main consequence of this principle in this book is a constant emphasis on program code.

The second insight of the literature view is that programs are read so they have to be readable. Style matters. But what are the elements of style?

Both writing and reading are sequential processes. Programs, however, involve loops, conditional blocks, function calls. When you're developing a program these structures grow in your hands and you know how to move around them. But a reader doesn't.

Rule 3

Make your code reveal its non-sequential structure to a sequential reader.

Following Rule 3 makes you think explicitly about readability and style. It's a high-level rule that automatically draws in low-level good practice if you follow it. Chapter 6 discusses ways of putting Rule 3 into practice.

We will also raise to the status of a rule the aspect of literary style that has most impact on programming:

Rule 4

Write no more than necessary.

Or, to put it several other ways: Be concise. Avoid needless repetition. Remove redundant code.

Programming languages provide facilities like subroutines and functions so that a procedure that is done repeatedly, perhaps with different parameters, can be abstracted into a single piece of code. We might even say that a corollary of the first, mathematical, principle is that a key part of representation is structuring repeated, parameterized, operations. But from the *textual* point of view we want to strengthen that idea. Don't repeat yourself in your program. Wherever you find yourself writing the same few lines of code, replace both occurrences with a call to a function. Whenever you have a reader and a writer, an encoder and a decoder, a data collector and a simulator, look for ways to share as much code as possible. Reduce program size.

Measuring program size

A line of code is an often-used but ambiguous measure given the variability in the expressiveness of programming languages, programming style and vocabulary. The number of Non-Commentary Source Statements (NCSS) has been widely used to indicate program size, but the issue of what constitutes a statement is ambiguous. In C and C++, it has become conventional to count the semicolons; that is, the number of lines of code is defined to be the number of semicolons in the program. So

```
for (i = 0; i < MAX; *p++ = *q++, i++);
```

is three lines of code, and

```
i = 0;
while(i++ < MAX)
   {
   *p++ = *q++;
   }
```

is two.

But there's more to not repeating yourself than condensing code. The code is the canonical representation, but alongside it there is usually other documentation, from the comments in the program file to interface specifications and user guides. The problem with these different representations is keeping consistency when things change, and the best way to avoid inconsistency is to say things only once. Now this is often impossible: the whole point of design documentation is to provide a bridge between high-level requirements and the implementation so it has to explain the same thing in a different way. A user wants a guide to program use, not to maintenance. But be ruthless: avoid resaying things that don't need to be resaid. The place to begin is comments. Always try to make your code so clear that it doesn't need commentary. Always keep comments tight up against the code to which they refer, so that if you modify one, you're likely to see the other. See Chapter 6 for more on this.

2.5 Organic software

The organic growth view	A program is a large, complicated, invisible organism, whose development and growth is determined by 'programming'.

Seeing a program as a mathematical statement puts software in the realm of pure mathematics. As far as it goes, this approach seems to be closest to what a program really is. But it does not go far enough. It cannot cope efficiently or reliably enough with the large programs that are of most practical importance. Seeing a program as a piece of

Fred Brooks (1931–) was educated at Duke and Harvard Universities, then worked for IBM. In 1965 he founded the Department of Computer Science at University of North Carolina and has remained there since. His web-page is www.cs.unc.edu/~brooks.

literature encourages careful writing, and emphasizes the value of the code itself. But it also gives the impression that you've understood what you've written. Interesting programs encounter situations that even the best author has never anticipated. Later in the chapter, when the engineering view is outlined, programs will be compared to other engineered artifacts. But what of the possibility that maths, literature and physical engineering can't handle the complexity of software – that programs are so different from all other human creations, that they need a different kind of model? What about thinking of software as if it were alive?

Fred Brooks is Kenan Professor of Computer Science at the University of North Carolina, a central figure in the development of the IBM System/360 mainframe computers in the 1960s, and author of *The Mythical Man Month*, a classic collection of essays on software engineering. In the late 1980s Brooks chaired a committee of the Department of Defense in the USA, looking into prospects for improving software productivity. The results of that committee's work were distilled in Brooks' article 'No silver bullet' for *IEEE Computer* magazine. The silver bullet in question is the one that will slay the werewolf of monstrous software, and Brooks' conclusion is that there is no such thing.

In the course of the article Brooks gives his definition of software. That is, he identifies the things which make software what it is – its essence. (As mentioned in Chapter 1, Brooks' search for essence was the inspiration for this chapter.)

Brooks sees three irreducible qualities:

- Software's complexity.
- Its changeability.
- Its invisibility.

Brooks says that software is complex in a way that no other artifact is – no two parts are alike. More important, software elements interact in a non-linear fashion, so that as a program grows, its complexity increases much more than linearly. Unlike physical science, where simplified models of phenomena have been developed to abstract away the complexity, software complexity is inherent. Part of the complexity comes from the need to conform to standards, which, from the point of view of the program as a whole, are arbitrary. They detract from the unity of the program, because different aspects – e.g. user interface and file format – all have different kinds of complexity imposed on them by conformity.

A contrary view

The assertion that as a program grows, its complexity increases much more than linearly has been disputed by Edsger Dijkstra. In his 1972 ACM Turing Award Lecture, Dijkstra said: 'I tend to the assumption – up till now not disproved by experience – that by suitable application of our powers of abstraction, the intellectual effort required to conceive or to understand a program need not grow more than proportional to program length.' The whole lecture is well worth reading. It was given more than 30 years ago, yet it is has not dated. It makes astute comments on the state of programming both then and now. On the other hand, Dijkstra's prophecies about the future were off-base. His confidence in the mathematical view of software has not been vindicated. As for the comment about abstraction, even today he would have a hard time convincing authors of 100 000 line programs that complexity was merely proportional to length.

Brooks' second property, the changeability of software, is not a statement that other artifacts are immutable, but that the pressures for software change are so much stronger than for other objects. Successful software is pressured into new variants; 'small' adaptations are proposed which extend the reach of an existing product. At the same time, computer hardware so quickly becomes obsolete, that programs must adapt to survive.

Brooks did not pursue the changeability property as far as he could. Perhaps this was because most of his readers would already be familiar with its implications. As much as 75% of a project's costs can be absorbed by 'maintenance'. This is not just work to fix bugs, but is exactly the kind of changes Brooks was talking about – changes to meet new technology, addition of features, and so on. Many programmers spend their careers maintaining software. Design for maintainability is a good maxim.

To some degree, the changeability of software is simply a reflection of its nature as 'infinitely malleable thought-stuff', as Brooks calls it. This leads to the third of his properties – invisibility.

Software has no spatial existence. Where there are geometrical abstractions to describe software (for example, in visual programming environments), they are really just metaphors. The program itself is unvisualizable. The essence of software is that as data structures and algorithms, it represents conceptual, not physical, objects.

Brooks goes on to insist that many of software engineering's hot topics – object-oriented programming, artificial intelligence, environments and tools, etc. – attack only the *accidents* of software, not its *essentials*. He then gives four proposed attacks on the essence:

- Buy, don't build.
- Use rapid prototyping to refine requirements.
- Grow, don't build.
- Nurture great designers.

The first of these is very much the conventional engineering mindset – no reinvention. Brooks is simply emphasizing a traditional value that has perhaps been too often neglected in software. The second attack also appears in the other engineering disciplines, though it does not have the same importance as in programming. In this text, prototyping will form a part of the discussion of design methodology. Skipping the third point for a moment, Brooks' final attack is really a management issue; recognizing that the best programmers are far better than the average, he recommends rewarding them appropriately. See the sidebar on page 26 for more discussion.

Brooks' third attack on software essence is the most interesting. In a way, it is nothing new. He is talking about incremental development, top-down design, stepwise refinement – all buzzwords since the early 1970s. We have already seen echoes of the idea in Van Dyke's literacy principles. But Brooks' argument that a system should first be made to run, even if all it does is call dummy 'stubs' (empty subroutines), then fleshed out gradually, with stubs being replaced with useful code and perhaps a lower level of stubs, has an almost biological feel. The program is born, alive but helpless, then as each function is added, it develops organically. As soon as there is a working system, no matter how limited, the development has already taken its first leap towards maturity. If each subsequent step keeps the program alive, progress may be slow, but at least it is happening. Here is a principle derived from Brooks' argument:

Principle 3

A program grows along the framework of its first version.

What does this mean for practical design? At first sight it looks rather intimidating: if you don't get the first version right you may be inviting trouble later. It's certainly true that the high-level structure needs to be thought about carefully. But the principle is more optimistic than that: it's really saying that the first version can be just a framework, an exoskeleton, and not do very much.

Rule 5

Make the first version durable not functional, and get it running early.

Though it does little more than nothing, concentrate on making the first version a good framework for the final program, and make it run. The better the framework, the more likely that the first version will grow into a well-organized final version.

Now there can be arguments about this kind of incremental program growth. Sometimes staying with the framework of the first version might be in conflict with another design principle, as we shall see shortly. But even when this happens, rapid prototyping of a framework is valuable. As the case studies (Chapters 15 to 17) demonstrate, an early version that has to be completely discarded is not wasted: it provides a reference for testing as well as experience for the design of its replacement.

The picture of software as a growing, organic entity explains why many programs become sprawling monsters – they just keep on growing. What can be done about this? Assuming that a program is going to sprawl, the best that can be done is to limit the lengths of its tentacles. In other words:

Rule 6

Localize information.

In this book, Rule 6 trumps Rule 5. You may have got a framework that you believe is durable, running and supporting initial development. But if the next piece you add to a growing program will increase the complexity of interaction between its many parts, and you see how an alternative structuring would make the whole system more modular, with fewer, cleaner, interactions, then, although it will be easier just to add the new piece, you should almost certainly go for the harder, longer, restructuring. Redo the framework. In the long run, you'll count the benefit.

Rule 6, Localize information, is very important and powerful. Even so, as a program grows, the number of combinations of inputs, states and histories it can encounter grows much faster. Before long, you will not be able to monitor all the different situations that your program might sometime have to face. So it could and probably will have idiosyncrasies that you haven't anticipated. Just like a living organism. Part of the response will be extensive testing, but first it's important to forearm yourself against the fact that your program will develop behaviours you hadn't anticipated. This happens much smaller and earlier in a program's life than you might expect. Indeed, a moderately complicated written algorithm is likely to have subtleties that are not obvious (see page 221 for the story of binary search). So prevalent is this phenomenon, and so important to remember, that it deserves to be a principle:

Principle 4

Every program has surprises.

Perhaps that principle is overstated. Some programs are simple and transparent. But its spirit is true: there are surprises lurking where you don't expect, so you have to be watchful, sceptical about the correctness of your own code, and, above all, you have to understand the methods and algorithms you are using.

In scientific and engineering programming, it is common to use software modules developed by others. You have to make judgements about where you need deep understanding. Few engineers using Matlab need to understand how the program chooses between different methods for finding eigenvalues. But if your product has to find principal components of incoming multidimensional data several times a second, and use the result to control a helicopter, then even if you are using someone else's eigenvalue library, you had better understand exactly what it is doing.

How do you develop understanding of programs and reduce the surprise count? The answer is by experimentation. Later we'll have much to say about testing. A whole chapter is devoted to it. Suffice to say now that testing is not just about finding straightforward faults, it is also about subjecting programs to diverse and unusual situations and observing their reaction. This is standard empiricism, and it is necessary because programs *are* large, complicated organisms.

Rule 7

Test to discover information.

Your own code should be written to help you discover surprises. Ask, how can I make my program help me visualize what it is doing? Although a debugger might be the standard way of tracking the states of internals, you as designer know most about what will help reveal the operation of your code. This leads to a rule which couples the best way of localizing information with the need to support visualization.

Rule 8

Hide information in public; reveal it in private.

Information hiding is discussed at length in Chapter 12 and, in the context of C++, in Chapter 5. It says that as you localize information, you keep implementation details secret from other modules. This is a very good thing. But the flip side is that when you are concentrating on the internals of a module or a class, you want it to reveal its workings to you. Chapter 16's case study illustrates the importance of being able to visualize internals and expose surprises.

There is one more lesson from the model of organic software. Related to Brooks' identification of the changeability of software as part of its essence, it might be said that a changing specification is a fact of most software projects. That is, customers, users, *and* programmers often do not know exactly what they want at the start of a project, and their requirements change while the work is in progress. A good development methodology should be able to accommodate changes in the direction of program growth in response to changes in requirements. Although a specification change is often seen as the thing to be avoided at all costs, it is more realistic to assume that it will happen many times. Therefore, software designers should assume, as a fact of life, that:

Principle 5

The requirements will change.

So, however formally or informally requirements are specified,

Rule 9

**Design software to minimize the damage
caused by changes to requirements.**

2.6 Software design as engineering

The engineering view	A program is an artifact, designed systematically, applying science and mathematics.

Engineering has been defined many ways. Probably the following five points sum it up:

- Engineering is applying scientific knowledge and mathematical analysis to the solution of practical problems.
- It usually involves designing and building artifacts.
- It seeks good and, if possible, optimum solutions, according to well-defined criteria.
- It uses reliable components and tools; it invents only where necessary, by synthesis from existing 'building-blocks'.
- Engineering is practised according to well-defined principles and methods.

Notice that the size of the project is not mentioned. One can talk of engineering a coupling as legitimately as engineering an aircraft. Also, methodology, which is certainly part of the total picture, does not on its own define engineering.

The five-point list above immediately suggests an equivalent summary for software design:

- Software design is applying scientific knowledge and mathematical analysis to the solution of practical problems.
- It usually involves designing and building programs.
- It seeks good and, if possible, optimum solutions, according to well-defined criteria.
- It uses reliable components and tools; it invents only where necessary, by synthesis from existing 'building-blocks'.
- Software design is practised according to well-defined principles and methods.

The engineer stands, metaphorically, with one foot in the technology and one foot in the problem. (In civil engineering this can be literally true.) The technology is the components, tools and processes for the job in hand; it is also the engineer's scientific and mathematical skill set. The problem is the real-world situation, the environment, the resources, the need that has to be addressed. The engineer must handle the technology competently, but equally must have a deep understanding of the problem domain. In software design, the practitioner must stand on the same two feet. Yes, understanding of theory and skilful use of tools are essential. But a good program can only be written if the problem domain is properly understood. Chapter 10 is about analysing and understanding problems.

Principle 6

Good design relies on understanding the problem and its context.

Principle 6 and Principle 5 (The requirements will change) are in conflict, so the rules we derive must recognize that.

Part of problem definition is deciding on evaluation criteria. Criteria are measurable answers to the question 'How will I know I have succeeded?' (or, more generally, 'How will I know *how well* I have succeeded?'). The question must be asked and answered at the start, then revised as necessary. If it's true that 'The requirements will change',

---

Some might be tempted to summarize software design more cynically. Here is a state of affairs you might recognize, but which I hope you will not adopt!

- Software design is patching together preliminary solutions that are destined to become final products.
- It involves successive modifications of a program, converging via fragility towards terminal collapse.
- It seeks a solution that works.
- It uses earlier programs, which we now realize were poor, but which like many bad programs, worked at the time (thus reinforcing bad habits).
- It is characterized by tricks, corner-cutting and undocumented features.

then evaluation criteria will change too, though usually only incrementally. At every stage, criteria enable you to assess how good your solution is. A solution that works – a kind of existence proof – isn't enough. You are looking for a *good* solution.

Rule 10

Steer design by clear criteria.

Formulating success criteria at the outset demands that you form a clear understanding of the problem and its context. Adapting them as you go along ensures they are up to date with your current understanding of what the project has to do. Making them explicit and prominent marks a course for design. Criteria are beacons. The most successful projects are those that steer by the beacons, where distractions don't cloud the clarity of what will count in the end.

The engineering perspective yields another principle, already anticipated in Brooks' 'Buy don't build', but more far reaching:

Principle 7

Something similar has already been done.

A fact that leads immediately to our next rule:

Rule 11

Exploit existing solutions.

Use existing code; existing techniques; existing methodologies. Where possible design by synthesis from reliable existing building blocks.

There is a corollary to Rule 11, so important that it is almost a rule in its own right:

The software corollary (Rule 11a)

The algorithm is in a book.

Practising software designers need to know how important algorithms work, where to find information on particular types of algorithms, and how to understand algorithm descriptions and performance analyses. It is very unusual for a programming job to require the invention of a substantial algorithm.

What about the last part of the engineering definition for software design? It mentions principles and methods. Principles are being developed in this chapter. Methods and methodology are frameworks for design. The use of a methodology can be summed up as: Design systematically. Design according to a system. In other words, 'Get organized'. We are going to take this advice to heart. Chapters 9 to 14 are about design methodology, and Chapter 3 is about management of your skill as a programmer.

The engineering perspective and 'software engineering'

Software engineering was originally a response to the 'software crisis' of the late 1960s, when it became clear that conventional programming techniques could not scale up well to large projects. It was (and is) concerned with the search for techniques to address this problem. Structured programming, high-level languages and time-sharing were originally seen as software engineering issues. The collection of articles entitled Programming Methodology, edited by David Gries, is a good illustration of this early perspective. But in the 1970s and 1980s, software engineering became identified with higher-level methodology and process.

Consider the higher-level perspective for a moment. The development cost for the initial release of a middle-sized program like a personal computer word processor or spreadsheet (which may have, say, 200 000 lines of non-library code) might be as little as 20 person years. This could have no relationship to the eventual revenues. What is certain, though, is that the larger the program, the higher the cost *per line* of code. Development cost increases much faster than program size. Very large programs, such as those for electronic telephone switches, are longer than 20 million lines. They never stop growing and changing, and keep hundreds of programmers perpetually busy.

Several studies have shown that, as well as cost per line, two other factors change as program size increases: the difference in productivity between the best and the worst programmers decreases, and the proportion of development time spent in defect removal increases. The details of one study (Capers Jones, for IBM, 1977) are summarized in Table 2.1. For these results, programming was in assembler language; a line of code corresponded to every non-commentary source statement. The lengths of time implied by this table seem staggering at first. One example Capers Jones gives in detail is of a 750 kline program consisting of 75 components, each with a team of five programmers (375 technical programmers in all). There were 30 user reference documents maintained (6000 pages) plus 75 component and maintenance documents (7200 pages).

Capers Jones' results suggest a simple management strategy: for small programs, invest in the best programmers because their productivity is substantially greater than average programmers; for large programs invest in organizational solutions, aimed particularly at reducing the defect removal time.

What are some of these organizational solutions? Certainly methods are needed for keeping track of the huge amount of information embodied in a program. Documentation, version and change control are software engineering issues that apply here. More fundamentally, all developers want to produce programs that are correct, robust, efficient, maintainable. Software engineering techniques for specification, testing, inspection are aimed at this.

Again, particularly for large systems, it is important that the cost and schedule be kept under control – software engineering is concerned with planning, estimating and tracking a project. Furthermore, there is increasing emphasis on speeding up development time. The need for new products quickly (and frequently) has driven the introduction of design methodologies geared to fast productization. For medium-sized software products, where cost is less of a concern than making the right product for the market, companies are prepared to spend more to achieve the same results faster. As a result, the emphases in software engineering are changing. The 'Mythical Man Month' is around – software development cannot be speeded up by adding people to a lagging project. But starting with a larger team might make sense, if there is confidence that it will speed up development. Subcontracting software development, using toolkits and object libraries, overlapping phases of the development process – all of these are things a company might try to reduce time to product. So the old emphases of software

Table 2.1 *Capers Jones' results for IBM programmer productivity*

Size of job in klines of code	Best programmer (months/kline)	Worst programmer (months/kline)	% time spent in defect removal
1	1	6	40
8	2.5	7	60
64	6.5	11	60
512	17.5	21	70
2048	30	32	75

engineering – getting the specification right and rigorous, maintainability, proper testing – remain; but they are joined by a new urgency that development should be *fast*.

So far as this book is concerned, the traditional software engineering aims motivate many of its techniques and methods. In particular, the book emphasizes rapid development in the context of a pessimistic view of software specification – 'the requirements will change'.

Despite the similarities with the aims of conventional software engineering, the treatment in this book diverges in several ways. First, it maintains that the distance between a reasonable definition of engineering and 'software engineering' is too great, and it aims for the first rather than the second. For example, it tries to apply some of the principles and attitudes that characterize the practice of engineering to day-to-day software design. Second, it places considerable emphasis on the detail of handling the technology – writing programs. This is a consequence of the definition of software as literature. In general, the anchoring of the book in specific principles derived from various definitions of software (other than the engineering perspective) puts the book somewhere between traditional software engineering and more radical Extreme Programming perspectives.

2.7 Putting the program in its place

> The system view   A program is an organized set of control components in a complex system.

The systems perspective recognizes the complexity of the physical environment where a program operates. Rather than being concerned with the mystique of software itself, it looks at the properties of the whole system in which a particular program is embedded.

One way of thinking about this perspective is to say that it is hardware oriented. Engineers concerned with embedded controllers, for example, bring a kind of bottom-up view to software: they know how semiconductor devices work, then how they are used to make logic gates, then the principles of combinatorial logic design, then sequential digital systems. They then see programming as the top layer of control of a particular flexible architecture. Figure 2.1 illustrates the layered model that an electronics designer has of a computer.

The hardware-oriented perspective is important to the relatively small group of programmers who are working 'close to the machine', perhaps designing control circuits, perhaps programming in assembly language. But the system view also has layers above the program (such as a human user) and other components around it which use it, feed it information and are controlled by it in a multitude of ways. Any application program may be insulated from the hardware by several layers of operating system, device drivers and so on, so there are many interfaces. Figure 2.2 illustrates the situation for your module (i.e. your part of a program) in a typical workstation application. For other systems (e.g. in an embedded controller, avionics system, or robot) the block diagram will be different. But in each case your module will be surrounded by others. Systems theory, and, more generally, systems thinking, analyses the whole picture.

The system view provides a series of insights that we will turn into three principles. The first is obvious – when you look from the system perspective.

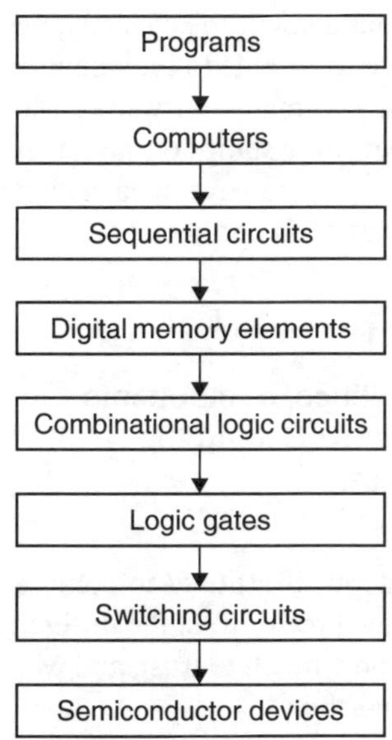

Figure 2.1 *Programs from a hardware perspective*

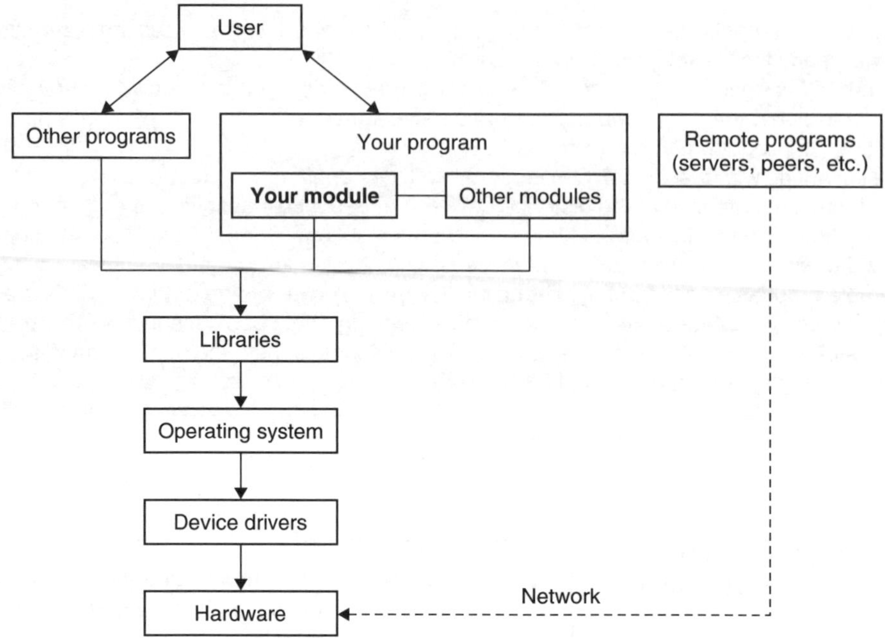

Figure 2.2 *Programs from a system perspective*

The words 'module', 'component' and 'subsystem' are synonymous here.

Principle 8

Every program is a component in a bigger system.

No program is an island. The system perspective makes us consider the implications.

The good designer will know enough about the capacity, performance and costs of a system as a whole to be able to include it in quantitative analysis of possible designs. Too many software projects get too far before someone realizes that an assumption about the platform's capabilities was incorrect – and could have been shown incorrect by a simple calculation at the start of the project.

Rule 12

Understand the capabilities, components and mechanisms surrounding the program.

Practical design often requires a sceptical attitude to some components of the system. You need to know how to trust the different parts of the system. That is, what to trust, how much to trust, and what to do when a component appears untrustworthy.

When you ask a component to do something, there are several things that can happen. What you hope is that the component will be able to do the thing, it will do it, then tell you that it's done. That's one possibility and usually the best. But there are others (assuming the component is supposed to tell you of its success or failure), and we can characterize what the component does as follows:

The silence	If the component never gets back to you, you are stymied. You don't know if it was successful or not. The component is faulty and a danger to the whole system.
The lie	If the component was successful and tells you it wasn't, or if it was unsuccessful but tells you it was, you have something that is worse than useless: your program state and the component's state are mismatched and disaster could result. The component is faulty and a danger to the whole system.
The wimp	If the component was able to do what you asked, but didn't (and told you it didn't – so no lie), the component is faulty, but can perhaps be worked around. The system as a whole is suboptimal, so the component should be repaired.
The polite refusal	If the component is not able to do as asked and tells you so, it is working correctly, but you may have a higher-level problem with the inability to perform.

Components are often other pieces of software, and it is important to be able to gauge their trustworthiness.

As a rule of thumb on a typical workstation:

- The hardware is not responsible for silences (program hangs, functions that don't return) because they would manifest across all programs; it does not lie (same reason); it never wimps out (because being able to do something is effectively the same as doing it at any particular time); and it often politely refuses, but almost all refusals are hidden from an application programmer by drivers and the operating system.
- The operating system is usually not responsible for silences; it usually does not lie or wimp out; and it sometimes passes polite refusals up from the hardware or generates its own.
- The compiler is very trustworthy. It has been so heavily tested that silences are all but eliminated, it won't lie, it will never wimp out, but it will make you many refusals if your program isn't perfect syntax (and they won't be all that polite either!). If you suspect your compiler of causing a problem, you are probably (but not certainly) wrong.
- Software libraries probably won't lie or cause silences (hangs) and they almost certainly won't wimp out. However, they are not as trustworthy as system or development software, so you might be

justified in suspecting a lie if you can't find the problem in your own code.

- Code that is under development is by far the most untrustworthy. Not only is this because it is new, it is also because it is most likely to suffer from poor error handling. That is, code under development might call reliable system code for a task that the system is not able to do. The system responds saying that is was unable to perform the task, but the new code doesn't properly check. Instead it carries on as if there had been success and passes on that message to other components. In other words, it lies on another, innocent, component's behalf.

However:

- Hardware under development is far less trustworthy than a regular workstation. Part of fulfilling Rule 12 includes knowing how to assess the platform's reliability.
- Old operating systems are more likely to be responsible for silences than new ones.
- Combining software libraries can expose errors in them that their developers missed, so you would be right to be more sceptical about combining open source libraries from many different places, for example.

You do not want to waste effort mistrusting reliable parts of the system. But it is far worse to assume something works that doesn't!

Rule 13

Program defensively.

Programming defensively means checking for errors, reporting on unexpected situations, providing means for debugging and monitoring. In particular, you should not trust parts of the system that are being developed or combined in new ways. And of which part of the system is that most true? That's right! Yours!

Well, how can you mistrust your own component? That sounds slightly schizophrenic. It is not something you can accomplish in the code itself, beyond prudent, careful programming. But there is something you can do to prevent the embarrassment of your component being detected as untrustworthy by other modules. It's time to codify a vital practice:

Rule 14

Test to break.

Or, to put it another way, 'Test *very* hard'. Do everything you can to make your software go wrong, to uncover obscure faults, to surprise your program so it doesn't surprise you! Testing is a hugely important stage in software design. Chapter 14 explains how to do it.

So how trustworthy are compilers?

A compiler is a program that converts source code into an executable. You have to trust it with your program if you're ever going to see it run. If you've written without syntax errors, the compiler will do its job. It won't protect you against errors in the logic of your program, but it will translate your code faithfully. Or will it?

Compilers are big and complicated. They use elaborate algorithms and maintain complex data. So there's considerable potential for bugs inside the compiler. But they are developed by elite programmers, tested exhaustively, and often used to compile themselves, so bugs should have been shaken out before the compiler gets to your program.

What won't get shaken out is a Trojan horse like the one Ken Thompson described in his 1983 Turing Award lecture. In designing the Unix C compiler, Thompson included a special test for the string 'login:'. This would detect when the Unix login program was being compiled. The compiler would then insert code that would allow Thompson to log on to any Unix system using a special user name. So the compiler covertly inserted a backdoor into any Unix system. Devious enough? But then, Thompson covered his tracks. He also added to the compiler a test to see if it was being used to compile itself. If so, it inserted the code for detecting login strings *and* for detecting compiler compilation in the generated executable. He only had to compile this version of the compiler once. Then he could remove the two special tests from his source code. Why? Because the next time the compiler was compiled, the two tests would be reinserted in the executable. And so on through all generations of a seemingly innocent compiler.

The moral is that compilers are among the most trustworthy software. It's compiler writers you have to be careful of.

So far we've talked about the system perspective as a static thing: components interfacing, requesting things of each other, exchanging messages. But the definition of program as an organized set of control components in a complex system takes us further. Systems theory, process control and cybernetics are concerned with describing how components interact in a system in terms of their states, inputs and outputs. Often they deal with the design of a control component that must act to stabilize the system after some event. In developing analysis methods, systems theory has adopted a principle of parsimony (simplicity) that has direct relevance to software.

Principle 9

Every system can be analysed and designed in terms of the interaction of a small number of concepts.

In digital computation, we can also characterize what goes on in the whole system with just a few concepts. Let's go straight to a rule for software design inspired by this perspective and see what it means. This is perhaps the most unusual and controversial rule in this chapter.

Rule 15

Design your program around these six sufficient concepts: states, events, conditions, actions, objects and types.

This is not a rule that can stand alone without further explanation. But it turns out to be one of the most powerful, as Chapter 11 shows. So stick with it!

Systems, Wiener and Cybernetics

The systems view has many strands. *Cybernetics* is the name coined by Norbert Wiener (1884–1964) to describe a particular approach that is either a central strand, or the whole subject, depending on your interpretation of Wiener's work. He argued that control systems, assembly lines, bureaucracies, biological systems (plants and animals), the whole universe, are feedback systems reliant on the exchange of messages. A learning component is one that changes its responses based on feedback from the rest of the system. Wiener's 1948 book, *Cybernetics: or Control and Communication in the Animal and the Machine*, was highly influential, launching cybernetics as a discipline and providing theoretic tools. Through the 1950s and 1960s cybernetics rose swiftly then declined, becoming reinterpreted later in the twentieth century as including neural networks, evolutionary computing, and other modern adaptive processing techniques. The IEEE still has a Systems, Man and Cybernetics society and prints a journal of that name.

The first thing the rule says is that the raw 'thought stuff' of a program is of six kinds. Objects and types are to do with nouns – an object is a particular thing, and a type is a class of things, so 'human' is a type and you are an object of that type. 23 is an object of type integer. Events and actions are verbs. Events are things that happen to the program, actions are things that it does. States are states of affairs – collections of adjectives if you like, that describe the current situation of the program (or subprogram). Conditions are like specialized adjectives – the way particular individual objects are at a particular time. The whole operation of the program is characterized by it being in a particular state, an event happening, the program doing actions in response and perhaps moving to a different state. The response to an event may also depend on an internal condition.

The second thing the rule says is that these six concepts – states, events, conditions, actions, objects and types – are sufficient for describing any program. However, they are not all necessary concepts. For many programs, everything that goes on can be expressed with just some. For example, a typical algorithm is a standalone procedure that just does things, one after the other, on some data. So it can be described in terms of objects and their types, actions, and, perhaps, conditions.

But the third thing the rule says is to shape your program around all six concepts. Despite their not all being *necessary* for expressing what a program is to do, the principle's standpoint is that using all is *beneficial*. We will develop this idea in Chapter 11, illustrating the use of state-oriented thinking even for standalone algorithms. This provides a different perspective on procedures that can help catch errors in thinking.

The system perspective has been fruitful in yielding design rules (even if Rule 15 brings in ideas specific to software). But it has one more gem of insight that we should capture. Systems can be described with simple concepts, but the interconnection of components makes them complex. A system designer or a control engineer will select components that have continuous, bounded, and, if possible linear responses. Any component that has an abrupt change in behaviour, dependent on an input crossing a particular threshold, is an additional complication in an already difficult problem. The best components are those that don't have parameters to be adjusted or tweaked – their behaviour is simple, smooth and predictable. Software systems are directly analogous. Here is a rule that may be difficult or impossible to adopt in many circumstances; but when you can,

Rule 16

**Remove magic numbers and thresholds.
Reduce the number of parameters.**

A magic number is a constant such as the length of a buffer that different parts of your code may share. If possible, make the buffer of variable length then you don't need the constant. Some constants are relatively benign: the stopping condition for an algorithm might be a number of iterations, but even here it's often possible to remove the dependence on a fixed number. Thresholds are typically part of an algorithm, included as a heuristic to deal with special cases or to speed things up. But where they can be removed or replaced with a smoother kind of transition, do so. You program will behave more 'linearly' and predictably from the outside. Parameters in general are values that specialize a general piece of code for a particular occasion. They are therefore essential to writing flexible modules. But whenever a method relies on tweaking a parameter for calibration, the product quality and the likelihood of errors are improved if there is a way of dispensing with that tweak.

2.8 User-centred design

> The user-centred view   A program is an interface that makes a computer usable for a particular task.

The last 40 years have seen the introduction of a huge number of unusable consumer products. Almost everyone has experience of difficult-to-use watches, clocks, VCRs, microwave ovens, mobile phones. Yet the problems of bad design in user interfaces have been around much longer. The psychologist Donald Norman includes ovens, doors, light switches in his list of poor interfaces. People's everyday exposure to frustrating, inefficient and just plain bad interfaces has led to increased interest in 'User-centered design' (Norman's phrase).

The user-centred designer would look at the other definitions in this chapter as missing the whole point – that programs are things that enable *people* to do things. Unless users are in the picture, then all this talk of essence is really accident – the program will not fulfil the practical purpose that people want to use it for. The software designer therefore needs to have a user-oriented perspective too. The problem is this:

Don Norman has teamed up with another usability guru, Jakob Neilson, to form the Neilson Norman group. He is also affiliated with Northwestern University. Information about him can be found at www.jnd.org and there are still old pages at cosci.ucsd.edu/~norman.

Principle 10

Users are complex.

There are hundreds of usability companies, consultants and conferences. The huge literature in Human Computer Interaction (HCI) ranges from physical ergonomics to the effect of emotion on computer users. And all this is just a part of the world of psychology research and application, proving that humans, and therefore users, are complex.

However, there are two things you can do.

Rule 17

Determine what users need.

It is essential to talk to your users and, if possible, work with them in developing a product for them. Your users have their own expertise.

Programming for scientific and engineering use

There are a huge number of scientific and engineering software applications. Many of these involve large pieces of software, commercially developed, for data analysis, modelling, simulation, graphics or numerical analysis. If you are working on one of these systems, you will probably be contributing a module to a comprehensive application which does many things. However, if you are in a small team developing custom software, perhaps for research, you are likely to be programming with a software tools approach.

In the software tools approach to programming, you follow Kernighan and Plauger's dictim:

> Make each program do one thing well.

I would have loved to have included that as one of my principles in this chapter, but time has moved on, and many programs today are large integrated applications with a graphical user interface. For a word processor, photo editor or CAD package, that is the appropriate structure. However, for many applications, the software tools approach is still the right one.

Programming with software tools involves the creation of relatively small programs that are then used in combination to achieve big effects. A software tools environment is typically a command-line environment, like a Unix shell or the Command Prompt under Windows (which emulates the old DOS prompt), where individual programs are invoked by naming, parameters are passed to programs via arguments on the command line, and the output from one program can be piped to the input of another. The command-line interface will support a scripting language, so that programs can be combined in sophisticated ways.

For some of the programs later in this book a software tools approach is assumed. The programs take arguments and run with minimal interaction. Part of the reason for this is that there is no space to develop graphical user interface versions of the programs, and in any case GUI code is platform dependent. But a deeper reason is that scientific and engineering programming of the type exemplified in these programs is best done under a command-line interface.

One of the biggest mistakes a developer of scientific and engineering software can make today is to believe that the command-line, scripting, and software tools approaches are dead. Because all common office applications are integrated and have GUIs, it becomes easy to think that a data-analysis, science or research package should look and feel similar. For work that involves a sequence of structured operations – like doing scientific experiments – the non-linearity of a GUI is an impediment. A command-line interface is far more efficient and flexible. It is no accident that the focus of Matlab's GUI is a command line. A nicely ironic twist to the old battle between MSDOS (command line) and the Macintosh (graphical interface) is that some university departments are today making buying decisions in favour of the Mac, because it now supports a powerful command-line interface.

See Jerry Peek, *Why use a command line instead of windows*, linux.oreillynet.com/pub/a/linux/2001/11/15/learnunixos.html, for more on the trade-offs between command-line interfaces and GUIs.

Sometimes they are experts in a particular field and their domain knowledge is as valuable to getting the program right as your software expertise. Almost always, they will know how they will get maximum benefit out of what your program could do, and so can advise you on what it should do. Sometimes it is possible to make thoughtless assumptions (particularly in respect of user interface, as the sidebar discusses) that saddle your users with difficult ways of doing the very actions that are central to their purpose. Often they are better able to identify such pitfalls long before you do. Just possibly, through working on the program, you might be able to suggest a way that they could change their working practices to be more productive, and embody support for the change in your software. In that case, designing alongside your users is invaluable: they are much more likely to heed your advice and perhaps take up your bright idea.

Of course, sometimes your user will be you and you alone. In the user interface, treat yourself well. In six months' time you want to be recommending the program to colleagues, not blaming your past self for an impenetrable user interface. This brings us to the more general rule,

Rule 18

**Understand how users will understand
the program.**

While it is impossible to give a complete picture of even the main issues in human–computer interaction in a book like this, we can make some comments about a 'generic user', i.e. a human. Chapter 10 provides some qualitative guidelines, chiefly concerned with pattern, visibility and mental models. The intent is at least to convey the necessity for humility on the part of the designer. Inefficient, error-prone, unhappy use of a program by a user is a fault of the design, and it is arrogant to bludgeon the user with commands like 'Do not attempt to use this product before reading the manual', when they just paid for it.

2.9 The craft of program construction

The artisan view	A program is a hand-made artifact, fashioned by a craftsperson, who applies creative judgement and experience to achieve beauty and utility simultaneously.

Programming has often been seen as an art or a craft. This is a valuable perspective because it helps in understanding programming and programmers.

The craftsperson embraces all the artistic virtues appropriate to their medium that do not conflict with utility. Clarity, style and beauty are all important in creating good artifacts. They are sought not for their own sakes, but because things work better if they are well crafted. Through formal learning, apprenticeship and practice, a craftsperson develops skills in choosing and handling materials and tools, adapting to – indeed exploiting – the peculiarities of each design situation, and consistently achieving a polished finish to their products.

Principle 11

**A good artisan has ability, knowledge,
skill and good judgement.**

Principle 11 summarizes a common, and probably accurate, belief about the constitutive parts of good craftsmanship. Native ability is a factor, but knowledge and skill can be acquired. Good judgement relies on coordination of all three.

Good programs are carefully shaped, elegant and economical. They display the work of good craftsmanship. How do you become a good artisan? Your native ability is a gift – you're stuck with it. But the other attributes are things that enable you to use whatever ability you have to greatest effect.

Rule 19

**Add to your knowledge and skills as you design.
Manage your knowledge and skills to build
good judgement.**

This rule has more to do with attitude than process. Chapter 3 is devoted to it, looking at the implications of seeing yourself as a craftsperson in the practice of software design. Because the *practice* of craft is fundamental to it, this book goes out of its way to encourage a few 'small' habits. The rationale is that if you, the reader, can be persuaded to practise them, they will be more valuable than wider-ranging knowledge that rarely gets applied. Procedural knowledge – skill – that is acquired by practising design needs to be guarded and exploited.

2.10 Programmers' programming

The organizational behaviour view	A program is the consequence of programmers programming.

The artisan view has a flipside. For the point of view of organizational behaviour, we shouldn't start with a definition of 'program' or even 'programming'. Rather, the program*mer* is the key. A program is the consequence of programmers' programming. Yes we need to understand what makes a program good, but our efforts should then be put into getting programmers to do what they do as well as possible. In these terms it doesn't even matter if the artisan (or any other) point of view is somewhat misguided – the point is whether, thinking of themselves as craftspeople makes programmers more or less productive.

Let's suppose we are software project managers. The program is what will eventually pay the bills, but our prime concern now is what can we do for programmers, so that their work on the program is effective, efficient, excellent. We may be aware of the design principles that they will work by. But is there a principle for getting the best out of them? Can we provide processes that leverage mathematical skills, develop stylistic fluency, promote demanding testing at every level?

We have already touched on the issue of individual ability. Brooks advocated hiring and rewarding great programmers; Capers-Jones showed that for small projects, individual ability strongly influences productivity. The first thing the manager has to face is that programmer ability does vary significantly. But then, given a team, what procedures and practices will accelerate development? Some candidates will be discussed in Chapters 9 and 10, though it must be admitted that it isn't always certain what works best. What does seem to be universally agreed, though, is that programmers' psychology plays a significant role in short-term productivity. Managers need to structure the project so that individuals can achieve, enjoy each other's help and encouragement, see progress towards goals, and think.

Principle 12

**Productivity depends on ability, morale
and environment.**

A manager should foster morale and provide a supporting environment. The way a team is organized has a significant effect. A practical rule that appears to apply is:

Rule 20

**Foster the sense of ownership that
programmers have of a project.**

A traditional group structure is a 'chief programmer team' which uses a surgical analogy to divide up jobs. In this structure it is not at all clear that those who are not the chief programmer will feel such strong ownership, or commitment, as the one who is. In contrast, a team organized along the lines of promoting individual ownership of parts of the program may be more democratic, but perhaps the solitary artisans do not give us as much as they would if they worked together. Another alternative is pair programming, where all coding is done by two programmers sitting side by side at a computer. The pairings change regularly so that it is impossible for anyone to feel ownership of any particular piece of the project. Yet proponents of pair programming claim that it engenders a much stronger corporate sense of ownership of the whole thing. The jury is still out on the best way to accomplish Rule 20.

Some of the other principles can be harnessed in a team setting to sustain morale. Incremental development enables programmers to demonstrate their achievement by advancing the working system. Designing with users who are not immersed in the minutiae of the software can inject fresh perspectives on a team's activity. Promoting a positive attitude towards defensive programming and hard testing can replace the frustration of later maintenance with collaborative removal of problems as pieces of software are integrated.

2.11 Living with ambiguity

> The multiple analogy view   A program is an X $\{X = $ movie, research paper, communiqué, ...$\}$.

In the end we could be forgiven for admitting that we don't know what software really is. Not only do the multiplicity of perspectives show that it's very difficult to get a handle on, but we can also cite other ongoing puzzles. For example, should software be patented or copyrighted? What does the answer say about the kind of stuff it is?

In an article in the May 2002 edition of *IEEE Computer*, Kyle Eischen, a sociologist, suggests that the analogies we use to describe a method can lead to confusion but, if spelt out, can also lead to better understanding. He features 'Software as university', 'Software as Hollywood' and 'Software as construction' as examples of analogies that are only partly accurate but provide models for key questions and assumptions. For example, if software development is seen as the interaction of independent, creative, professional thinkers (as a university), then peer review, subjectivity of learning and research, apprenticeship and mentorship might map over from one domain to the other to explain the social activity of software design. Similarly, if software is Hollywood, and a program is a movie, does that give us any insights into design team organization, quality evaluation, and struggles over content and distribution?

Almost all analogies of what a program is give a model for how different people interact in the preparation and use of software. That there are multiple analogies shows that software means different things to different people, and many attitudes and goals will be brought to bear on a particular programming project. This multiplicity is (almost) an essential feature of software design, and it manifests itself socially as well as in the range of programming types and styles that you will encounter. Anyone who has tried to integrate code from a gifted young hacker with the textbook examples of a computer science teacher will know that the gulf is wide, yet each is offering a valuable resource to be harnessed. Software's people are eclectic, and the cross-disciplinarity of so many projects means software products are rich in diversity. That's a problem for maintenance and standardization, but it also means that the software designer's richest resource is other people. A principle and a rule, even for the solitary coder:

Principle 13

Software design is a social activity.

Rule 21

**Develop the ability to apply other
people's ability.**

2.12 Summary

This chapter has looked at software design from nine different perspectives, each offering a different definition of program. From these, we have drawn 13 principles and 21 rules that will be used to guide the rest of the text. For convenience they are repeated here.

THE MATHEMATICAL VIEW

Principle 1: Software design is representation.

Rule 1: Begin every step in software design by asking: *How* should this information, this activity, this state, this relationship be represented?

Rule 2: Don't expect to get the representation right first time. Look for opportunities to improve code by *re*-representation.

THE LITERATURE VIEW

Principle 2: The text of the program itself – the code – is the canonical expression of the design.

Rule 3: Make your code reveal its non-sequential structure to a sequential reader.

Rule 4: Write no more than necessary.

THE ORGANIC GROWTH VIEW

Principle 3: A program grows along the framework of its first version.

Rule 5: Make the first version durable not functional, and get it running early.

Rule 6: Localize information

Principle 4: Every program has surprises.

Rule 7: Test to discover information.

Rule 8: Hide information in public; reveal it in private.

Principle 5: The requirements will change.

Rule 9: Design software to minimize the damage caused by changes to requirements.

THE ENGINEERING VIEW

Principle 6: Good design relies on understanding the problem and its context.

Rule 10: Steer design by clear criteria.

Principle 7: Something similar has already been done.

Rule 11: Exploit existing solutions.

THE SYSTEM VIEW

Principle 8: Every program is a component in a bigger system.

Rule 12: Understand the capabilities, components and mechanisms of the system surrounding the program.

Rule 13: Program defensively.

Rule 14: Test to break.

Principle 9: Every system can be analysed and designed in terms of the interaction of a small number of concepts.

Rule 15: Shape your program around these six sufficient concepts: states, events, conditions, actions, objects and types.

Rule 16: Remove magic numbers and thresholds. Reduce the number of parameters.

THE USER-CENTRED VIEW

Principle 10: Users are complex.

Rule 17: Determine what users need.

Rule 18: Understand how users will understand the program.

THE ARTISAN VIEW

Principle 11: **A good artisan has ability, knowledge, skill and good judgement.**

Rule 19: **Add to your knowledge and skills as you design. Manage your knowledge and skills to build good judgement.**

THE ORGANIZATION BEHAVIOUR VIEW

Principle 12: **Productivity depends on ability, morale and environment.**

Rule 20: **Foster the sense of ownership that programmers have of a project.**

THE MULTIPLE ANALOGY VIEW

Principle 13: **Software design is a social activity.**

Rule 21: **Develop the ability to apply other people's ability.**

2.13 Chapter end material

Bibliography

The references given at the end of this, and every, chapter tell where to find the things explicitly referred to in the chapter, plus some other (usually introductory) references that are directly relevant. Although there may be many significant papers and books in a particular area, in general only one or two are given. These almost always have good lists of references themselves.

The mathematical view

Hofstadter, D. R. (1979). *Gödel, Escher, Bach: An Eternal Golden Braid*, New York: Basic Books Inc. A readable, and very popular, introduction to ideas about mathematical systems, theorem proving, autonoma, and much else.

Gries, D. (1981). *The Science of Programming*, Springer-Verlag. One perspective on the mathematical view (scientific view?) of programming, which references much of the important work in the field. Highly recommended.

The literature view

Kernighan, B. W. and Plauger, P. J. (1978). *The Elements of Programming Style*, 2nd edn, McGraw-Hill.

(An updated treatment with a similar theme is

Kernighan, B. W. and Pike, R. (1999). *The Practice of Programming*, Addison-Wesley.)

Kernighan, B. W. and Plauger, P. J. (1981). *Software Tools in Pascal*, Addison-Wesley.

Van Dyke, C. (1987). 'Taking "computer literacy" literally', *Communications of the ACM*, **20**, pp. 366–374.

Knuth, D. E. (1984). 'Literature programming', *Computer Journal*, **27**, pp. 97–111.

Bentley, J. (1986). 'Programming pearls: literate programming', *Communications of the ACM*, **29**, pp. 364–369. (See also the follow-ups in 1986, pp. 471–483 and 1987, pp. 284–290.)

Horowitz, E. (1983). *Fundamentals of Programming Languages*, Computer Science Press.

Schneider, B. R. (1985). 'Programs as essays', *Datamation*, **30**, pp. 162–168.

The announcement for the 'Literate Programming' column in *Communications of the ACM* was on page 593 of the July 1987 issue. The first article in the series (pp. 594–599 of that issue) has D. Hanson's solution to the problem that Knuth solved in the Bentley reference above. Hanson's error was pointed out in the December 1987 issue, page 1000.

The organic growth view

Brooks, F. P. (1987). 'No silver bullet: essences and accidents of software engineering', *IEEE Computer*, **20**, pp. 10–19.

Much of the literature on Extreme Programming could be said to derive from the organic growth view. The focal point of online information is Ward Cunningham's WikiWiki. Cunningham is the leading figure in Extreme Programming, and his pages are a delight. Advocacy and debate (as well as an explanation of Wikis) are at c2.com/cgi/wiki?ExtremeProgrammingRoadmap.

Some good print references are:

Beck, B. (2000). *Extreme Programming Explained*, Addison-Wesley.

McBreen, P. (2003). *Questioning Extreme Programming*, Addison-Wesley. It has to be said that McBreen's questioning is not particularly aggressive.

Stephens, M. and Rosenberg, D. (2003). *Extreme Programming Refactored*, Apress. Gives a genuine, funny, critique of extreme programming.

The engineering view

The two big 'standard texts' are:

Pressman, R. (2000). *Software Engineering: A Practitioner's Approach*, 5th edn, McGraw-Hill.

Sommerville, I. (2000). *Software Engineering*, 6th edn, Addison-Wesley.

Others I have found helpful include:

McDermid, J. A. (ed.) (1993). *Software Engineer's Reference Book*, CRC Press. Has pointers to many other sources.

Mynatt, B. T. (1990). *Software Engineering with Student Project Guidance*, Prentice-Hall.

The references mentioned in this Engineering Perspective sidebar are:

Capers Jones, T. (1977). 'Program quality and programmer productivity', IBM Technical Report TR02.764.

Capers Jones, T. (1986). *Programming Productivity*, McGraw-Hill.

The system view

Tanenbaum, A. S. (1998). *Structured Computer Organization*. Illustrative of the hardware-oriented perspective on computers.

The system view is broad, ranging from control theory and cybernetics to so-called 'soft systems'. The Principia Cybernetica Web gives a good overview and collections of links: http://pespmc1. vub.ac.be/CYBSYSTH.html

Wiener not only initiated cybernetics, but also wrote other influential books on natural and physical systems.

Wiener, N. (1961). *Cybernetics or Control and Communication in the Animal and the Machine*, MIT Press.

Wiener, N. (1988). *The Human Use of Human Beings: Cybernetics and Society*, Da Capo Press.

A book, now out of print, that gives a good physical science perspective on systems:

Koenig, H., Tokad, Y., Kesavan, H. and Hedges, H. (1967). *Analysis of Discrete Physical Systems*, McGraw-Hill.

A popular and successful presentation with a software slant is

Weinberg, G. M. (2001). *An Introduction to General Systems Thinking*, Dorset House.

Many books on control reflect the system perspective. For example,

Franklin, G. F. (2002). *Feedback Control of Dynamic Systems*, Prentice-Hall.

The user-centred view

Norman, D. A. (2002). *The Design of Everyday Things*, Basic Books. (Originally appeared as *The Psychology of Everyday Things*, Basic Books, 1988.)

The artisan view

The view of software design as a craft is reflected in many programming books. It figures prominently in

McBreen, P. (2002). *Software Craftsmanship*, Addison-Wesley.

The organizational behaviour view

DeMarco, T. and Lister, T. (1999). *Peopleware*, Dorset House.

The multiple analogy view

Eischen, K. (2002). 'Software development: an outsider's view', *IEEE Computer*, **35**, pp. 36–44.

3 The craft of software design

The artisan perspective of Chapter 2 – seeing yourself as a craftsperson – encourages the development of a particular mindset. We've already seen the tension between traditional software engineering and craft-based approaches to design. Although we are trying to draw out common advice from both, you may have guessed, from its high view of the code, that this book leans towards a craft-based view. The purpose of this chapter is to anchor that bias in some down-to-earth practicalities and tell you how to fulfil:

Rule 19

Add to your knowledge and skills as you design. Manage your knowledge and skills to build good judgement.

The advice given here is based on observation of proficient designers. Although students sometimes see the practice of 'good habits' as optional – just for the very keen – and certainly lacking in glamour, such habits are *sure* ways of improving any designer's understanding and productivity.

This chapter outlines six simple practices which, if adopted as habits, will accelerate your development as a programmer and continue to inform your practice of software design. These are collaboration, imitation, finishing, tool building, maintaining a logbook, and maintaining a personal library.

3.2 Collaboration and imitation

For a software apprentice, the ideal environment for learning good design habits is working in a team with more experienced programmers. Apprenticeship is about learning by example and learning by doing. Experienced colleagues can function as mentors, their programs are examples to imitate, and they can be consulted about dealing with unfamiliar situations. They can give feedback on the apprentice's programs, and quickly spot problem areas.

Rule 21

Develop the ability to apply other people's ability.

Unfortunately most programmers do not learn to program by apprenticeship. Often, though, they can consult peers: when there is no expert, another novice might be able to provide different insights on problems and advice about something that they themselves have recently discovered. It pays to collaborate with other designers, and collaboration is a good habit to adopt right at the start.

Principle 13

Software design is a social activity.

In an academic context, many teachers are prepared to accept genuine collaboration (as opposed to alternating assignment preparation) as a legitimate part of learning. It goes without saying that students must be careful to observe whatever rules about collaboration are laid down for them to follow.

Principle 7

Something similar has already been done.

Many software problems are analogous to other, solved, problems. Later chapters will have much to say about exploiting existing solutions. Try to develop skill at finding analogies between problems: this will allow reuse of ideas from one solution to another. At the least, it is often possible to imitate the structure of an example program closely, changing those parts that need it, but preserving as much of the known, tested, original as possible. The exercises in Chapters 4 and 5 of this book are often problems similar to those just dealt with in the text. Practise recognizing this, and exploit the similarities.

Rule 11

Exploit existing solutions.

Imitation can, of course, be subject to legal constraints – there is no right to copy material when the author has not given permission, and close imitation might amount to copying. Be especially careful in an academic context. Don't imitate your fellow students' assignments because that *does* count as copying in most rules of assessment.

Legal copying

Free software (www.gnu.org/philosophy/free-sw.html) and Open Source software (osdn.com) is source code that you are free to copy and adapt. The huge amount available is a wonderful resource for programmers. Not only does it allow you to exploit existing solutions, it also provides good examples from which to learn. There is a catch, at least for certain kinds of free software. When you have adopted and extended a program, you may be obliged to make it available under the same terms as the original – that is, anyone else can freely apply what you have done. There are a number of different licensing mechanisms, so you need to check what you are and are not allowed to do with free software if you intend to use it in a program that may not itself be free. An often used licence is the GNU public licence, available at the GNU website www.gnu.org, which also gives lots of information about the various kinds of free software (www.gnu.org/philosophy/categories.html).

3.3 Finishing

If software design is a craft, then programs are the objects of that craft. When a designer has finished with a program, what state should it be left in? The craftperson's answer is, if a thing is finished with, it is either complete or rejected. Surprisingly many programs are left in a kind of limbo state, where they work, but not with sufficient help, diagnostics or documentation, to be used again easily in a few years' time. The world is full of old programs that have to be studied in depth to get the first idea of what they do (or did).

The craft analogy is exact. A program might be put together all right, but does it have a high-quality finish? Polishing a program may seem cosmetic, but it reflects a designer's concern that their product be durable and presentable.

In programming, finishing includes:

- Making sure the program works right.
- Making the executable as self-documenting as appropriate. That is, cue the user with informative prompts, and provide help where necessary.
- Making the source as self-documenting as appropriate. In particular, each program file should contain a header giving the program's purpose, author, outline usage information, any known problems or hazards (things that *could* go wrong), and the version history (initially just the date, but with dated comments appended when the program is subsequently revised). See page 185.
- Removal of all unused code.

Rule 4

Write no more than necessary.

- Clear identification of the final version.
- A makefile, or clear documentation about how to compile, link and run the program.

For the want of these few things, too many programs have been reinvented.

The designer who practises finishing soon earns their colleagues' respect, by being able to relocate and reuse even the smallest of tools developed during earlier projects.

3.4 Tool building

Craftspeople choose their tools carefully, and many make their own. The software designer will usually be supplied with the major tools of compiler, editor, debugger (perhaps in an integrated environment) for a particular platform. But a good strategy for getting to know that platform is to develop software tools for it.

Practising skills by developing tools therefore fulfils a rule:

Rule 12

Understand the capabilities, components and mechanisms surrounding the program.

Unix has often been regarded as the ideal tool-building environment, because of the way it gives two levels of control: programs are built as simple tools, each designed to do one thing well, then combined in 'pipelines' to do complicated things. (Unix allows the 'piping' of one program's output into the input of another; pipelines of several programs can be constructed easily.) Some of the most elegant examples of software tools are those written for Unix, as, for example, in Kernigan and Plauger's book *Software Tools*. A developer using Unix as a platform should become familiar with the tried and elegant methods of Unix tool building. Although a huge number of tools are widely available, and although the non-interactive, serial mechanism for which Unix is famous is often inappropriate for end-user applications, learning to tool-build with Unix is still a good way of learning design.

Because of the increasing dominance of the graphical user interface, and the demand for interactive, non-serial user control, a simple tool for a windowed operating system often ends up looking like a huge application. The smallest reasonable Microsoft Windows program is several pages long. Even so, in learning such a system, it is wise to invest practice time in developing tools that will be used later. For example, you might start with an inspector tool that looks inside applications and their documents, revealing their structure and giving their contents in some textual form. Such a tool is sure to exercise the system in a demanding way, helping you truly to understand how it works, and will continue to be useful as work on the system progresses.

Learning by tool building is an excellent way to fulfil Rule 12 (understanding the mechanisms undergirding the program). Where a software project will use interprocess communication, a tool that monitors messages moving between processes will be a worthwhile development. If a particular piece of hardware must be controlled, an ideal place to begin is with a tool that interrogates the hardware's current status and prints out the result. In each case the tool builder learns about the system, while creating a valuable debugging tool.

3.5 Logbooks

Rule 19

Add to your knowledge and skills as you design. Manage your knowledge and skills to build good judgement.

A good software designer keeps a logbook. Why? Mainly because it is the single most effective way to empower your thinking.

Here are the three conventional roles of the engineering logbook:

1. As a personal journal, it is a store of knowledge and experience.
2. As a commentary for others, it can be used by colleagues to understand and reconstruct the work.
3. As a legal document, it can be used as evidence to support claims about dates of inventions, developments, etc. (In this case, the logbook is usually attested to by witnesses, and more formal rules are followed about what does and does not go in the book.)

Of the three, only the first is certain; the other roles are provisional. The logbook will be used every day by the writer; but it may never be read by anyone else, and probably it will never act as legal evidence. The other two roles cannot be neglected, but it is in the certain role of journal that the logbook will be of most benefit.

The perfect logbooks?

Leonardo da Vinci's notebooks famously record his diverse interests and deep insights in beautifully organized words and pictures. They are a model to aspire to (except you don't need mirror writing). See www.museoscienza.org/english/leonardo/Default.htm for examples.

Most people have at some time kept a journal (that is, a diary of what has been, not what is to be). Even if done unwillingly (typically as a school project), in retrospect the record takes on unusual value. A single sentence unearths a large body of almost-forgotten memories. Events simply related are lenses that focus whole scenes. And when it comes to specifics, the journal records what choices there were to make and why one was chosen, of how the particular differed from the regular.

Like any other journal, an engineering logbook is dated but otherwise unstructured, written once, not revised, and gives a personal perspective. Like any other, its value is that it provides for the memory a framework and a gallery of snapshots. A designer needs their logbook to do this as effectively as possible. But the amazing thing about a journal is that it is almost impossible to fail! Even if the wrong things are recorded; even if what then seemed important, now looks trivial; even if tantalizing gaps abound; the journal will probably still succeed. In human psychology, reasoning and memory work together to recover vital memories from inadequate cues. That the journal provides any cues at all is often sufficient to stimulate the mind into recovering what matters. The logbook is, then, a surprisingly robust partner to your memory.

The logbook is a kind of user interface to the designer's knowledge, some of which is in the book and some of which is in the head. The most reliable way to create such an interface is to assume that the knowledge that will stay in the head is whatever a reasonably proficient person in the field would have in their head anyway. This means, the way to write a good logbook is to write it for someone else. The best answer to the question, what should go in the book? is, whatever commentary someone else would need to follow what is going on. That neatly addresses roles 2 and 3 of the logbook too.

It could be argued, fairly tritely, that in six months' time you will *be* someone else (from the point of view of understanding what you did today)! But the main reason for writing as if for someone else is that it gives the opportunity to practise 'public' writing without cost. The writing need not be in prose – almost certainly it will be in point form, with plenty of diagrams – but it will be written to communicate. There is no overwhelming reason to suppose that careful composition of the logbook is going to be slower in the long run. One who practises writing carefully will learn to write carefully quickly. That combination is a most valuable communication skill.

So write the logbook as a commentary for someone else, providing enough information for them to understand your decisions, follow your reasoning, reconstruct your work. The someone else is an informed colleague, not an ignoramus, and not a genius. Write it once. Do not go back and hone the record. Write in the first person singular, and name names when referring to other people's work. Date all the entries, and if necessary, get colleagues to witness entries (for legal purposes). Stick things into the book where appropriate, but do not leave loose sheets slipped between pages.

A logbook should be a real book, not soft documentation stored on a computer. There are benefits in keeping a soft logbook: it is in the same medium as programs; most programmers type better than they handwrite; automatic date stamping is possible; programs and logbook can be effectively cross-referenced – perhaps even linked with hypertext; it is easy to share materials with colleagues. But against these, there are strong counterarguments:

No hardware platform yet combines the portability and economy of a book. Portability is vital because the logbook is used in the problem domain (perhaps during meetings with customers, perhaps gathering information on-site), in the library, on the bus, and even (horrors) at home.

No software base yet is as convenient as pencil/pen and paper for multimodal input involving text, graphics and annotations. Drawing with a fine instrument on a good-sized canvas is an important facility.

No hardware/software combination yet gives such a powerful range of browsing, searching, multiple-view facilities as a book.

Logbook material stored in a computer is likely to suffer fates similar to other soft documentation. It will not be kept up to date, old material will be mistrusted, it will get accidently deleted, then abandoned in the belief that what it contains was not important enough to justify recovery from backup. Because of its flexibility, a soft logbook is also susceptible to tinkering: some authors find it difficult not to spend time improving it and bringing it up to date after the fact.

> I have not discussed security issues and ownership of material here. Obviously, if your employer owns your work (as they normally do), it may not be possible to take your logbook home. But if inspiration strikes in the middle of the night, it is obviously easier to paste in your home notepaper, than to type it in, when you get to work.

3.6 The personal library

Rule 21

Develop the ability to apply other people's ability.

An apprentice may learn from a local mentor, but also tries to understand the masters and keep abreast of current happenings in their area. In software design, which is both an intellectual discipline and a discipline concerned with information, the major resource for the works of the masters is books. The major resource for current developments is technical journals. An expert designer will build a library of the most valuable writing on the subject. This section gives advice on selecting books to go in a personal library and makes some recommendations. Note that here the concern is with material about software technology – that is, about programming – and with problem solving *in general*. Chapter 10 discusses gathering and sieving information about a particular application or problem domain.

There are four levels on which it is useful for you to have reference material about software technology. The lowest is the platform – information about the hardware, language and tools that are being used. The next is the level of 'Programming in the Small' – issues of program design, style, testing. Above this is the level of data structures and algorithms – the abstract components of programs. Finally, the top level is 'Programming in the Large' – software engineering of large systems.

A language reference manual is essential, and hardware and operating system manuals are often valuable. Most compilers for personal computers come with good language documentation. On the other hand, it is usually possible to do better than the manufacturer's manual for hardware and operating system information. Because there are many books on popular platforms, operating systems and languages, it is worth reviewing reader comments on amazon.com or books.slashdot.org before deciding what to buy. Network newsgroups can also help (see Chapter 10) – programmers can seek advice from experts about what the current best references are.

Then, there are books that deal with the craft of writing code. Kernighan and Plauger's books on software tools and programming style are classics, so much so that their influence permeates the C++ examples later in this book. But even better are Jon Bentley's Programming Pearls books. These are not systematic treatments. Indeed they depend largely on examples. But they are so good at illustrating the craft of software design, that they shouldn't be missed. If I could have only one programming book on my bookshelf, it would have to be the first *Programming Pearls* – beautifully written, fun, and an unexpected source of good applicable ideas.

At the next level are references on data structures and algorithms. Donald Knuth's *Art of Computer Programming* books are classics, but expensive. They are comprehensive, detailed and usually give the best advice around. Except that it might be out of date. In some respects, at least, Knuth's books have been superseded by newer treatments like Sedgewick's *Algorithms*. This is now available in C, C++ and Java language editions. So far as a numerical techniques book goes, a strong contender is probably *Numerical Algorithms*, available with code in C, C++, Pascal and FORTRAN. Some important algorithms are given in C++ code in this text. In using any prewritten version of an algorithm, remember:

Principle 4

Every program has surprises

and many important algorithms are tricky. Don't put your trust in the recipe. Understand the method!

So far as programming in the large is concerned, there are many books on software engineering. Perhaps the only one that has attained the status of 'classic' is Brooks' *The Mythical Man Month*. It is a bit dated, but everyone still looks back to it as the seminal work on pragmatic software engineering. However, it is not a reference book.

Practitioners keep up to date with developments in technology. The way to do this is to subscribe to technical journals. There are several, fairly general magazines and journals that provide good coverage at a particular level. The key professional ones are the *Communications of the ACM* (which members of the ACM get automatically) and *IEEE Computer* (for members of the IEEE Computer Society). A periodical like *Dr Dobbs' Journal* is good on the application of software design.

An important resource for problem solving is quantitative data on particular platforms. After all…

Principle 8

Every program is a component in a bigger system.

This is technology information that the designer will use in analysis while problem solving. Programmers working on a particular computer platform might save benchmark listings and performance figures and keep them somewhere accessible (perhaps a logbook).

3.7 Chapter end material

Bibliography

For C++ books suitable for a personal library, see Chapter 4. For data structures and algorithms books, see Chapter 8.

Programming in the Small

Bentley, J. L. (1999). *Programming Pearls*, Addison-Wesley.
Bentley, J. L. (1988). *More Programming Pearls*, Addison-Wesley.
See also the Kernighan and McConnell books referenced in Chapter 6.

Programming in the Large

Brooks, F. P. (1975). *The Mythical Man Month: Essays on Software Engineering*, Addison-Wesley.

Keeping a logbook

Roush, G. E. (1989). 'Documenting one's work', *IEEE Potentials*, May 1989, pp. 24–26.

4 Beginning programming in C++

Software design is about representing information.

Principle 1

Software design is representation.

A crucial issue is the choice of programming language, because the language defines the kind of representations that can be used in the program. Different languages provide different kinds of abstractions, so are suited to different goals. This book's position is that use of a real programming language is essential for discussion of software design:

Principle 2

The text of the program itself – the code – is the canonical expression of the design.

So we need a language, preferably one that is standard.

Arguably there are many standards in programming languages. FORTRAN, LISP, C, Ada, COBOL, BASIC, Pascal, Java, Perl, C# … are all standard according to some criteria. C++ also has a good claim, given its high profile in personal computer application development, and its status, with C, of being the language with the broadest support across major computing platforms. As an extensible language, it can fill roles previously held by 'higher-level' languages. It can be as fast as FORTRAN for numerical computation. As the oldest thriving sibling to Java and C# in the family descended from C and Simula (the earliest object-oriented language), C++ is the best choice for a book like this. Both historically and practically, it will be important for a long time to come.

This chapter and the next provide an introduction to C++. They cover the main features of the language concisely, and their approach is consistent with the design principles of this book. Readers new to C++ can get started here and move to more in-depth treatments for details. Experienced C programmers should read this chapter, because several non-C ideas are introduced here, then concentrate on Chapter 5. Experienced C++ programmers are invited to the sidebar.

A language you *won't* want to use is Intercal. Intercal's virtues include being the only language with the statements COMEFROM (see page 16) and PLEASE (it expects programmers to be reasonably polite). But it also has a downside. Intercal is documented at catb.org/~esr/intercal.

The idiom of the book

This C++ tutorial provides a 'traditional' rather than 'contemporary' introduction to the language. Much of the present chapter is applicable to C programs as well as C++, and object-oriented programming isn't properly introduced until the next. Experienced C++ programmers will also note some omissions:

- There is minimal reference to the standard library for collections. Instead we freely use arrays, pointers, and home-made data structures.
- C-strings appear extensively.
- Very little is said about namespaces, exceptions, and multiple inheritance.

Apart from the benefit of slipping in an introduction to C alongside C++, what is the thinking behind these choices? There are three main reasons for the book's C++ idiom:

- C++ was once thought of as 'a better C'. It still is. C++ can do everything C can. Although its creator refuses to make the claim (Stroustrup, 2002), *C++ supersedes C*. At least it does if you allow C++ to be learned and used in a C-like way. That means being aware of the low-level features that make it good for system programming, as well as the high-level features that support object orientation, respecting the heritage of the C standard library (which is incorporated in the C++ standard library), being aware of the possible program-size consequences of using exceptions, and familiar with the mechanics of constructing data structures out of arrays and pointers. For some teachers of C++ this seems like a retrograde step, but I believe the opposite. Here's why:
- In the 2010s and 2020s there will be two major types of program: those compiled for the single- or multiple-processor machines we're familiar with today, and those that compile into reconfigurable hardware. The second type of programming is moving up from complete hardware orientation in languages like VHDL to intermediates like Handel-C, suffering a performance hit in the process, but not a crippling one. It seems likely that the potential of speed, resilience and flexibility of reconfigurable computers will eventually be accessible to all languages. But for the foreseeable future, it will be programs written in a (parallelized) procedural style that compile most efficiently. By introducing C++ via procedural programming, I hope that my readers will be equipped for these newer computers and their applications.
- My coding idiom is calibrated to the software I most admire. Having learned a lot in using and applying Open Source software, it is natural that I advocate programming in a way that mirrors the best. In particular, I am a great admirer of the Trolltech QT library. I aspire to produce code as good as theirs, and am unembarrassed to suggest you too should program in a similar idiom. (I would also recommend a study of the Trolltech white papers comparing QT with Java and on the Slots and Signals paradigm which provides a very elegant object-oriented framework. See www.trolltech.com.)

So far as the three deficiencies I pointed out above go, I'll add:

- As this book is about the whole software design process, I set up the language in a way that allows discussion of data structures and algorithms using basic constructs like arrays and linked lists. Don't worry: I'll point out all the standard library support for collections later on!
- Writing a string class is perhaps the best learning exercise available to a novice. It makes the dangers of C-strings very clear, highlighting the problems of array access in general. It also raises interesting design issues: speed versus simplicity in reference counting versus copying on assignment; understanding users' understanding in deciding whether to overload '+' for concatenation. Yes the standard library string class is the final solution we should adopt, but we need C-strings anyway (see, for example, page 168), so until the String class exercises in Chapter 5, they will be enough.
- Every program has surprises. Exceptions are for dealing with surprises, but discovering the surprises is harder than coding the exceptions! Where I've left things out, it's usually to emphasize something more fundamental.

4.2 The programming environment

The first requirement for software development is a programming environment. This includes:

- The hardware – the computer or workstation: PC, Mac, Unix machine, etc.

- The operating system – the software that directly controls the hardware, and interacts with the user – Windows, OS X, GNU/Linux, etc.
- The compiler that generates programs from source code. On a personal computer, this is usually an integrated system that includes a program editor and a debugger – Borland C++ or Visual C++ on a Windows machine, CodeWarrier on a Mac, Kdevelop on Linux. Old time Unix (and Linux) programmers still sometimes run the compiler separately from the editor and debugger; programmers write and edit using one of the many editors that Unix provides, such as vi or emacs, and debug with one of the debuggers, such as dbx.
- Libraries of classes and functions, which programs can include to do particular things.
- Other programming tools such as User Interface Builders, Source Code Control Systems, etc.

In general, the programming environment may be different from the machine on which the program is to run. For example, code for an embedded microcontroller might be developed and cross-compiled on a workstation. In this book, though, it is assumed that the development and target environments are the same.

To begin, a programmer must know how to edit, compile and run programs in the programming environment. The documentation that compiler manufacturers provide is usually very extensive, and includes tutorials for getting new users started. Regrettably, students often do not have guaranteed access to hard-copy documentation. In these cases it is usually best to get someone knowledgeable to provide a tutorial introduction to the programming environment.

In all cases, the process of creating a single-file program involves the following steps:

1. Enter the source text of the program, either using a text editor, or the built-in editor of an integrated development environment (IDE) like Visual-C++. In many IDEs, you must first create a project which will include all the files that make up your program. Your environment may create a code framework or skeleton for you once you've answered the questions you're asked by a Setup Wizard. If you are not working in an IDE, you should save the program in a file whose name ends with **.cpp**.
2. Compile the program. The compiler will syntax-check the program, and if there are no errors it will generate an executable program as output. On a Windows system, this will be a file with the same name as the source, except with the extension **.exe** instead of **.cpp**. On a Unix system, the executable program will be in the file **a.out**. (In the Unix case, if everything works, it is a good idea to tidy up with **mv a.out program**, renaming the executable to have the same name as the source, but without the **.cpp** suffix.)
3. Test the program by running it, either in its own right by simply invoking it (entering its name without an extension), or from within the integrated environment.

Step 2 really has two parts. For simple, one-file, programs these are invisible to the programmer, but understanding both is important when writing larger programs. The two parts are both handled by the compiler, but the first is called Compilation and the second Linking.

A resource for getting started in some of the popular PC development environments is the Deitel downloads webpage. Deitel provide short introductions to using Visual C++, Borland C++ and other systems at http://www.deitel.com/books/downloads.html. This page also supports their book on Learning C++ (see references at the end of the chapter).

2a. During Compilation, the program source is translated into machine code, or 'object code'. If the compiler was stopped after this stage, it would output a file with a **.obj** or a **.o** extension. This file is almost ready to be run on the computer, and it also contains a 'symbol table' of where important things (like program variables) are in the object code. However, it is not complete, because it only has the code directly generated by the source file. Almost every program (including all those in this book) needs some other object code too, most of which comes from already-compiled libraries.

2b. During Linking, all the object files that go to make up a program are linked together into a single executable. If the program has several files (as larger programs do), each can be developed and compiled independently, then left as **.obj** or **.o** files until all are ready to be linked. Even with a single-file program, there is a need for a separate linking stage, because that program will call functions, or use classes, from 'The Standard Library'. These are stored in special object files called libraries that the linker part of the compiler knows about. At link time, these libraries are searched to pull in any of the functions that the program uses. Sometimes a programmer will link in non-standard libraries too, and then the compiler has to be told explicitly what these are.

Of course, most programs demand more than one iteration around the write–compile–run loop, as programs grow and errors are identified and corrected.

Although you will certainly want to learn how to use other features of your environment (such as the debugger and help facilities), so long as you can edit, compile and run successfully, you can enter any of the programs in this chapter.

4.3 Program shape, output, and the basic types

Here is a simple C++ program:

```cpp
#include <iostream>
using namespace std;
int main()
    {
    cout << "Hello World\n";
    return 0;
    }
```

Enter this program in your chosen environment. Compile and run it. If you are in an IDE like Visual C++ you will need to create this program as a *console application*. All programs in the next two chapters are console applications.

After making sure you can go through the edit–compile–run cycle successfully, introduce some errors into the program, like dropping a bracket or misspelling **cout**, and observe the compiler's response.

C++ is based on the older language C. For this example, we are going to consider the equivalent program written in C. The 'classic' C equivalent is short and sweet:

```
main()
  {
  printf("Hello World\n");
  }
```

Today's C is more like C++. Here's a version in C99 (C according to the 1999 ISO standard):

```
#include <stdio.h>
int main()
  {
  printf("Hello World\n");
  }
```

We'll begin with the classic C version and work out to the C++ program. The C program contains one function: **main()**. **main()** appears in all C and C++ programs and is where execution begins. **main()** can call other functions by naming them, as can any function. So here **main** calls **printf**, which is a C standard library function to format and print output on the display. The C standard library is a collection of functions provided with the C/C++ compiler that can be linked into anyone's programs. The compiler automatically links in just those functions the program calls. C's standard formatted output function **printf()** is superseded in C++, but it is worth knowing of its existence, because it is very common in legacy programs and **printf()**-style formatting even pops up in parts of the C++ standard library.

Data are passed between functions using 'arguments'. The two brackets () delimit the argument list for a particular function, while the two brackets {} delimit the statements (or instructions) of the function. So the function **main** in this example is passed no arguments, or at least chooses to ignore any arguments it is passed. **main**'s one and only statement, the call to **printf**, does involve an argument: the string **"Hello World\n"** is the argument that **main** passes to **printf**.

The system knows that **"Hello World\n"** is a string because it is included in double quotes. The **'\n'** at the end of the string is an example of an *escape sequence character constant*. Its meaning is New Line. So a line break is output after "Hello World" when the program is run. There are several escape sequence character constants, all represented by a backslash (\) (called the escape character) followed by a letter or a number. Table 4.1 summarizes these.

A subtlety to be aware of in dealing with characters is that strings are written enclosed in double quotes, but individual constant characters are written enclosed in single quotes. Single characters and strings are different data types, as explained on page 75.

C++ and modern C are stricter than classic C about defining things before they are used. Nowadays, no program may call a function, even a library function, like **printf()**, before *prototyping* it. A function prototype specifies the number of arguments to a function, their types, and

Table 4.1 *Escape sequence character constants*

Escape sequence	ASCII character represented	Interpretation	Example
\"	" (Double quote)	Display actual double quote	"\"Hi!\", he said" prints as: "Hi!", he said
\'	' (Single quote)	Display actual single quote	"It\'s good" prints as: It's good
\?	? (Query)	Display actual query	"What\?" prints as: What?
\\	\ (Backslash)	Display actual backslash	"\\ is the escape char" prints as: \ is the escape char
\a	BEL (Bell)	Sound beep on terminal	"\aDing dong" beeps the computer before printing Ding Dong
\b	BS (Backspace)	Go back one position	" 1\b\b2\b\b3" prints as: 321
\f	FF (Formfeed)	Start new screen	
⇒ \n	LF (Linefeed)	Move to next line. On most systems, move to start of next line	"one\nword\nlines" prints as: one word lines
\r	CR (Carriage return)	Move to start of this line	" 1\r2\r3" prints as: 321
⇒ \t	HT (Horizontal tab)	Go to next horizontal tab stop	"Won\tLost\tTied" prints as: Won Lost Tied
\v	VT (Vertical tab)	Go to next vertical tab stop	
⇒ \num	num is an octal number of up to 3 digits. \num represents the char with octal value num		if (c == '\0') is a test for a null character at the end of a string
\xnum	num is a hexadecimal number of up to 2 digits. \xnum represents the char with hex value num		

the return type of the function. We will see how to write prototypes shortly. For all the standard library functions, the prototypes are given in system 'header' files, so programmers simply **#include** the appropriate ones of these for the library functions they use. **#include** is a directive to the compiler preprocessor to include the text of another file at this point. The angle brackets around the filename <> tell the compiler to look in all the usual include directories for the named file. In the modern C example, the include file **stdio.h** has the prototype for **printf()**; in the C++ example the include file **iostream** has the required prototypes.

C99 and C++ also insist that every function return a value (or be explicitly declared **void**, meaning it does not return a value). As a special case, the function **main** has to return an integer, which in C and C++ is denoted by the keyword **int**.

Returning now to the original C++ program (now annotated with line numbers):

```
1:   #include <iostream>
2:   using namespace std;
```

Some of the programs in this book have line numbers, some don't. You *never* put line numbers in a real program. They are used here only so the text can refer to the program easily.

```
3:    int main()
4:    {
5:        cout << "Hello World\n";
6:        return 0;
7:    }
```

Line 1 includes the system-wide header file **iostream**, which has the necessary prototypes and definitions for this program. Everything in C++ that involves basic user input/output is defined in this header file. The file **iostream** is usually hidden away in some compiler directory – programmers rarely need to look at it (and you are not advised to do so yet).

Line 2 is a shortcut that we will use a lot in this book. Everything in the C++ standard library – which includes the i/o streams, support for strings and many collection classes – is in *namespace* **std** (short for standard). Namespaces are a way of dividing code up so that if two libraries use the same name for a function or a class, they won't get confused. You can refer to a function **blog()** from namespace **og** by the writing "og::blog()", while function **blog()** from namespace **nog** will be "nog::blog()". Since we're writing small programs at present, we won't be including many libraries. Indeed, the *only* library that we'll use at all is **std**. The real name of **cout** in line 5 is **std::cout**. As we'll see in a moment, **std::cout** is very useful, but it's tedious always writing the namespace then the name. Instead, all our programs will feature the line

using namespace std;

which says that every library function, class or object, that isn't recognized, should be prepended by **std::** because it's a member of the standard library namespace.

Line 3 now prefaces **main()** with the **int** keyword, saying that **main()** will return an integer to whatever called it. Since **main()** is where the program begins execution, the caller is the operating system shell from which the program was invoked. Usually the return value from **main()** is ignored by the shell, but sometimes it can be used, so in this book, **main()** always returns a value. 0 is returned if the program terminates normally, and -1 if there was an error.

Lines 4 and 7 are the brace delimiters of the function **main()**.

Line 5 is the one that arranges for the phrase 'Hello World', followed by a new line, to be output. Its shape is rather different from the **printf()** call used in the standard C program, and it reflects the object-oriented nature of C++. **cout** is an object (console outputter) which knows how to print things out on the console. (Remember, it's really **std::cout**, defined in the standard library.) **<<** is an operator, called 'put to', that **cout** and objects of the same type have 'overloaded'. **cout** on the left-hand side of a **<<** operator actually calls a function associated with **cout**, sending the thing on the right of the operator to that function. So the outcome is similar to the **printf()** function call, but accomplished with an object and an operator. Don't worry if this explanation doesn't make much sense yet. Chapter 5 has much more to say about the mechanism behind 'put to'. For now, we simply need to use it to print things.

Line 6 makes **main** return an integer value of 0. The system running the program might interpret **main()**'s return value (although it probably won't). Returning 0 signals successful completion; anything else signals failure. **main()** returns 0 by default, so this line could be omitted.

Here is a slightly more complicated program:

```
#include <iostream>
using namespace std;
int main()
   {
   cout << "3 divided by 7 equals " << 3/7 << '\n';
   cout << "3.0 divided by 7.0 equals " << 3.0/7.0 << '\n';
   cout << "Go figure!\n";
   return 0;
   }
```

Enter this program, compile and run it. Again you may want to experiment by introducing errors.

Every statement ends with a semicolon. This is C++'s way of delimiting statements. Spaces, tabs and carriage returns ('whitespace') can be inserted anywhere except in the middle of names and operators. It is the semicolons that tell the compiler where to split sentences. But code should be organized neatly, with whitespace used to make it readable. For example, groups of statements should be indented to show structure in the code.

Although this program is only two lines longer than the first, it introduces three more important ideas:

1. The program includes constants of four types:
    *   Strings, as before, are text enclosed in double quotes: e.g. **"Go figure!\n"**.
    *   Integers are whole numbers and arithmetic done on them results in whole numbers: e.g. **3/7** causes 3 to be divided by 7 and rounded down to the nearest integer – 0.
    *   Floating point numbers, are numbers with decimal points, and arithmetic done on them results in floats: e.g. **3.0/7.0** gives the result 0.428571.
    *   Single character constants are single characters enclosed in single quotes: e.g. **'\n'**.
2. **cout** and **<<** can handle these four types consistently.
3. A sequence of **<<**s can be cascaded on a single command line.

Again, the mechanism by which **cout** and **<<** achieve things is discussed in Chapter 5. For now, it is necessary only to know how to use them to output results. There is an analogous method for input, using **cin** and **>>** ('get from') which is illustrated in the first program of the following section.

> ## Bonjour tout le monde! (Hello World in other languages)
>
> Here's the Hello World program in Java:
>
> ```java
> public class HelloWorld {
>   public static void main( String args[] ) {
>     System.out.println("Hello World\n");
>    }
>  }
> ```
>
> You can find versions of Hello World in over 200 different programming languages at http://www2.latech.edu/~acm/HelloWorld.shtml.

4.4 Variables and their types

Here is a C++ program to print the sum of three integers:

```cpp
#include <iostream>
using namespace std;
int main()
  {
  int a, b, c, sum;
  cout << "Enter three integers, \n";
  cin >> a >> b >> c;
  sum = a + b + c;
  cout << "Their sum is " << sum << '\n';
  return 0;
  }
```

> Enter this program, compile and run it. Test your program with a variety of inputs. What does **cin** do when the input is not what it expects (three integers separated by spaces, followed by return)? **cin** will keep looking at the input until it has found three things that can be interpreted as integers, to put successively into a, b and c. If you type in things that can't be interpreted as integers, or if you don't put in a return (**cin** only receives things from the keyboard input when return is pressed), it might seem that your program has stuck, or crashed. If your program *has* crashed, on most systems you can get out by typing Control-C.

Any variables used, in this case **a**, **b**, **c**, and **sum**, must be declared before or at the place of first use in the function. Here they are declared at the beginning of the function. **a**, **b**, **c**, and **sum** are declared as **int**s – 16-bit signed quantities on old machines, 32-bit signed quantities on most current computers, and 64-bit or some other length signed quantities on other systems. An **int** is an integer of the 'natural word length' of the computer on which you're working. A single statement is used to declare all four variables. Four separate statements

An **int** in Java is *always* 32 bits. Good for Java!

could have been used instead. Other possible data type declarations include **char** (8-bit signed), **float** (single precision floating point) and **double** (double precision floating point). All the possibilities for arithmetic types are summarized in Table 4.2. Remember that the arrow in the left-hand column indicates those you're likely to use most.

Table 4.2 *C++ arithmetic types*

	Type	Typical size (bytes)	Range	Typical use
$\Rightarrow$	char	1	-128 to 127	ASCII characters, very small integers
	unsigned char	1	0 to 255	ASCII characters, very small integers
	bool	Implementation dependent, typically 4 (i.e. the same as int), but logically just one bit (1/8 byte)	false or true (corresponding to values 0 and 1)	Flags, results of conditional statements
$\Rightarrow$	int	4 (Depends on machine. int is nominally the natural word length of the processor)	$-2\,147\,483\,648$ to $2\,147\,483\,647$	Counting (e.g. in loops), integers
	unsigned int	4 (Depends on machine)	0 to $4\,294\,967\,295$	Large magnitudes
	short	2	$-32\,768$ to $32\,767$	Counting, integers
	unsigned short	2	0 to $65\,535$	Counting, magnitudes
	long	4	$-2\,147\,483\,648$ to $2\,147\,483\,647$	Large integers
	unsigned long	4	0 to $4\,294\,967\,295$	Large magnitudes
	enum	2 or 4	(As int)	Ordered sets of values
	float	4	3.4×10^{-38} to 3.4×10^{38}	Scientific arithmetic (7-digit precision)
$\Rightarrow$	double	8	1.7×10^{-308} to 1.7×10^{308}	Scientific arithmetic (15-digit precision)
	long double	10	3.4×10^{-4932} to 1.1×10^{4932}	Scientific arithmetic (19-digit precision)

Values are assigned to variables using the equals sign, and arithmetic operators, +, -, *, /, are used to combine values, together with brackets () to show precedence. + and - are addition and subtraction; these have lower precedence than * and /, multiplication and division, which means that

```
d = a + b * c;
```

sets **d** to **a** plus the product of **b** and **c**. To get the other grouping, parentheses must be used:

```
d = (a + b)*c;
```

Apart from the four standard arithmetic operators, C++ also provides the modulus operator %. This divides two integers and gives the remainder. For example, **x** = 10%7 sets **x** to the remainder after 10 is divided by 7, which is 3. There is no exponentiation operator in C++; to raise floating point numbers to a power, the math function **pow** may be used. A power function for positive integers is given on page 71.

Another operator that C++ provides is **sizeof()**. This 'returns' the size of a type in bytes. 'return' is a misnomer, because although **sizeof()** looks like a function call, it is really an operator. That is, it's part of the language, not part of some library. On the other hand, you do have to include a header file, **cstddef**, in a program that uses **sizeof()**. Why? Well, the type that **sizeof()** 'returns' is defined there with a line something like:

typedef unsigned int size_t;

size_t is usually an **unsigned int**, but in some implementations it's an **unsigned long**. Because this return type of **sizeof()** is used extensively for manipulating chunks of data, and because it is implementation dependent, it is given a type name of its own. By convention, some programmers use **size_t** to declare the type of variables that will be used for describing chunks of memory. You will see it extensively in the table of library functions at the end of this chapter.

In Table 4.2, the third column gives the typical size of each type in bytes. This looks useful, for helping to calculate how much data memory the program will use. However, some type sizes are machine/implementation dependent. What is more, standards for C and C++ guarantee very little about the sizes of things. Therefore, the reliable way to find the size of something is to use **sizeof()**. This always yields the correct sizes for the machine on which the program is running. This brings us to our first example program which is worth keeping as a useful tool.

```
//
// Program to print out the sizes of the basic types
//
#include <iostream>
#include <cstddef>
using namespace std;
int main()
  {
  cout << "Basic type sizes on this implementation are \n";
  cout << "char\t" << sizeof(char) << '\n';
  cout << "int\t" << sizeof(int) << '\n';
  cout << "float\t" << sizeof(float) << '\n';
  cout << "double\t" << sizeof(double) << '\n';
  return 0;
  }
```

> Enter this program, compile, run and save it. Later on, you will probably want to expand it to include pointer size.

Apart from the use of **sizeof()**, and \t for inserting tabs (see Table 4.1), the only new thing this program introduces is comments. The C++

comment mechanism is line-by-line, where the start of a comment is signalled by two slashes //, and the end of the line terminates the comment. This program starts with a comment, describing its function. This is always a good policy.

C++ also supports C-style comments. In C, the beginning of a comment is flagged by the pair of symbols /* and the end of a comment is flagged by */. Text between these two delimiters is completely freeform, except that comments cannot be nested. In general, the double-slash C++ style is preferable because single-line comments are easier to do and multi-line comments are more clearly set off from non-comment material. Most programmers reserve the old style for 'commenting out' sections of the code to remove a chunk of code temporarily, just put /* at the beginning and */ at the end, and the compiler will think that it is a comment.

4.5 Conditionals and compound statements

The format of the if conditional in C and C++ is:

if (*expression*) *statement*

The expression is evaluated and if it is true (which in C++ is equivalent to saying if it is non-zero), the statement is executed. Otherwise the statement is skipped. The statement can just be a single instruction, or a compound statement (sometimes called a 'code block') which is a sequence of statements all enclosed between a pair of curly brackets {}. In general in C++, wherever a simple statement may appear, a compound statement may appear instead. Furthermore, compound statements may be nested. For example:

```cpp
#include <iostream>
using namespace std;
int main()
   {
   int a, b, swap;
   cout << "Enter two integers: ";
   cin >> a >> b;
   if (a > b)
      {
      swap = a;
      a = b;
      b = swap;
      }
   if (a == b)
      cout << a << " = " << b << '\n';
   else
      cout << b << " > " << a << '\n';
   return 0;
   }
```

The **if** conditions show two of the C++ relational operators:

```
>    greater than
==   equal to
```

Note the difference between == equal to, and = assignment (i.e. *make equal to*). It is a common error to confuse these. Beware!

The other relational operators are:

```
<      less than
<=     less than or equal to
>=     greater than or equal to
!=     not equal to
```

All these operators yield results of non-zero if true and 0 if false. They can be logically combined with the operators

```
&&     logical AND
||     logical OR
!      logical NOT
```

For example, here is a test for **a** between 5 and 10:

```
if ((a >= 5)&&(a <= 10))
```

Operator precedence is such that this could be written:

```
if (a >= 5 && a <= 10)
```

but there is never harm in using parentheses. They help show structure.

The **&&** and **||** operators are intelligent enough only to test as many of the conditions as necessary. They work left to right, so in the above test for **a** between 5 and 10, if **a** is less than 5, it fails the first half (**a >= 5**), and it cannot therefore ever pass the test as a whole, so the second half (**a <= 10**) is never even tried. Similarly, in an OR combination of tests using **||**, as soon as the first succeeds, then that is sufficent for the whole test to succeed, and the remainder are never done.

An **if** statement can take an optional **else** clause which is executed if the condition is false. In fact **if**s and **else**s can be used to do multi-way branching:

```
if (...)
  {
  ...
  }
else if (...)
  {
  ...
  }
else
  {
  ...
  }
```

Write a program to read a character followed by two decimal numbers from **cin**. If the character is a +, print the sum of the two, if it is a −, print the difference. Once you have this working, add * and / operators.

There are two other ways to do conditional branches in C++. There is a shorthand version of **if ... else** that uses the characters ? and :. Writing

sign = (value < 0) ? '-' : '+';

(where **sign** is a **char**) is equivalent to

```
if (value < 0)
   sign = '-';
else
   sign = '+';
```

The third mechanism for conditionals is the **switch** statement. **switch** is good for multiway branching. Instead of writing

```
if (c == 'a')
   ...;
else if (c == 'b')
   ...;
else if (c == 'c')
   ...;
else
   ...;
```

switch provides a compact form:

```
switch(c)    // Have to have a char or int here. You cannot
             // switch on a string, for example.
   {
   case 'a':
      ...;
      break;
   case 'b':
      ...;
      break;
   case 'c':
      ...;
      break;
   default:
      ...;
      break;
   }
```

The **case** statements label the various actions, and **default** is done if none of the others are satisfied. Any number of statements can be listed for each case. The **break** statement at the end of each case means jump to the end of the **switch** block. If the **breaks** were not there, execution would just continue sequentially through the statements for lower **cases**. Sometimes this is desired, but usually it is best to avoid by putting a **break** after the statements for each **case**. Sometimes a **return** appears at the end of a case, rather than a **break**; this is when the **case** is the last thing that has to be done in the current function, and it can then return directly to its caller.

4.6 Loops

There are four ways to do loops in C++. Three of them, **while, for,** and **do-while,** are closely equivalent. The fourth, using **goto** and labels, is frowned upon, and is mentioned here just for completeness.

The general form of a **while** statement is

while (*expression*) *statement*

The expression is evaluated and if its value is non-zero the statement is executed. The same thing is done again and again until the expression evaluates to zero. Then the loop terminates and the program continues after the loop. Again, the statement can be a compound of several simple statements. Furthermore, assignments can be incorporated in the conditional expression.

A simple piece of code to print the squares of all the integers from 1 to 10 would be

```
cnt = 1;
while(cnt <= 10)
   {
   cout << cnt << " squared = " << cnt*cnt << '\n';
   cnt = cnt + 1;
   }
```

The statement **cnt = cnt + 1;** takes care of the incrementing of **cnt.** Adding numbers to variables is such a common operation that C++ provides a special operator **+=** so that

```
x += 8;
```

has exactly the same meaning as

```
x = x + 8;
```

The incrementing statement above could therefore be written as

```
cnt += 1;
```

There are **-=, *=** and **/=** operators which work analogously. In fact, incrementing and decrementing are so common, that C++ goes even further. It provides **++** and **--** to increment and decrement variables. That is,

```
cnt++;
```

is equivalent to

```
cnt += 1;
```

If the **++** is appended to the right-hand side of a variable, the increment will be done after the variable has been used. So

```
b = 3; a = b++;
```

leaves **a** equal to 3 and **b** equal to 4. In this case the **++** (or **--**) is called a post-increment (or post-decrement) operator, because it is written after the variable and takes effect after the variable is used. Similarly there is pre-increment and pre-decrement:

```
b = 3; a = ++b;
```

leaves both **a** and **b** equal to 4.

C++ means add one to C

The **++** idiom is where the name of C++ comes from. In *The C++ Programming Language*, special edition, Chapter 1, Bjarne Stroustrup, C++'s inventor, relates the history of the language. Early versions were known as 'C with Classes'. 'C++' was coined by Rick Mascitti, signifying the evolutionary nature of the changes from C.

Another way to do a loop is with the **for** statement. The general form of a **for** statement is

```
for (initialization; expression; increment)
  statement;
```

Its meaning is exactly the same as

```
initialization;
while(expression) {statement; increment;}
```

So **for** is a handy way of doing loops compactly. For example, the printing of squares example given above could now be done:

```
for (cnt = 1; cnt <= 10; cnt++)
  cout << cnt << " squared = " << cnt*cnt << '\n';
```

Here is a slightly longer example involving looping.

```
//
// Print Fahrenheit - Celsius table for f = -40, -30, ... 200
//
#include <iostream>
using namespace std;
int main()
  {
  const int lower = -40;
  const int upper = 200;
  const int step = 10;
  double fahrenheit, celsius;

  fahrenheit = lower;
  while (fahrenheit <= upper)
    {
    celsius = (5.0/9.0)*(fahrenheit - 32.0);
    cout << fahrenheit << '\t' << celsius << '\n';
    fahrenheit += step;
    }
  return 0;
  }
```

> Modify this program to use the **for** statement. Compile and run it. Make a further modification to allow the user to input the range of the table to be printed (i.e. the start and finish values).

Observe the use of the keyword **const**. This is a signal to the compiler that the value initially put into a variable is to stay there through the life of the variable. **const** statements are not essential, but they provide protection. If a new value were assigned to something that had been declared **const**, the compiler would flag this as an error. This might uncover a bug that would otherwise have been missed.

The third mechanism C++ provides for loops is the **do-while** loop.

do *statement*
while (*expression*)

The **do-while** loop is almost the same as an inverted **while** loop. Whereas **while** performs the test first and executes the enclosed statements only if the result is true, the **do-while** executes the statements and then performs the test. This means that the statements will be done at least once whatever the value of *expression*.

4.7 Random numbers, timing and an arithmetic game

Here is a division tester program that incorporates simple interaction, random number generation, and timing. As such, it includes some new functions from the C/C++ standard library, but the language constructs are almost all ones introduced already.

```
 1:  //
 2:  // Division tester program
 3:  //
 4:  #include <iostream>
 5:  #include <cstdlib>        // For random number functions
 6:  #include <ctime>          // For time functions
 7:  using namespace std;
 8:  int main()
 9:  {
10:      int dividend, divisor, answer;   // answer = dividend/
                                          // divisor
11:      int users_answer;
12:      long start_time;      // long because return value from
                              // time() is long
13:      int duration;
14:      int sum_correct = 0;
15:      int sum_durations = 0;
16:      int i;                // Counter
17:      srand(time(NULL)); // Use the current time to "seed"
18:                          // the random number generator
19:      for (i = 0; i<10; i++)
20:        {
21:          answer = rand()/(RAND_MAX/8) + 2;
22:          divisor = rand()/(RAND_MAX/8) + 2;
23:          dividend = answer*divisor;
                // Generate answer and divisor, then work out
                // dividend from them. See text for explanation
                // of random number generation.
24:          cout << "What is " << dividend << " divided by " <<
                    divisor << "?\n";
25:          start_time = time(NULL);
26:          cin >> users_answer;
```

```
27:        if (users_answer != answer)
28:           cout << "Wrong! The answer is "<< answer << '\n';
29:        else
30:           {
31:           cout << "Right! (in " << (duration =
                   time (NULL)-start_time) << " seconds)\n";
32:           sum_correct++;
33:           sum_durations += duration;
34:           }
35:        }
36:     cout << sum_correct << " out of 10 correct.\n";
37:     if (sum_correct)
38:        cout << "Average time = "<< (double)
                sum_durations/sum_correct << " seconds\n";
39:     return 0;
40:     }
```

> Enter this program (without line numbers), compile and run it.
> It consolidates many of the earlier sections. It will make a good
> base for the Rock–Scissors–Paper exercise later.

This program has three **#include** directives. Again, all the files mentioned are supplied by the compiler manufacturer. They are required because they contain prototypes for the library functions used lower down. How do we know which header files to include for which functions? The library documentation gives the information, and the summary of important library functions at the end of this chapter also specifies appropriate headers to include.

Again all the variables have been declared at the top of the **main** function (lines 10 to 16). This time there are several **ints** and a **long**. Because this program is required to produce a different division problem every time, it uses library functions to generate random numbers. Consider the statement **srand(time(NULL));** (line 17). **srand** and **time** are both standard library functions, while **NULL** is a constant defined in **cstdlib**. **time**, with an argument of **NULL** returns the number of seconds since Midnight, GMT, on January 1st, 1970, and **srand** uses its argument to initialize the random number generator with a 'seed'. This is so that the random number generator starts at a different place every time the program is run. Line 17 is our first example of using a function's return value by slotting it into an expression. Similarly lines 21 and 22 use the return value from **rand**. **rand** returns a random integer, between 0 and **RAND_MAX** (a value defined in **cstdlib**). So the statements

answer = rand()/(RAND_MAX/8) + 2;
divisor = rand()/(RAND_MAX/8) + 2;

turn the return value of **rand** into a single digit number between 2 and 9.

Running out of time

In the late 1990s, people worried about the 'Y2K' bug. Many legacy programs had been written to maintain just two decimal digits of year information, and the fear was that the change from 99 to 00 would break computer systems all over the world. Millions of dollars were spent on making systems Y2K compliant and the millennium came in with hardly a ripple. time() returns the number of seconds since 1 January 1970 but on most systems that value is just 4 bytes wide. A signed, 32-bit quantity can count up to 2 147 483 647. That's about 68 years worth of seconds, meaning time will run out in 2038. If there are any systems still using time() with a 4-byte integer type in 2038, they will face the equivalent of the Y2K bug.

The general way to get a random integer from 0 to $n - 1$ is:

```
answer = rand()/(RAND_MAX/n);
```

An alternative method might be:

```
answer = rand()%n;
```

Remember that the modulus operator % produces the remainder after division, so in this case that is a number between 0 and $n - 1$. But the first way is better. Many current random number generators are not all that random in their least significant bits; it is better to use more significant bits. Even though this problem may be fixed in the standard C/C++ library you are using, you should always regard random number generators with scepticism.

The number that the user has to work out is the one called **answer**. Because it is a single digit, only one keystroke is required from the user. However, 'get from' will accept more; it does not return to the program until a carriage return has been pressed. The lack of immediate response to keystrokes is one problem with **cin >>** as we have covered it so far; real-time interaction is impossible. There are ways to fix this, and the standard libraries are often augmented with other libraries for interactive programs. Unfortunately, those other libraries differ from platform to platform.

The final **cout** statement

```
cout << "Average time = " << (double)
        sum_durations/sum_correct << " seconds\n";
```

does introduce a new feature: the use of (**double**). This construct, where a type name is put, in brackets, before a variable (or a constant, or an expression), is called a 'cast'. Here, **sum_durations** is being cast into a **double**. That is, a double-precision floating point value, equal to the integer value of **sum_durations** is to be used in this statement. The reason for doing this here is to make **sum_durations** into a **double** before it gets divided by **sum_correct**. That will ensure that the result is a **double**, so that a decimal fraction is printed out. If everything had been left as integers, the print statement would have been

```
cout << "Average time = "<< sum_durations/sum_correct <<
        " seconds\n";
```

and the printed result would have been rounded to the nearest whole number.

C++ will do type conversion automatically within arithmetic expressions. In fact, this happens in the above example. **sum_durations** is first cast into a **double**. Then, because the fraction **sum_durations/sum_correct** is required, **sum_correct** is 'promoted' to a **double** automatically before the division is done. In general, C++ promotes smaller types to larger ones according to the rules given in Table 4.3. The top half of Table 4.3 shows the conversions that are done every time a variable is involved in an arithmetic expression, then the bottom half shows how the smaller operand of an expression is cast to the type of the larger operand.

More on casts

The process of converting one type to another (in this case, an int to a double) can be accomplished, as here, with an explicit cast. C++ also provides more nuanced control over type conversions with **static_cast**, **reinterpret_cast**, **const_cast**, and, for handling derived types, **dynamic_cast**. These are not covered in this book. Because they cover the different uses of casting, the conventional cast is sometimes deprecated, but its use is still widespread, and for converting between built-in types, perfectly acceptable.

Table 4.3 *Type conversion*

Type	Converts to
First, C does these conversions:	
char	int
unsigned char	int
signed char	int
short	int
unsigned short	unsigned int
enum	int
float	double

Then it goes through the following list, and if either of the operands has the specified type, it converts the other operand to the same type.

long double
double
float
unsigned long
long
unsigned int

4.8 Functions

For anything larger than the division tester program (page 67), a program should be divided into functions. A function has a name, arguments and a return type, and it appears in three ways:

1. Before it can be called, a function must be *prototyped*. This is analogous to declaring a variable before using it. Prototyping is accomplished by a line of the form:

   return_type function_name(arg1_type arg1_name, ... argn_type argn_name);

   (The names of arguments are not strictly necessary in a prototype: including them is sometimes helpful for debugging.)
   Prototyping is usually done at the top of the program, or in a header file.

2. The function must be *defined*, that is, its statements must be written out. The function definition is of the form:

   return_type function_name(arg1_type arg1_name, ...
   argn_type argn_name)
   {
   statements of function
   }

   (If the function is defined before its first call, then it does not have to be prototyped. Even so, as a rule, prototyping everything is good practice.)

3. Function calls (of which there may be many) are typically of the form:

   variable = function_name(value1, value2, ... valuen);

   though assignment of the return value to a variable is not necessary.

Here is an example that uses a function to raise an integer to a positive integer power.

```
1:    // Raise an integer to a positive integer power
2:    #include <iostream>
3:    using namespace std;
4:    int ipower(const int num, const int exp); // Prototype
5:
6:    int main()
7:      {
8:      int a, b;
9:      cout << "Enter two integers: ";
10:     cin >> a >> b;
11:     cout << a << " to the power of " << b << " is " <<
               ipower(a,b) << '\n';
12:     return 0;
13:     }
14:   int ipower(const int num, const int exp)
15:     {
16:     int value = 1;
17:     for (int cnt = 1; cnt <= exp; cnt++)
18:        value *= num;
19:     return value;
20:     }
```

Here **ipower** is the programmer's own function, being called from **main**. Before **ipower** is even used, right at the top of the program, on line 4, it is prototyped. The prototype shows how many arguments are passed, what their types are and what the return type is. **ipower** is a function of two integers and returns an integer.

int ipower(const int num, const int exp);   // Prototype

The **const**s in the prototype mean that the function **ipower()** will not change the values of its parameters.

Within **main**, on line 11, **ipower** is called by naming it:

ipower(a,b)

Note that the names used for arguments in a function do not have to agree with those used by the caller. But the types have to agree.

Below **main**, from line 14 to line 20, the actual definition of **ipower** is given. The first line:

int ipower(const int num, const int exp)

is almost identical to the prototype, but is followed by the function itself (between { and }) rather than a semicolon.

The value returned by a function is specified using the **return** statement as on line 19. A **return** may happen from anywhere within a function; it does not have to be the last statement. Of course, the returned type must agree with the prototype. In this case an **int** has to be returned.

Raising a number to a power is an example of a function that could be implemented recursively. A recursive alternative to the **ipower()**

defined above is:

```
int recursive_ipower (const int num, const int exp)
    {
    if (exp == 0)
        return 1;
    return num*recursive_ipower(num, exp-1);
    }
```

This is not as efficient as the first implementation, but it illustrates recursion. A function calls itself repeatedly, nibbling away at the problem, until some degenerate case (in this program, **exp** == 0) is reached.

> To see how the recursive example works, try hand executing the function with simple parameters; for example, use it to raise 3 to the power 4.

For some algorithms, recursion can be very efficient and elegant, but it has associated dangers, because every call to a function involves space and time overhead.

The **ipower()** examples pass arguments to the function, which the function then uses before returning a value. C, from which C++ is derived, is a call-by-value or pass-by-value language, which means that when arguments (parameters) are passed to a function, it is copies of those arguments which the called function actually gets to manipulate. When **a** and **b** are passed to **ipower()** by the **main** function, it is only their values that **ipower()** receives. So if **ipower()** tried to change **num** and **exp** (which are the names that it knows its arguments by), it would merely change the copies, not the original **a** and **b**. C++ has added to C the ability to pass arguments by reference (call-by-reference), which means that the called function can directly manipulate the original variables. (C itself has now added the same facility.) The arguments involved must be explicitly declared as *references* in the function prototype and the function definition. The way to do this is to put the character & after the type of the argument. The following simple program shows how a function can exchange the values of two variables in its caller, if they are passed as references.

```
#include <iostream>
using namespace std;
void put_bigger_into_first_param(int& a, int& b);
    // a and b are reference parameters
int main()
    {
    int bigger = 3;
    int smaller = 5;
    put_bigger_into_first_param(bigger, smaller);
    cout << bigger << " is bigger than " << smaller << '\n';
    return 0;
    }
void put_bigger_into_first_param(int& a, int& b)
        // Now a and b are aliases for the original
```

```
                              // bigger and smaller, and this function
                              // manipulates those parameters directly.
                          {
                          if (a >= b)
                             return;
                          int temp = a;
                          a = b;
                          b = temp;
                          }
```

The keyword **void** is used as the return type of the function here, because it does not return a value. Observe the **return** statement standing alone. Variables cannot be of type **void**, only functions.

Note that **const** is not used for the arguments this time, because they are altered by the function.

It is also possible for a function to have 'default parameters'. Suppose a function has n arguments. When it is prototyped, the last m arguments may be initialized. Then the function can be called with any number of arguments from $n - m$ to n. Those arguments that do not appear in the function call are filled in with their default values from the prototype. Here is an example. The function **print_complex** can be called with one or two arguments. When only one argument is used, the second parameter, **imaginary**, takes its default value of 0.0.

```
#include <iostream>
using namespace std;
void print_complex(const double real, const double
imaginary = 0.0);
int main()
   {
   print_complex(1.2,3.4);
   print_complex(5.6);
   cout << '\n';
   return 0;
   }
void print_complex(const double real, const double imaginary)
   {
   cout << real << " + " << imaginary << "j ";
   }
```

The program prints out:

```
1.2 + 3.4j   5.6 + 0j
```

4.9 Arrays and C-strings

Arrays are declared in the following form:

data_type array_name[size_of_array];

For example:

```
int x[100];
char name[20];
char table[100][20];
```

The first example declares an array of 100 integers, the second an array of 20 characters, and the third a two-dimensional array of 100 rows and 20 columns. In each case, space is allocated in memory to hold the number of items specified of the type specified.

In multidimensional arrays, the right-most subscript varies fastest, so the declaration of **table** might be appropriate to hold 100 names each of maximum length 20 letters.

When an array is declared, the number(s) in the square brackets gives the size of the array. Afterwards, when the array is used, the number in the square brackets is the subscript of the particular element being accessed. Subscripts run from 0 to $n - 1$ for an n element array.

Here is a program that reads in 10 integers from **cin** (console input), then prints them out in reverse order:

```cpp
#include <iostream>
using namespace std;
int main()
   {
   int stored_input[10];
   int i;
   cout << "Enter 10 integers:\n";
   for (i = 0; i <= 9; i++)    // Read in from 0 up to 9
      cin >> stored_input[i];
   for (i = 9; i >= 0; i--)    // Print out from 9 down to 0
      cout << stored_input[i] << ' ';   // Output list with space
                                        // ' ' separators.
   cout << '\n';
   return 0;
   }
```

If a function is to manipulate an array, the way to do it is to pass the *address* of the array (more accurately, the address of its first element) as a parameter to the function. In C++, the array name without square brackets is interpreted as the array address. So the caller can just name the array to pass its address. The function that is called may declare this parameter as an array of unknown size. To show how this works, here is a program that prints out the highest-valued of 10 integers, using a function that searches an array of integers for the highest value and returns that value.

```cpp
#include <iostream>
using namespace std;
int highest(const int array[], const int num_in_array);
      // Note [] - array of arbitrary size
int main()
   {
   int stored_input[10];
   int i;
   cout << "Enter 10 integers:\n";
   for (i = 0; i <= 9; i++)    // Read in from 0 up to 9
      cin >> stored_input[i];
```

```cpp
        cout << "The highest of these is " <<
            highest(stored_input, 10) << '\n';
    return 0;
}
int highest(const int array[], const int num_in_array)
{
    int highest;
    if (num_in_array <= 0)   // Error in parameters, can't have
                             // such an array
        return 0;    // Just return a 0 in this case - don't worry about
                     // signalling an error
    highest = array[0];
    for (int i = 1; i < num_in_array; i++)
      {
      if (array[i] > highest)
          highest = array[i];
      }
    return highest;
}
```

The declaration of the first argument to **highest()** as **array[]** means
that the argument is to be interpreted as an array. The size is not known
by **highest()** – it is the concern of the caller (in this case **main()**) to
ensure the array is big enough for whatever the called function (in this
case **highest()**) is asked to do.

Note how **highest()** has been defined to have more general behaviour than actually required for this program. It can find the highest
element in an array of arbitrary size. When designing functions, it is
often good practice to make them general purpose. They are then
more likely to be reusable without modification.

The difference between individual characters and strings was mentioned back at the first program example. In fact, all the strings we have
been dealing with so far are little more than arrays of characters. The
only subtlety is that they are 'null-terminated'. That is, the last character in a string is 0 (which, written as a character, is '\0' – i.e. not the
character '0' but the value 0). Functions to manipulate these strings
don't need to know the length of the array that the string is stored in;
they simply keep working until they hit a '\0'. So when 'Hello' is
stored as a constant string, it is really the array of characters:

'H'   'e'   'l'   'l'   'o'   '\0'

The compiler changes constant strings into null-terminated character
arrays for you. But in general, to use **cout <<** to output a variable
string, for example, the programmer is responsible for making sure
the array of characters has a '\0' at the end of the string.

Null-terminated strings are standard C-language strings. C++'s
standard library does, however, have its own string class. We will
defer discussion of that until late in the next chapter (after designing
our own, small, string class). For now, every reference to string means
an array of 8-bit characters terminated by a 0. C++ people often call
null-terminated strings, 'C-strings'.

Pros and cons of C-strings

C-strings are made up of 8-bit ASCII characters. Two reasons why it is worth learning about them:

- A null-terminated string is a simple data structure so you can manipulate it directly and efficiently, and, in a learning context, it's easy to see what's going on.
- A large amount of legacy code uses C-strings, including functions in the C++ standard library (such as opening a file given its name as a C-string).

Two reasons why it might be better to support the C++ **string** class:

- Functions that manipulate C-strings, like those in this section, often use pointer manipulation. That's efficient, but also a common source of bugs. Using a **string** class protects you from the opportunity to make those mistakes.
- ASCII characters in C-strings can only represent Romance languages. For internationalization strings should be able to support other character sets. The C++ **string** class does so.

We will get to C++ **strings** eventually, but in the process we'll build our own string class from C-strings (see page 139 onward).

Let us consider how we might write a function to copy a string from a source character array to a destination character array. Starting at index 0 of each, the function will copy elements over one by one. The '\0' at the end of the source string will signal when the copy is complete.

The heart of the string copy function could be:

```
for (i = 0; (dest[i] = source[i]) != '\0'; i++)
    ;
```

If this seems esoteric, remember that condition tests can include assignments. The condition part of the **for**, which is (**dest[i] = source[i]**) !='\0', is doing the copy from the source array to the destination array of element i, *and* the test for whether this character is a '\0' terminator. It is also possible to have a null statement (just a semicolon) as the body of the loop, just as in this example.

Adding code to turn this into a function definition gives:

```
int my_strcpy(char dest[], const char source[])
    // source[] is const because unchanged,
    // dest[] is changed however.
{
int i;
for (i = 0; dest[i] = source[i]; i++)
    ;
return(i);  // Length of string
}
```

Here the start addresses of the arrays will be passed from the caller. Again, the declarations within **my_strcpy** do not include the size of the array because that is defined in the caller. Now any other function can use **my_strcpy**, provided it passes the addresses correctly. As a bonus, the length of the copied string is returned, if the caller wants to use it.

Note what has happened in the condition part of the **for** loop. Instead of

(dest[i] = source[i]) != '\0';

there is just

dest[i] = source[i];

This works because zero and non-zero are equivalent to FALSE and TRUE respectively in C++.

So while any character but '\0' is being copied, the condition is satisfied. When '\0' is reached, then

dest[i] = source[i] = '\0'

and the condition fails so the loop terminates. Some compilers flag a warning when **dest[i] = source[i]** is used within a condition test – they are detecting a possible confusion between = and ==, but in this case the warning may be ignored.

Copying strings is so common that there is a standard C++ library function to do it, **strcpy()**, which is sure to be at least as efficient as the one above.

Here is a better string copy function:

```
int my_strncpy (char dest[], const char source[], const int
      dest_length)
// dest_length is the length of the dest[] array
   {
   int i;
   for (i = 0; (i < dest_length)&&(dest[i] = source[i]); i++)
      ;
   if (i == dest_length)  // Special case - have to replace
                          // last char with '\0' so string
      dest[i-1] = '\0';   // manipulation works correctly
                          // hereafter.
   return(i);   // Length of string
   }
```

This function copies a string from a source array to a destination array of length **dest_length**, to a maximum of **dest_length-1** characters. If the full **dest_length-1** characters are copied (so the **source** has not yet included a '\0'), then a '\0' is appended to the truncated string in **dest[]**.

Why is this better than the straightforward copy?

The answer is this: *One of the most dangerous sources of bugs in C and C++ is writing to unallocated memory.*

Consider an example.

```
char  mytext[22];
```

The compiler makes sure there are 22 characters of space available for this array. Now, suppose the string copy function defined above is used to write a string into this array:

```
my_strcpy(mytext, "This is my text string");
```

FALSE and TRUE here refer to logical categories. **bool** variables take the equivalent keyword values **false** and **true**. Many people rely on non-**bool** types to carry the meanings FALSE and TRUE because they can also be used for other things, as here, where non-zero characters are part of the string itself, and the zero character signifies termination.

The best defence in *C and C++* against buffer overrun is to be very careful when copying things of variable length. But in Java you don't need to worry! Java does bounds checking on array accesses and doesn't even allow you to make this kind of bug. The penalty is time overhead every time you access an array.

This will compile without problems, but it is very wrong! There are 22 letters in **"This is my text string"** and 22 **char**s of space. But an extra **char** is always needed for the null character at the end. So **my_strcpy()** will actually write 23 characters to a 22 character array. This extra character is going to overwrite something else in memory, next to the space allocated for **mytext[]**. Perhaps it will overwrite another variable in your program and cause intermittent malfunction; perhaps it will overwrite some variable in the operating system, and later cause a very mysterious crash. Such bugs are very hard to track down. The best defence against them is to be very careful when copying things that are of variable length (like strings).

Rule 13

Program defensively.

Using **my_strncpy** provides a safeguard:

```
char mytext[22];
my_strncpy(mytext, "This is my text string", 22);
```

This code copies as much of the source string as will fit in the first 21 array slots, and ensures that '\0' terminates the string.

In a system that does a lot of string manipulation, where most of the strings have the same length (say 256 **char**s, for example), **my_strncpy()** might be slightly modified to have a default parameter:

```
int my_strncpy(char dest[], const char source[], const
    int dest_length = 256)
// dest_length is the length of the dest[] array - defaults to 256
// Rest of the program is unchanged.
  {
  int i;
  for (i = 0; (i < dest_length)&&(dest[i] = source[i]); i++)
    ;
  if (i == dest_length)    // Special case - have to replace last
                           // char with '\0' so string
    dest[i-1] = '\0';      // manipulation works correctly
                           // hereafter.
  return(i);  // Length of string
  }
```

so that the programmer could use it with just two arguments, except where the destination string lengths were non-standard.

4.10 Program example: A dice-rolling simulation

The Central Limit Theorem is a result in probability theory that states that the probability distribution of summed independent random variables converges to a normal curve as the number of random variables increases. A normal (or Gaussian) distribution is bell shaped, and is of considerable significance in probability and statistics.

The Central Limit Theorem is a theorem, so it can be proven. But it is helpful to visualize what is happening as random variables are added.

One way is to roll a dice, and note that the frequency distribution of the six possible outcomes is uniform (if the dice is not loaded) – that is, all six values are equally likely. But roll two dice, and the sum of their values is more likely to be 7 than any other number (because 7 can be made in the most ways from 2 dice: (1,6), (2,5), (3,4), (4,3), (5,2), (6,1)). Increase the number of dice and the frequency distribution of different results tends to a normal curve.

Suppose we are asked to design a program that illustrates the Central Limit Theorem with simulated dice rolling. Between 1 and 12 dice are to be rolled simultaneously and their total value calculated. Between 1 and 10 000 rolls of the dice are to be simulated, and the histogram of totals displayed. On any roll of the dice, the total will be between 1*number_of_dice, and 6*number_of_dice, meaning the highest total possible will be 6*12 = 72. Thus, a histogram with 72 bins will be required. Because a standard computer console display or window has 80 columns it is possible to display the histogram using (ASCII) standard text characters. Fixing the height of the histogram at 20 will allow it to be displayed on just about any console window. A constraint for the examples in this book is that they should not be platform specific, so having a textual histogram is almost essential. Having said all this, the histogram display part of the program could be modified for graphical displays. Furthermore, being able to display histograms is something other programs might wish to do. It is therefore sensible to separate the histogram display from the rest of the code. This is done in the code presented below.

Rule 6

Localize information.

```cpp
#include <iostream>
#include <cstdlib>
#include <ctime>
using namespace std;
int roll(int dice[], const int num_dice);
int print_histogram(const int num_bins, const int bin_array[]);
int main()
    {
    const int max_dice = 12;
    int dice[max_dice];
    const int possible_outcomes = max_dice*6;
    int freq_counters[possible_outcomes+1];
    // Extra 1 is because outcomes will go from 1 to
    // possible_outcomes, not from 0 to possible_outcomes-1
    int num_dice;
    int num_rolls;
    int roll_cnt;
    int i;
    cout << "Enter number of dice (1 - " << max_dice << "):";
    cin >> num_dice;
    if ((num_dice<1)||(num_dice > max_dice))
        {
```

```cpp
                     cout << "Sorry. Can only handle between 1 and "
                          << max_dice << " dice\n";
                     return 0;
                 }
            cout << "Enter number of rolls: ";
            cin >> num_rolls;
            if (num_rolls < 1)
                {
                cout << "Cannot do less than 1 roll!\n";
                return 0;
                }
            srand((int) time(NULL));    // Seed rand number generator
            for (i = 0; i < possible_outcomes+1; i++)  // Initialize range
                                                       //of cnts
                freq_counters[i] = 0;
            for (roll_cnt = 0; roll_cnt < num_rolls; roll_cnt++)
                {
                roll(dice, num_dice);
                int sum = 0;
                for (i = 0; i < num_dice; i++)
                    sum += dice[i];
                freq_counters[sum]++;
                }
            print_histogram(num_dice*6,freq_counters);
            return 1;
            }
        int roll(int dice[], const int num_dice)
            {
            for (int i = 0; i < num_dice; i++)
                {
                dice[i] = rand()/(RAND_MAX/6) + 1;
                }
            return 0;
            }
        int print_histogram(const int num_bins, const int bin_array[])
            {
            const int max_height = 25;    // Desired height of histogram
            const int line_length = 79;    // Length of a text line
            char outline[line_length+1];
            int i;
            // First, clear outline
            for (i = 0; i < line_length; i++)
                outline[i] = ' ';
            outline[line_length] = '\0';    // Null termination
            // Find highest peak in histogram
            int height = -1;
            for (i = 0; i < num_bins; i++)
                {
                if (bin_array[i] > height)
                    height = bin_array[i];
                }
```

```cpp
                    // Now make the histogram fit in max_height:
                    // On each iteration decrement the height threshold by
                    // height_dec - set so histogram will occupy max_height lines
                    int height_dec = height/max_height;
                    if (height_dec < 1)
                        height_dec = 1;
                    while(height > 0)
                        {
                        for (i = 0; i <= num_bins; i++)
                            {
                            if (bin_array[i] >= height)
                                outline[i] = '*';
                            }
                        cout << outline << '\n';
                        height -= height_dec;
                        }
                    return 0;
                    }
// Example run of the program:
// (Distribution may appear skewed when printed with
// variable-width font)
//
//  Enter number of dice (1 - 12): 8
//  Enter number of rolls: 9999
//                 *
//                ***
//               *****
//               *****
//              *******
//              *******
//              *******
//             *********
//             *********
//             *********
//            **********
//            **********
//            **********
//           ************
//           ************
//           ************
//          **************
//          **************
//          **************
//         ****************
//         ****************
//        ******************
//       ********************
//       ********************
//      ************************
//     **********************************
//
```

This is the first substantial program that we've seen without extra annotations. Trying to interpret an unannotated program is a common situation in real life, so this is an opportunity to practise. Here are some tips: Begin reading at the **main()** function and trace through the functions as they are called. 'Play computer' with possible input values if that helps in understanding, but, if possible, just think of variables as variables, not particular values. If you don't understand what a particular variable is doing, see if it is passed to a function. In the function it may be called something else which is more meaningful. You can also read from the point of view of a data structure, seeing what parts of the code manipulate it or convert it to another form. In this program the **freq_counters** array would be a good data structure to follow.

Write a computer version of the children's game 'Rock–Scissors–Paper'. The computer will play several rounds of the game against a human opponent. It will cue the user, for example by printing the string **"Enter r, s or p, followed by return"**, then wait for the user to enter a selection. Before the user has input their selection (no cheating!) the computer makes its selection, which is revealed to the user after they have input. The computer figures out the winner according to the rule Rock beats Scissors, Scissors beats Paper, Paper beats Rock. The computer also keeps score. You can consider how the computer could learn its opponent's patterns and predict their behaviour. For example, you could count choice frequencies to infer probabilities for the next human choice. You could go further and include higher-order prediction using previous user input sequences.

4.11 Bitwise operators

Here are the remaining C/C++ operators. These manipulate individual bits of **ints** and **chars**.

   &   Bitwise AND. This combines its two operands by doing a bit-by-bit AND operation; i.e. each bit in the result is set (equal to 1) if and only if the corresponding bit in both operands is set.

   | Bitwise OR. Each bit in the result is set if the corresponding bit in either of the operands is set.

   ^   Bitwise Exclusive OR. Each bit in the result is set if the corresponding bit in exactly one of the operands is set.

   >>   For a **signed int**, shift bits right the number of places given by the following number; fill in 1s at the left if the number is negative, 0s if the number is positive (this is called 2s complement sign extension). For an **unsigned int**, fill in 0s at the left.

   <<   Shift bits left the number of places given by the following number; fill in 0s at the right.

   ~   (Unary operator) Produces the 1s complement of its operand (i.e. every bit is inverted).

Table 4.4 *Precedence and associativity of operators*

Operator	Meaning for built-in types	Associativity
()	explicit parentheses function call	Left to right
{}	block of code	
[]	array subscript	
.	direct reference to structure or union member	
->	pointer reference to structure member	
!	logical NOT	Right to left
~	bitwise complement	
+	unary plus	
-	unary minus	
++	increment	
- -	decrement	
&	address of	
*	contents of	
(*typecast*)	cast to specified type	
sizeof	size in bytes of data structure	
*	multiply	Left to right
/	divide	
%	modulo (remainder after division)	
+	add	Left to right
-	subtract	
<<	bitwise shift left	Left to right
>>	bitwise shift right	
<	less than	Left to right
<=	less than or equal to	
>	greater than	
>=	greater than or equal to	
==	equal to (test for arithmetic equality)	Left to right
!=	not equal to	
&	bitwise AND	Left to right
^	bitwise Exclusive-OR	Left to right
|	bitwise OR	Left to right
&&	logical AND	Left to right
||	logical OR	Left to right
?:	conditional expression	Right to left
= *= /= %= += -= &= ^= | =<<= >>=	assignment	Right to left
,	sequential expression	Left to right

Each of these can be immediately followed by an equals sign (just as with +, -, etc.) to mean modify the single variable specified. The shift operators can be used to do fast multiplication or division by powers of 2. For example

```
a >>= 4;
```

(shift **a** right 4 places) is often faster than

```
a /= 16;
```

(assuming of course that **a** is an integer type for which shifting has this effect).

It is important not to confuse the bitwise **&** and **|** with the logical operators **&&** and **||** which interpret any non-zero number as true, zero as false, and produce a boolean result.

We have now introduced most of the standard C/C++ operators. For reference Table 4.4 lists all of them in order of precedence, with their associativity. The first line in the table indicates operators with highest precedence. Associativity tells the direction in which the compiler evaluates the operands.

There are a couple of unusual placings in the precedence table. For example,

```
x_times_seventeen = x<<4 + x;
```

does not parse as

```
x_times_seventeen = (x<<4) + x;
```

but as

```
x_times_seventeen = x<<(4 + x);
```

which is probably not what you want.

This turns out to be ideal for using **<<** and **>>** as put to and get from operators, their other well-known role in C++. How operators can come to have such different roles is discussed in Chapter 5, page 134.

If in doubt about operator precedence, use parentheses!

4.12 Pointers

A pointer in C/C++ is the address of something. We saw on page 74 how the addresses of arrays are passed between functions by giving the array name. Sometimes it is necessary to pass addresses of other types of data, and for this pointers are required. Pointers are much more important than this though. They are at the heart of C's low-level efficiency and C++ has inherited them. They can be tricky to use, which is why some programmers are wary of them. But they are the most natural way to link data items together. They are essential for low-level programming and for fast manipulation of large amounts of data. They are also very important to C++'s implementation of object-oriented programming.

The **&** operator gives the address of something.

```
b = &a;
```

sets **b** equal to the address of **a**. Note that this use of **&** is different to its placement at the end of an argument type to declare a reference parameter (page 72).

To get at the contents of an address, use the * operator.

```
a = *b;
```

sets **a** equal to the contents of **b**.

A pointer is just an address, so how should it be declared? C++ requires that a pointer is declared as a pointer to a particular type.

```
char *a;
int *b;
float *c;
```

a is a pointer to type **char** – the only things it may point to are **chars**. ***a** is always a **char**. Similarly, **b** is a pointer to type **int**, so ***b** is always of type **int**. **c** is a pointer to type **float**, and ***c** is always of type **float**.

Pointers can be used to implement a swapping function (modelled on the swap program given earlier):

```
#include <iostream>
using namespace std;
void swap(int *p, int *q);
int main()
  {
  int a,b;
  cout << "Enter two integers: ";
  cin >> a >> b;
  if (a > b)
     swap(&a, &b);   // Pass addresses of a and b to swap
  if (a == b)
     cout << a << " = " << b << '\n';
  else
     cout << b << " > " << a << '\n';
  return 0;
  }
void swap(int *p, int *q)
  {
  int temp;
  temp = *p;
  *p = *q;
  *q = temp;
  }
```

Here the addresses of the two items to be swapped are passed, and **swap** does its work by using the contents of those two addresses.

Pointers really come into their own in dealing with arrays. Suppose an array and a pointer are declared:

```
int line[256], *pline;
```

Now **pline** can be pointed to the zeroth member of the array (the start of the array) as illustrated in Figure 4.1.

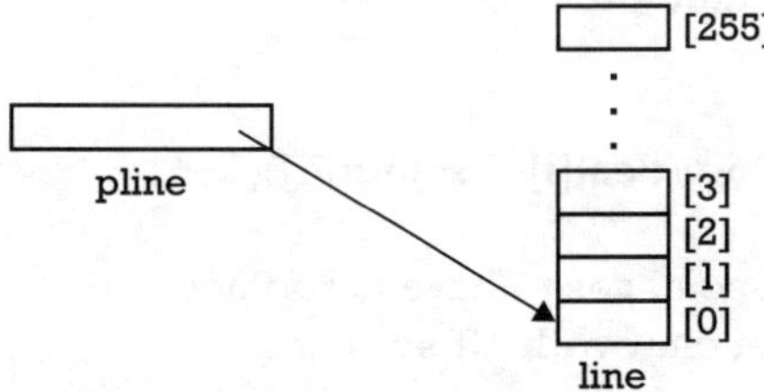

Figure 4.1 *Pointer **pline** set to the address of the zeroth element of array **line***

```
pline = &line[0];
```

As mentioned earlier, the name of the array represents the start address, so equivalently:

```
pline = line;
```

Now ***pline** gives the contents of the array start address, i.e. **line[0]**. It is important to realize:

- **pline** is just an address (possibly 2 bytes, probably 4, possibly 6 or 8, depending on the size of addresses on the computer – but *just* an address). Where **pline** is in memory is irrelevant.
- Somewhere in memory (it could be anywhere) are 256 contiguous ints that make up the array **line[256]**.
- **pline** has been made to point at the start of the **line[]** array, which means the address stored in **pline** is the address of the zeroth element of **line[]**. What this address actually is does not matter much – but having it stored in a pointer means the array can be accessed via the pointer.

Because **pline** was declared as a pointer to a particular type, the compiler knows just how big that data type is and therefore the space between elements of the array. This means that **pline + 1** points at line[1], and ***(pline + 1)** gives **line[1]**, ***(pline + 2)** gives **line[2]** and so on. This would work if the array was of **chars**, **ints**, **floats**, or whatever.

Even better, the value of a pointer can be changed so that it points to a different member of the array. For example

```
pline = line;
pline++;
```

means that **pline** is made to point at **line[1]**, and ***pline** is **line[1]**. See Figure 4.2.

Consider again the function designed for copying a string. Here is the form we have seen before:

```
int my_strncpy(char dest[], const char source[], const int
      dest_length = 256)
// dest_length is the length of the dest[] array - defaults to 256
// Rest of the program is unchanged.
   {
   int i;
   for (i = 0; (i < dest_length)&&(dest[i] = source[i]); i++)
      ;
   if (i == dest_length)  // Special case - have to replace
                          // last char with '\0' so string
      dest[i-1] = '\0';       // manipulation works correctly hereafter.
   return(i);  // Length of string
   }
```

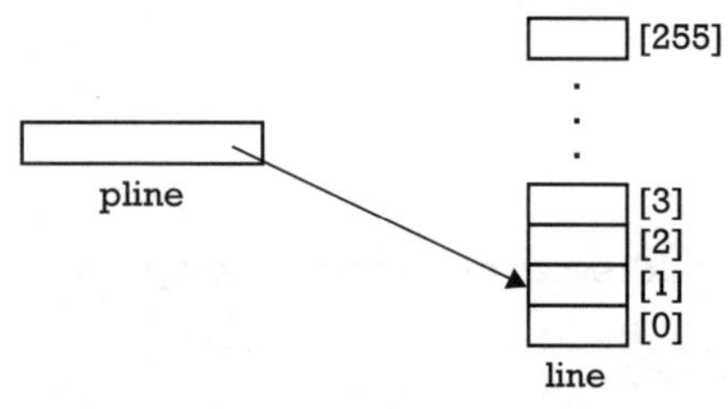

Figure 4.2 *Pointer* **pline** *incremented to point at the next element of array* **line**

Now, with pointers, the first and fifth lines can be replaced to make something more efficient:

```cpp
int my_strncpy(char *dest, char *source, const int
dest_length = 256)    // Replacement
// *dest means the same as dest[] - the start address of an array
  {
  int i;
  for (i = 0; (i < dest_length)&&(*dest++ = *source++); i++)
    // Replacement
    ;
  if (i == dest_length)    // Special case - have to replace last
                           // char with '\0' so string
    *(dest-1) = '\0';    // manipulation works correctly
                         // hereafter.
  return(i);  // Length of string
  }
```

Do not panic! The idiom ***dest++ = *source++** is hard to understand at first, but it is very common and useful.

dest and **source** have been declared as pointers to **chars**. **my_strncpy** can be called as before – the caller still has to supply the start addresses of the arrays; the only difference is that **my_strncpy** now uses these addresses as pointers. In the condition part of the **if** statement, the contents of what **dest** is pointing to are set to the contents of what **source** is pointing to; **source** and **dest** are then both incremented to point to the next item in their respective arrays. If the value that was copied is zero, the loop terminates. All this is accomplished by

(*dest++ = *source++)

Here is another example – a rewrite of the program on page 74 for reversing the order of 10 input integers. This version uses pointers.

```cpp
#include <iostream>
using namespace std;
int main()
  {
  int stored_input[10], *p;
  int i;
  cout << "Enter 10 integers:\n";
  p = stored_input;
  for (i=0; i < 10; i++)
    cin >> *p++;   // Put the integer into the array location
                   // currently pointed to by p, then move p
                   // to the next location
  for (i=0; i < 10; i++)
    cout << *-- p <<' ';  // Move p one location back, then get
                          // the value in that location
  cout << '\n';
  return 0;
  }
```

> Compare this with the version on page 74, and 'play computer', executing the code yourself, to grasp the function of pointers.

Confusion between pointers and arrays is a common problem. The examples above show that when a function's arguments are declared, the statements

```
char dest[]
char *dest
```

are completely equivalent. However, inside a function,

```
char dest[];
```

would be an invalid declaration because it does not give the array size, and if it did,

```
char dest[256];
```

for example, it would mean something very different from

```
char *dest;
```

When first learning C or C++, it is easy to forget just how different arrays and pointers are. Misuse of pointers rapidly leads to disaster. The story usually begins with something like this:

```
char *pstring;
strcpy(pstring, "word");  // Initialize pstring with "word"
```

At this point the programmer believes that the string **"word"** has been stored starting at location **pstring**. And that's right – it has! What's more, the string is properly null terminated, so that if the next line is:

```
cout << "This is the final " << pstring;
```

the program would probably print out:

This is the final word

just as expected. Yet this program has an insidious bug. Pause for a moment to see if you can find it.

The problem is that a pointer has been declared, but a pointer is just a variable to hold an address. No memory has actually been reserved for the string. In this program **pstring** could be pointing anywhere. When the characters are stored, they are put at the location in memory where **pstring** happens to be pointing, which could be a part of memory reserved for other variables, or perhaps even part of the operating system's working space. When those other variables are next accessed by their real owner, perhaps by a critical portion of the operating system, they have been corrupted. Bugs like this can cause very nasty crashes, and are difficult to track down. The thing to remember is that

```
char *pstring;
```

declares the memory location for the pointer (i.e. the address of something), but not for the thing(s) to which it points.

> Write a **my_memcopy** function that takes three arguments – a **source** address, a **destination** address and a **count** and copies **count** bytes from **source** to **destination**. So far, this is easy – it is like **my_strncpy** without the test for '\0'. But here's the hard part: make sure your function works properly even when the **source** and **destination** arrays overlap. If the address **destination** is within **count** bytes of **source**, then you have to ensure that you do not overwrite part of the **source** array before it is copied. Test your function by designing an appropriate **main()** with arrays and data to copy, which calls **my_memcopy**.

4.13 Arrays of pointers and program arguments

It is not possible to have an array of variable-length items, so the next best thing is to have an array of pointers (all of which have the same size, because they're all just addresses), and make them point to the items of interest. The most common example of this is with strings. One way to manipulate strings of different lengths – for example, sort them into alphabetical order – is to leave them where they are in memory, set up an array of pointers so that each points to one of the strings, then move the pointers around the array as required.

For example, Figure 4.3 shows an array **char *greet[4]** storing pointers to strings of different lengths. The strings could be located anywhere in memory.

A program to print the strings in alphabetical order could simply shuffle the values in the **greet** array without moving the strings to which they point. See Figure 4.4.

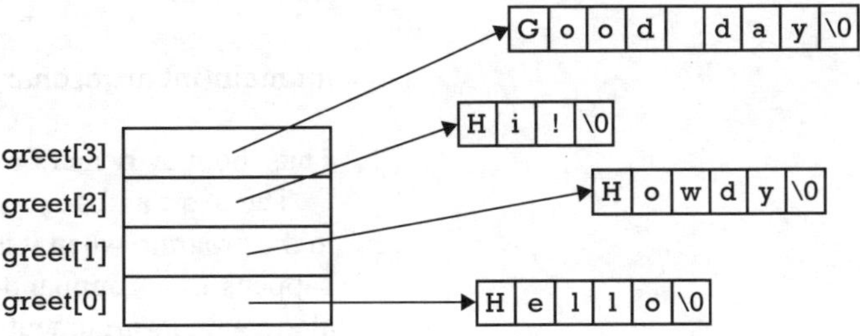

Figure 4.3 *An array of pointers to* **char** *with each entry pointing to a null-terminated string*

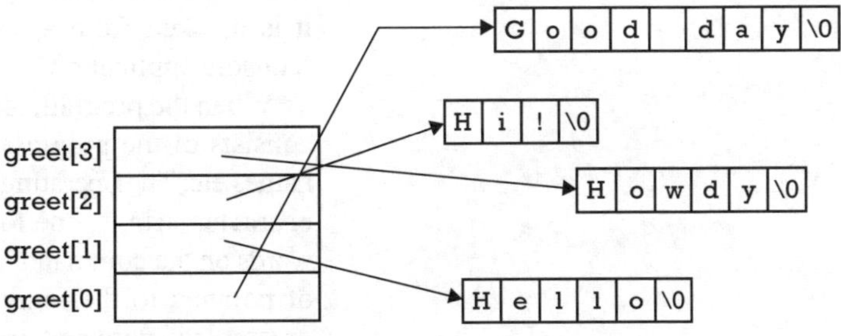

Figure 4.4 *The* **greet** *array for Figure 4.3 reordered so the strings are in alphabetical order*

Sometimes the address of an array of pointers must be passed from one function to another. This address is a pointer to a pointer. There is nothing special about pointers to pointers – they are just pointers. Every pointer is declared according to the thing pointed to at the end of the chain of pointer links. So, for example:

```
char c, d;
char *p;      // Pointer to a character
char **q;     // Pointer to a pointer to a character
q = &p;       // A legal assignment, making q point at the
              // pointer p
*q = &c;      // Also legal: makes the pointer pointed to by
              // q - in other words, p - point at c
*p = 'a';     // Puts character 'a' into char pointed to by p - in
              // this case the variable c
d = **q;      // Puts the character pointed to by the pointer
              // pointed to by q into d. So this sets d equal to 'a'
```

Usually things do not get this complicated. It is simply a case of passing an array address (such as **greet** in the example above) so that a function can manipulate an array of pointers.

One important use of an array of pointers is in the arguments to **main()**. At least, this is so in the software tools approach to programming (see page 35).

So far the programs in this chapter have opened with **main()** – no function prototype, no arguments. In fact, the operating system does pass three (or sometimes just two) arguments to **main** – these are commonly known as **argc**, **argv**, and **env**. The declaration below shows them:

```
int main(int argc, char *argv[], char *env[])
```

Thus, both **argv** and **env** are the addresses of arrays of **char** pointers.

The **argc** and **argv** arguments refer to parameters that are passed to the program when it is invoked. Traditionally, and still usefully, this happens in a command-line interface, where the program is named along with its arguments. A program with a GUI is unlikely to receive arguments when **main()** is called (i.e. when the program is started). Indeed, many GUI programming environments replace **main()** with an alternative that does not take arguments. In the following sections, we will assume that **main()** does take arguments conventionally. That is, it is invoked from a command-line interface, and is compiled as a 'console application'.

When the program is invoked, the user types a command line which consists of the program name with (perhaps) some parameters, filenames etc. The operating system converts the command line into a set of character strings, one for each word. The value **argc** is the number of words on the command line and the value **argv** is the address of an array of pointers to the individual word character strings. **argc** is always greater than zero and **argv[0]** points to the name of the program itself. **env** works in the same way as **argv** – it is an array of pointers to strings, but its meaning varies between operating systems. In general, the strings are of the form *ENVIRONMENTVARIABLE=value* where *ENVIRONMENTVARIABLE* is the name of some potentially relevant variable passed from the invoking program (the shell or command interpreter) to the program. In the Windows XP Command prompt, for example, the strings are those manipulated by the SET command, such as: PATH=C:\WINDOWS. Most simple programs ignore the **env** argument to main.

Here is a program that echoes its arguments in reverse order:

```cpp
#include <iostream>
using namespace std;
int main(int argc, char *argv[])
    {
    int i;
    for (i = argc-1; i > 0; i--)
        cout << argv[i] << ' ';
    cout << '\n';
    return(0);
    }
```

If this program's name was rev_arg, then the command line

rev_arg aaa bbbbb c dd

would produce the output

dd c bbbbb aaa

Recall that **argv** is a pointer to a pointer to character, so **argv[i]** (or ***(argv+i)**) is a pointer to a character – in fact, a null-terminated string of characters. Here is a program that finds the sum of its (integer) arguments:

```cpp
#include <iostream>
#include <cstdlib> // Has function prototype for atoi
using namespace std;
int main(int argc, char **argv)
    {
    int i, sum;
    sum = 0;
    for (i = 1; i < argc; i++)
        sum += atoi(argv[i]);   // Convert the ith argument string
                                // into an int type, then add it to sum
    cout << sum << '\n';
    return(0);
    }
```

atoi is a library function that converts a string to an integer.

> Rewrite the **ipower** program from page 71 to read its arguments (**a** and **b**) from the command line, rather than prompting for them.

4.14 Static and global variables

So far, all the variables used here have been 'local' and 'automatic'. These variables are declared within a particular function and exist only as long as the function is running – when it returns, the variables disappear. Sometimes a variable that exists, and holds its value, for the duration of the program is required. There are three reasons for using such variables:

1. A function is called repeatedly, with some data items that must retain their values between calls. (For example, there may be a variable which counts how many times the function has been called.)
2. A function manipulates a large number of variables (or, perhaps, a large array), and the platform has difficulty managing so much 'automatic data'.
3. Data are to be shared between several functions.

In almost all cases the solution to these issues is to create a class that stores the variables as well as the functions that manipulate them. See Chapter 5. However, there are older alternatives which are still sometimes useful and we take a quick look at them here.

A static variable is local, but exists for the program's duration. Such variables can be declared with the use of the keyword **static**. For example:

```
int num_times_called()
   {
   static int counter = 0;  // counter initialized to 0 at start of
                               // program run
   cout << "This function has been called " << ++counter <<
      " times\n";
   return counter;
   }
```

Each time the program enters the function **num_times_called, counter** will have kept its value from the previous time it was entered.

Static variables declared inside a function are only 'visible' within that function. That is, the 'scope' of the variable is the function in which it is declared. Suppose you want a variable that is visible and usable by several functions.

Declaring a variable outside of any function (and before all those that use it) automatically makes that variable last for the duration of the program and allows it to be shared. In the following swap program, the variables a and b are shared between **main()** and **swap()**:

```
#include <iostream>
using namespace std;
int a, b;
void swap();
int main()
   {
   cout << "Enter two integers: ";
   cin >> a >> b;
   if (a > b)
      swap();
   if (a == b)
      cout << a << " = " << b << '\n';
```

```
  else
     cout << b << " > " << a << '\n';
  return 0;
  }
void swap()
  {
  int temp;
  temp = a;
  a = b;
  b = temp;
  }
```

a and **b** in this program are called 'global variables'. At first glance use of global variables seems easier than passing arguments around. Why not have all the data declared outside of functions, so that all functions can access what they want when they want? But this apparent convenience causes many problems when writing larger systems. Since a global variable can be seen, accessed and *changed* by any function, there is a greater risk that the wrong function will change it at the wrong time. For good software engineering, it is important NOT to have everything globally visible and accessible.

Rule 8

Hide information in public; reveal it in private.

Data should be hidden, so far as possible, as a key step in achieving good modularity. Therefore it is good practice to avoid using global variables, except where really necessary.

There is also an **extern** keyword which flags a global variable declared in another file that will be used in this one.

4.15 File input and output

Nearly all useful programs have to read and write information from disk files. Just as we have been using standard input and output streams, despite their reliance on more advanced concepts, here we introduce file streams, and defer a full explanation of the mechanisms to Chapter 5.

Some programs must read and write arbitrary files without making assumptions about their structure. But there are many particular cases where a known structure can be assumed (perhaps because the program reading the file was responsible for creating it). Textual databases, for example, can be created automatically, assumed to be uncorrupted, and read directly into the program.

Here is a very small database file, together with a program for reading it in. The first line of the database file specifies the number of 'records' or lines in the file. The individual records then consist of five 'fields' each: a location name, the latitude in degrees followed by a character indicating north or south (of the equator), the longitude in degrees followed by a character indicating east or west (of the Greenwich meridian).

The database file:

```
5
Toronto    43.66 N 79.33 W
```

```
London     51.5 N 0.17 W
Delhi      28.63 N 77.23 E
Sydney     33.87 S 151.22 E
Lima       12.05 S 77.05 W
```

The program:

```
 1:   #include <fstream>       // Required for file i/o
 2:   #include <iostream>
 3:   using namespace std;
 4:   int main(int argc, char **argv)
 5:      {
 6:      if (argc != 2)
 7:         {
 8:         cout << "Usage: readlocs locationsfile\n";
 9:         return -1;
10:         }
11:      ifstream infile(argv[1]);  // Declares a stream object,
12:              // and makes it refer to the input file (the first and
13:              // only argument on the program command line)
14:      if (!infile)     // If the file could not be found, infile is 0
15:         {
16:         cout << "Can't open " << argv[1] << " for reading\n";
17:         return -1;
18:         }
19:      int numrecords;
20:              // First line of database file is number of
21:              // records, so...
22:
23:      infile >> numrecords;  // Using getfrom operator - just like
24:                             // standard input!
25:      for (int i = 0; i < numrecords; i++)
26:         {
27:         char name[20];
28:         float latitude, longitude;
29:         char ns, ew;
30:         // Get records one by one and print them out
31:         // on the screen
32:
33:         infile >> name >> latitude >> ns >> longitude >> ew;
34:         cout << "Location " << name << " is at latitude " <<
35:            latitude << ' ' << ns << ", longitude " << longitude
                << ' ' << ew << '\n';
36:         }
37:      infile.close();
38:      return 1;
39:      }
```

On lines 6 to 10, the program illustrates standard practice for a program that takes a particular number of arguments: check that the command line has the right number of words, and if not print usage information. Similarly, if an error in reading or writing a file occurs (lines 14 to 18), print an error message so the user can track down the problem.

The statement on line 11

```
ifstream infile(argv[1]);
```

uses one of the input types defined in **fstream** to declare the variable **infile**. The type **ifstream** is used for opening a disk file, prior to reading it as a stream (just as the standard input **cin** is read as a stream). This declaration of **infile** actually looks a bit like a function call, because a value (**argv[1]**) is involved. Later in the code **infile** seems to be working as a variable, a structure (see next section) and a function! It is, in fact, an *object* and we shall see in Chapter 5 how to declare and use other variables like this. The important thing to note though is that apart from the declaration of **infile** and the **close** line (37), it is being used with the get from (>>) operator (lines 23 and 33). Using file streams, the get from >> and put to << operators work exactly as they do with **cin** and **cout** for standard input and output.

Also note the use of error returns. When **infile** is declared (line 11), it opens the appropriate input file. If the opening operation fails, then the object **in** is set to zero. Zero is often used as an error return value when a function returns a pointer. For functions that return integers, an often-used convention is to return a negative value if an error occurs. This is what is done in the above program, in the return value from **main**. Many systems do not do anything with a program's return value. But nothing is lost by returning a value.

Enter and run this program. Why do some of the floating point values get changed? (A hint is given on page 60).

The program has a latent bug. Normally it will work fine, but the bug is waiting to pounce. Read carefully and see if you can find it. If you can't, a clue is coming up on page 105.

When dealing with arbitrary files, the get from and put to operators cannot be used because the structure of the file is unknown. In this case it is necessary to use other facilities of file streams. Here is a preview of a program explained fully in Chapter 5. It copies one file to another using the file stream facilities **get** and **put** to read and write single characters. It is included here so that readers can start using unstructured file input/output now, if they have need. The exercises in this chapter do not require it.

```
// Program to copy one file to another
#include <iostream>
#include <fstream>
using namespace std;
int main(int argc, char **argv)
    {
    if (argc != 3)
```

```
    {
    cerr << "Usage: copy source dest\n";
    return(-1);
    }
ifstream in(argv[1]);   // Opens file given by argv[1] (first
                        // program line argument)
                        // for reading
// If working on a PC e.g. Borland C++, have to use
// ifstream in(argv[1], ios::binary);
if (!in)
    {
    cerr << "Can't open " << argv[1] << " for reading\n";
    return(-1);
    }
ofstream out(argv[2]); // Opens file given by argv[2] for
// writing. ofstream is the mirror image of ifstream (just as
// cout is the mirror of cin)
// If working on a PC e.g. Borland C++, have to use
// ofstream out(argv[2], ios::binary);
if (!out)
    {
    cerr << "Can't open " << argv[2] << " for writing\n";
    return(-1);
    }
char c;
while(in.get(c))        // Gets a character from the in file into c
        // in.get(c) returns 0 when the end of file is reached.
    out.put(c);             // Puts the character in c into out file
in.close();
out.close();
return(0);
}
```

Note that in MS Windows, a file can be opened either in 'binary' mode, or in 'text' mode. Windows, and its predecessor MSDOS, differentiate between the two file types mainly because a newline '\n' actually involves two characters – carriage return followed by line feed. This means that text files are stored slightly differently to the way a program manipulates them. When dealing with files known to be text, it is appropriate to open them in text mode. Otherwise use binary mode. See the comments in the code above for how to do this.

Write a program to print out a text file on the console display. Open in text mode, but read a character at a time because you do not know the structure of the file.

For a text file of limited size – say 20 lines of at most 80 characters – write a program that will print out the lines of the file in reverse order.

4.16 Structures

Programmers make their own data types by putting together disparate basic types (**ints**, **chars**, etc.) into 'structures'. The basic form of a structure definition is:

struct *type_name*
{
*List of data items to be put together in the structure in the
    format:*
type member_name;
};

Once defined, variables with this structure can be declared simply with

type_name variable_name;

Suppose we wanted to define a structure for the locations database program discussed on page 94, or one that holds a student's name, I.D. number and mark, and then use an array of these structures for a marks reporting program. We could, for example, define structures like these:

**struct location
{
char name[20];
float latitude; char ns;
float longitude; char ew;
};
struct student
{
char name[30];
unsigned long id;   // unsigned long is used so that the full
                    // range of possible 8-digit
                    // I.D. numbers can be covered.
float mark;
};**

Variables with these structures can now be declared:

**location university_of_poppleton;
student the_ideal_student;
student syde321[200];**

This declares a single **location** structure's worth of storage for the variable **university_of_poppleton**, a single **student** structure's worth of storage for the variable **the_ideal_student**, and an array of 200 **student** structures for **syde321**, all with the format defined in the original **struct** statements. (None of the storage is initialized to contain meaningful values yet though.)

Access to particular members of a structure is through the dot '.' operator – writing *structure_name.member*. For example:

**university_of_poppleton.ns = 'N';
the_ideal_student.mark = 100;
syde321[14].id = 12345678;**

The last of these has set the **id** field of the 14th entry of the **syde321** array to 12345678. (14th counting from zero, remember!)

Here is a code segment that steps through the array of structures:

```
average = 0;
for (i = 0; i < 200; i++)
   average += syde321[i].mark;
average /= 200;
```

In each case the **mark** item is correctly retrieved from the appropriate structure.

The following program is a simple example of the use of structures. This program converts between Roman and Arabic numbers. Unfortunately it only deals with 'primitive' Roman numbers: these do not use the subtractive prefix convention that allows, for example, formation of 4 as IV (= 1 subtracted from 5) or 90 as XC (= 10 subtracted from 100). Instead the program recognizes IIII as 4, and LXXXX as 90.

```
// Simple Roman number convertor: Doesn't handle subtraction
// prefices
#include <iostream>
#include <cstdlib>
#include <cctype>   // Needed for isdigit()
using namespace std;
struct convertor       // A convertor stores an arabic integer
                       // and its equivalent Roman character.
                       // E.g. 10, 'X'
   {
   int value;
   char roman_symbol;
   };

// The array of convertors is declared globally and initialized
// to contain all relevant arabic/roman number pairs.
const convertor conv[7] =
      {
      1, 'I', 5, 'V', 10, 'X', 50, 'L', 100, 'C', 500, 'D', 1000, 'M'
      };

void arabic_to_roman(const char *in, char *out);
void roman_to_arabic(const char *in, long *value);

int main(int argc, char *argv[])
   {
   char outroman[20];
   long outarabic;
   if (argc != 2)
      {
      cerr << "Usage: romanize number (where number is in\
         Arabic or Roman numerals)\n";
      return -1;
      }
```

> When a string is too long for one line of program and has to be split, as the 'usage' string is here, the backslash character '\' is put at the end of each line to denote continuation onto the text.

```cpp
                        // Read the first character to see whether input is Arabic or
                        // Roman
                        if (isdigit(argv[1][0]))    // First character is a digit, i.e.
                                                    // an arabic numeral
                           {
                           arabic_to_roman(argv[1], outroman);
                           cout << outroman << '\n';
                           }
                        else
                           {
                           roman_to_arabic(argv[1], &outarabic);
                           cout << outarabic << '\n';
                           }
                        return 0;
                        }
                void arabic_to_roman(const char *in, char *out)
                   {
                   long value;
                   value = atoi(in);    // Convert input to an integer value
                   int i = 6;           // Start at largest numbers
                                        // (i indexes the convertor array from
                                        // large values to small)
                   while(value >0)
                      {
                      if ((value - conv[i].value) >= 0) // Keep outputting
                           // current symbol so long as
                           // its value can be subtracted from the residue
                           // of the input value
                           {
                           *out++ = conv[i].roman_symbol; // Output symbol
                           value -= conv[i].value;    // Reduce "input" value
                                                      // appropriately
                           }
                      else
                           i--;   // Move to next convertor down (next lowest value)
                      }
                      *out = '\0'; // Terminating null
                   }
                void roman_to_arabic(const char *in, long *value)
                   {
                   *value = 0;
                   char c;
                   int i = 6;   // Assume Roman number is correctly formed:
                                // Starts with highest value symbol and moves
                                // down values monotonically through string
                   c = *in++;
                   while ((c != '\0')&&(i >= 0))
                      {
                      // Find character
                      if (c == conv[i].roman_symbol)    // Matches current
                                                        // symbol
                         {
```

```
                        *value += conv[i].value;    // Add symbol value to
                                                    // output value
                    c = *in++;
                    }
                else
                    i--;    // Move to next convertor down
                }
            }
```

> Study the above program to see how it converts numbers both
> from Arabic to Roman and from Roman to Arabic. Modify the
> program so it works with input Roman numbers that use prefix
> subtraction. That is, a smaller value symbol on the left of a larger
> value symbol subtracts the smaller from the larger. For example:
> MCMLXXXIV = 1984. Think carefully about your algorithm!

4.17 Pointers to structures

Pointers to structures are declared just like pointers to any other data
type. Pointer arithmetic works, so by incrementing a pointer value it is
possible to step through an array of structures, no matter how big each
struct is. For example, with the **student** structure definition as above,

```
student syde321[200];    // Set up an array of 200 student
                         // structures
student *pstudent;       // pstudent is a pointer variable so will
                         // contain the address of a
                         // particular student structure.
pstudent = syde321;      // Pointing at syde321[0]
pstudent++;              // Now pointing at syde321[1]
```

Just as the dot operator is used to look inside a named structure, the '->'
operator is used to look inside a structure from a pointer. The syntax
is *structure_pointer->member*. For example:

```
pstudent = syde321;          // Pointing at start of array
pstudent->mark = 73.4;       // Sets syde321[0].mark to 73.4
strncpy(pstudent->name, "Alice Nonesuch",30);
pstudent->name[29] = '\0';
```

(where **strncpy** is a library function that does much the same thing as
the **my_strncpy** that we wrote earlier, except that when all 30 characters are copied it doesn't overwrite the last with a null terminator.
Hence the extra line to ensure the last character is a terminator).

Pointers to structures can be passed down to functions to manipulate data declared in the calling function.

Here is an example of a program using pointers to structures. This
is the first prototype of a program for reading a database of student
marks and printing out the top mark.

```
#include <iostream>
#include <cstring>    // To manipulate C-style strings
```

```cpp
using namespace std;
struct student {
  char name[30];
  unsigned long id;
  float mark;
  };
float top_mark(const student *pstudent);
void load_data(student *pstudent);

int main() {
  static student syde321[200];
  load_data(syde321);
  cout << "Top mark was " << top_mark(syde321) << "\n";
  return(0);
}

float top_mark(const student *pstudent) {
  int i;
  float top;
  top = -1.0;                        // Initialized so that any real
                                     // student mark will overwrite

  for (i = 0; i < 200; i++)
    {
    if (top < pstudent->mark)   // If this student's mark is
                                // better than current top

       top = pstudent->mark;    // Replace top with its value
    pstudent++;                 // Move to next student
    }
  return(top);
}
void load_data(student *pstudent) {   // Presently a STUB
  for (int i = 0; i < 200; i++)
    {
    strcpy(pstudent->name, "Stubbs");    // A pun
    pstudent->id = 12345678;
    pstudent->mark = 75.5;
    pstudent++;
    }
}
```

top_mark returns a **float** and its single argument is a pointer to type **student**. **top_mark** does its stuff by stepping through all 200 array entries and finding the highest value of **pstudent->mark**.

The function **load_data** is defined to set up specific values – the same ones in each of the **syde321** array entries! Eventually, this function will be modified to read input data from a student database file (in a few pages' time). Right now, **load_data** is a stub – a kind of minimal implementation just to get the program going. Remember:

Principle 3

**A program grows along the framework of
its first version.**

and

Rule 5

**Make the first version durable not functional,
and get it running early.**

so using stubs helps you to concentrate on getting a good framework running early.

Clearly, because load_data puts a mark of 75.5 into every student, the calculated top mark will be (should be) 75.5. A stub is therefore useful not just to get a first implementation working, but also as a vehicle for preliminary testing – if the program does not calculate 75.5, then something is wrong.

A structure can include a pointer to itself. For example:

```
struct link {
  int value;
  link *next;
  };
```

Structures like this are the building blocks of linked lists, discussed in detail in Chapter 7.

4.18 Making the program more general

The code shown in the student database example above works, but it assumes that the class size will be exactly 200 students. If the size is substantially less, then the program wastes a lot of space (and might go wrong because unused student structures are tested for 'marks' which have never been entered); if the size is 201, then the program fails. In any case, what is so special about 200?

There are times when a constant used in a program really is a special number. Perhaps all classes are known to contain exactly 200 students. Then it is good practice to flag the constant (in this case 200) as a 'magic number'. Declaring a variable and using the **const** keyword is one way to do that. An older way, still sometimes useful, is the #define keyword.

#define *name something*

is a macro definition statement in C and C++ that tells the compiler wherever you see *name* substitute *something*. In the above example an appropriate definition would be

#define MAXSTUDENTS 200

It is usual to **#define** constants at the beginning of the program, so that array sizes, parameters, etc., can be altered just by changing the initial **#define** statement.

What if 200 were a maximum, but there was the possibility of having fewer students? One way of dealing with this is to add a flag item to the **student** structure, then have **load_data** set the flags of structures as it fills them. When the list of students is complete, **load_data** should ensure that the flag fields of all the other (unfilled) entries in the table are zero. Another solution could be to have a variable which stores the number of students, and loop for that many times. A third solution is the use of a 'sentinel'. A sentinel is just a value that normally cannot occur for a particular item, stored in the array element just after the last used element. To use a sentinel the code needs to be rewritten:

```cpp
#include <iostream>
using namespace std;
#define   MAXSTUDENTS   200
// An alternative, perhaps better, would be:
// const int MAXSTUDENTS = 200;
#define MAXNAMELENGTH        30
// Similarly,
// const int MAXNAMELENGTH = 30;
struct student {
   char name[MAXNAMELENGTH];
   unsigned long id;
   float mark;
   };

float top_mark(const student *pstudent);
void load_data(student *pstudent);

int main()
   {
   static student syde321[MAXSTUDENTS];
   load_data(syde321);
   cout << "Top mark was " << top_mark(syde321) << "\n";
   return(0);
   }
float top_mark(const student *pstudent)
   {
   float top;
   top = -1;
   for (int i = 0; (i < MAXSTUDENTS) && pstudent->id; i++)
      {
      if (top < pstudent->mark)
         top = pstudent->mark;
      pstudent++;
      }
   return(top);
   }
```

Here it is assumed that the I.D. number 0 can never occur, and that **load_data** has set the **id** field of the array item just after all the student items to 0. In each iteration of the loop,

```cpp
for (int i = 0; (i < MAXSTUDENTS) && pstudent->id; i++)
```

the **id** number is checked to see whether the 0 sentinel has been reached – if it has, the loop is terminated, even though i has not yet reached **MAXSTUDENTS**. Recall that when **&&** (AND) and **||** (OR) occur in condition tests, they are evaluated left to right, and as soon as sufficient information is available to know whether the test will succeed or fail, evaluation stops. So in

(i < MAXSTUDENTS) && pstudent->id;

if i is greater than or equal to **MAXSTUDENTS**, the first part of the test yields 0 or false. Logically, then, the whole condition must be false whatever the value of the second condition. C++ exploits this and avoids the second test in this situation.

Why is this property of **&&** useful in the above program? Most of the time the array is not full, so **pstudent->idnumber** equals zero before i goes greater than or equal to **MAXSTUDENTS**. You might even think that

pstudent->id && (i < MAXSTUDENTS);

would be better, because on the last time through it saves the comparison of i against **MAXSTUDENTS**. But see what happens when the array *is* full: **pstudent** is incremented at the end of the previous loop to point just after the end of the array, and then it is tested. Now, although everything will still work (probably), the last array access is beyond the upper limit of the array. This is a bad thing: you shouldn't access data outside the ranges of your variables. The earlier formulation,

(i < MAXSTUDENTS) && pstudent->id;

is therefore better. When the array is full, the test of i against **MAXSTUDENTS** fails and the (now invalid) **pstudent** is never used.

4.19 Loading structured data

Now, what should **load_data** look like? That depends on the format in which the data are stored. It is often helpful to store databases as ASCII text files with descriptive tags included. This means that humans can read the database with just a text editor; it often helps in debugging too. It also means **load_data** can be written easily, using **>>** operators, just as in the locations database example earlier. The database will be stored in a file **students.dat** in the following way. The first line of the file contains a number in ASCII characters, giving the number of student records in the file. The remainder of the database file just consists of lines of text each giving the name, id number and grade of a student. Student names are in the form **alice_nonesuch** – i.e. underscores are used as word separators. This is because get from operators read strings up to white space characters (i.e. up to a space, a tab or a newline). Using the underscore ensures the full name is loaded into a single character string buffer. (To read strings with spaces, see the streams details at the end of Chapter 5.)

```
#include  <fstream>
using namespace std;
#define DATAFILE    "students.dat"    // Strings can be #defined
                                      // just like any other
                                      // constant
int load_data(student *pstudent)
    {
    char c;
    unsigned int numstudents;
```

```
    ifstream in(DATAFILE);
    if (!in)
       {
       cerr << "Can't open " << DATAFILE <<" for reading\n";
       return(-1);
       }
    in >> numstudents;    // Using the >> operator - again, just
                          // like standard i/o
    if (numstudents > MAXSTUDENTS)
       {
       cerr << "Too many database entries\n";
       return(-1);
       }
    for(int i = 0; i < numstudents; i++)
       {
       in >> pstudent[i].name;
       in >> pstudent[i].id;
       in >> pstudent[i].mark;
       }
    in.close();
    pstudent[numstudents].id = 0; // Sentinel
    return(0);
    }
```

Note that **main** will have to be modified to take account of the return
value from **load_data** and abort the program if that value is less than 0.

> Review the second exercise on page 95. The above function has
> a similar bug to the one lurking there. You need to think about
> this with the aid of the **struct student** definition on page 103.
> See what the bug is? More clues on page 129.

4.20 Memory allocation

We often don't know how much memory will be required until runtime. It might depend on the size of a database, the number of iterations
of an algorithm, the volume of incoming data. We need a mechanism
to ask for memory from the operating system as and when needed. In
C++, the program requests memory using the **new** operator which
'returns' a pointer to the start of allocated memory. For example, to
allocate a new student structure:

```
student *pstudent;
pstudent = new student;
```

pstudent points to a newly allocated **student** structure, which can be
used as required. When no longer needed, the memory would be
returned to free store with:

```
delete pstudent;
```

Allocating arrays of objects is easy.

```
student *pstudent;
pstudent = new student[numstudents];
```

What if **new** fails?

If **new** cannot get the memory you asked for, it throws an exception. We haven't discussed exceptions yet (see page 160), but, in the meantime, there are alternatives. First, you can rely on the default exception handling mechanism – if an exception isn't caught, the program terminates. This could well be the right thing for a failure with **new**: exhausting the memory is very unusual, and making your program work around it could be impossible. Alternatively you could use the **nothrow** version of **new** like this:

```
pstudent = new(nothrow) student[numstudents];
if (pstudent == 0) {
   cerr << "Allocation failed! Panic!" << endl;
   return -1;
   }
```

When the **nothrow** version of **new** cannot allocate the required memory it returns value 0. This can be tested as above, and be handled in the calling code.

And now **pstudent** is the start address of a block of memory that can be used as if it were a **numstudents** long array of **student** structures.

The memory would be released with

delete [] pstudent;

The use of the array brackets in the **delete** statement ensures that the whole array of items is deleted and not just the first.

Every pointer value used in a **delete** statement must have been returned by an earlier use of **new**. When a program terminates, all allocated memory is returned to the operating system anyway, but it is standard practice explicitly to **delete** everything allocated with **new**.

Here is a complete program to read in a student database and calculate the average mark. Some of the functions have been changed from those given above.

```
#include <iostream>
#include <fstream>
#include <cstdlib>
using namespace std;
#define    MAXSTUDENTS 80
#define    MAXNAMELENGTH 30

struct student
   {
   char name[MAXNAMELENGTH];
   long id;
   float mark;
   };

float average_mark(int num_students, student *p);
int  load_data(char *fname, student **pp);

int main(int argc, char **argv)
   {
   student *syde321;   // Will point to the base of the array of
                       // structures
```

```cpp
int num_students;
if (argc != 2)
   {
   cerr << "Usage: average filename\n";
   return(-1);
   }
if ((num_students = load_data(*(argv+1),&syde321)) < 0)
   return(-1);

cout << "Average was " << average_mark(num_students,
      syde321) << '\n';
return(0);
}

float average_mark(int num_students, student *p)
   {
   int i;
   float sum_marks;
   sum_marks = 0.0;
   for (i = 0; i < num_students; i++)
      sum_marks += p[i].mark;
      // Remember, a pointer passed to a function can
      // be treated like the start address of an array.
   return(sum_marks/num_students);
   }

int load_data(char *fname, student **pp)
   {
   char c;
   student *memory;
   unsigned int numstudents;
   ifstream in(fname);
   if (!in)
      {
      cerr << "Can't open " << fname << " for reading\n";
      return(-1);
      }
   in >> numstudents;
   if (numstudents > MAXSTUDENTS)
      {
      cerr << "Too many database entries\n";
      return(-1);
      }
   memory = new student[numstudents];
   // Memory is allocated here
   // Don't catch bad_alloc exception from new: Just exit if
   // memory exhausted. Write into student pointer
   *pp = memory;
   for(int i = 0; i < numstudents; i++)
      {
      in >> memory[i].name;
      in >> memory[i].id;
```

```
      in >> memory[i].mark;
    }
  in.close();
  memory[numstudents].id = 0;
  // Sentinel
  return(numstudents);
}
```

Read and interpret the program to understand how it works. You may wish to enter it and extend it by incorporating other calculations and the ability to add and remove records, and write out a modified array.

Write a program that prompts the user for input of **student** records and from them creates and writes out a **student** database file.

Using an array of **location** structures, read in a locations database (maximum of 100 locations), then write out a text file stating for each city which of the earth's zones it is in. The earth has five zones: North of the Arctic Circle (lat 66.6 deg North) is the North Arctic Zone; between the Tropic of Cancer (lat 23.5 deg North) and the Arctic Circle is the North Temperate Zone; between the Tropics is the Torrid Zone; between the Topic of Capricorn and the Antarctic Circle is the South Temperate Zone, and South of the Antarctic Circle is the South Frigid Zone.

4.21 typedef

C++ has a keyword, **typedef**, that allows the renaming of data types. It can be used with any of the basic types to make declarations more descriptive. For example

typedef char BYTE;

allows the declaration:

BYTE codeword;

typedefs may make the code more descriptive and readable. They can also help you to reconfigure code quickly for a different data type just by adjusting a single **typedef** statement.

4.22 enum

enum is a handy way of providing names for a set of integer values. It gives much the same facility as a series of **#defines**, but is more compact. Here is an example:

enum months { Jan=1, Feb, Mar, Apr, May, Jun, Jul, Aug, Sep, Oct, Nov, Dec } thismonth;

This defines **Jan** equal to 1, **Feb** equal to 2 and so on. **months** is a tag for the **enum** (just like a structure tag), and **thismonth** is an integer variable that can take one of the enumerated values. **enum** values normally start at 0; explicitly making **Jan** equal to 1 makes it possible to have code like:

```
int numdays[13];
for (thismonth = Jan; thismonth <= Dec; thismonth++)
  numdays[thismonth] = 31;
numdays[Feb] = 28;
numdays[Apr] = 30;
...
printf("The last day of the year is %d/%d\n",numdays[Dec],
    Dec);
```

4.23 Mechanisms that underlie the program

Rule 12

Understand the capabilities, components and mechanisms surrounding the program.

What is going on in the computer when a program executes? Here is a small C++ program:

```
1:   #include <iostream>
2:   using namespace std;
3:   void put_bigger_into_first(int &a, int &b);
4:   int counter;
5:   int main()
6:     {
7:     int x, y;
8:     counter = 0;
9:     for (int i = 0; i < 4; i++)
10:      {
11:      cout << "Enter two integers:\n";
12:      cin >> x >> y;
13:      put_bigger_into_first(x,y);
14:      cout << "Bigger is " << x << '\n';
15:      }
16:      cout << "Number of times bigger was already first: "
                << counter << '\n';
17:      return 0;
18:      }
19:
20:   void put_bigger_into_first(int &a, int &b)
21:     {
22:     if (b > a)
23:       {
24:       int swap = a;
25:       a = b;
26:       b = swap;
27:       }
28:     else
29:       counter++;
30:     }
31:
```

When this is compiled, a sequence of machine instructions is generated called the CODE or TEXT of the object file. The instructions include some that manipulate the processor stack where automatic variables will be stored when the program runs. All static data, including initialized strings such as **"Enter two integers:\n"** (line 11) in the above example, are listed, and space is provided for them in the DATA of the object file. All the references to static data locations, and the destinations of jump instructions, in the code, are marked as relocatable addresses. That is, the machine addresses are not yet filled in. The object file also contains a symbol table listing all the symbols and their relocatable addresses, marking which symbols in the code are as yet unaccounted for. For example, the program compiles because **cin** and **cout** are declared in **iostream**, but the code that implements them is not included in this file, so they are as yet unresolved references.

The linking step merges all the object files of a program (in this case there is only one) and pulls from libraries whatever standard code is required to account for the program's use of library functions and classes. In this case the standard C++ streams library contains the implementation of **cin** and **cout** and this is linked to the program code, resolving the previously unaccounted-for symbols. The output of the linker is an executable program. This still has relocatable addresses: the decision about what actual memory addresses to allocate to the program is left to the operating system which 'fixes up' the executable immediately before running it. So a particular program does not always run at the same address in memory. The executable file may include a symbol table: this is not needed for the program to run, but it is for symbolic debugging. Thus the linker will have a flag that allows the programmer to choose whether or not to include a symbol table in the executable. With the table, the executable will be bigger, but it will be easier to debug.

When the program runs there are four areas of memory (not contiguous) that have a role. These are:

- The area in which the text of the program is stored. The computer fetches instructions from this area for execution. Once the program is running, it has only read access in this area of memory – this is so that programs cannot rewrite themselves.
- The area in which the program's static data items are stored. In the case of the above program, this will include 4 (or 2, or 8, …) bytes for the global variable counter (line 4), 21 bytes for the string **"Enter two integers:\n"** (line 11), 11 bytes for the string **"Bigger is "** (line 14), 1 byte for the first single-character '\n' (line 14), 42 bytes for the string **"Number of times bigger was already first: "** (line 16), and 1 byte for the second '\n' (line 16).
- The system stack. When automatic variables are declared, they are put on the stack. The stack grows from high memory downwards, with a fixed top location and a processor pointer (called the stack pointer) pointing at the current bottom of the stack. When the declaration int **x, y;** appears in the code (line 7), the stack is grown downwards by 8 bytes (4 for each integer) by moving the stack pointer down eight locations. Those 8 bytes are then used by the code to store the values of the symbols x and y. When the variables

go out of scope (when the function returns) the stack pointer moves back up to reclaim their space. Many other things are put on the stack too. Just before a function is called, the address of the next sequential machine instruction is pushed onto the stack – this will be the return address for the processor to move back to when it has finished executing the function. Arguments to functions are also pushed onto the stack before a function call. In the example above, references to **x** and **y** are pushed onto the stack at line 13; that is, their addresses are pushed. Of course, those addresses happen to be locations just a little higher up the stack.

- The system heap. This is free memory which the operating system can allocate. The program gets chunks of this memory using **new**, and returns them using **delete**. There is no example of heap use in the above program.

4.24 More on the C/C++ standard library

The standard C library is part of all C and C++ compilers. In addition, C++ compilers include the standard C++ library, discussed in Chapter 5. The standard library is divided into modules according to which header file a particular function's prototype is given in. The really useful ones are **cstdlib** and **cstdio**. There are some handy things for dealing with characters and strings in **cctype** and **cstring**, and for mathematics, **cmath** and maybe **cfloat** are needed. Table 4.5 lists the most useful library functions, together with their header files, and an explanation. This is not a comprehensive list, but will be sufficient for many programs. For a full list, see your compiler documentation, which in many cases (e.g. Borland C++) will be available online.

> If you read a C program, you'll see these common library functions just as here in C++. The only difference is the name of the header file that's included. In C, header files have names like **stdlib.h** and **math.h**. The C++ header names are derived from the C ones by removing the '.h' and putting 'c' on the front. Any C++ header that starts with a 'c' is using the standard library functions shared by the two languages.

Table 4.5 *Useful standard library functions derived from the C standard library*

Function prototype	Header file	Operation
int isalpha(int c)	cctype	Returns nonzero if c is any alphabetical character (lower or upper case) (a-z, A-Z)
int isdigit(int c)	cctype	Returns nonzero if c is any decimal digit (0-9)
int islower(int c)	cctype	Returns nonzero if c is any lowercase letter (a-z)
int isprint(int c)	cctype	Returns nonzero if c is any printing character (including space)
int isspace(int c)	cctype	Returns nonzero if c is space, tab or newline
int isupper	cctype	Returns nonzero if c is any uppercase letter (A-Z)
int tolower(int c)	cctype	Returns corresponding lower case letter for c if c is an upper case letter; otherwise, returns c

(Continued)

Table 4.5 (*Continued*)

Function prototype	Header file	Operation
int tolower(int c)	cctype	Returns corresponding lower case letter for **c** if **c** is an upper case letter; otherwise, returns **c**
int toupper(int c)	cctype	Returns corresponding uppercase letter for **c** if **c** is a lowercase letter; otherwise, returns **c**
double acos(double x);	cmath	Returns arccos of **x** in range $[0,\pi]$ in radians
double asin(double x);	cmath	Returns arcsin of **x** in range $[-\pi/2,+\pi/2]$
double atan(double x);	cmath	Returns arctan of **x** in range $[-\pi/2,+\pi/2]$
double ceil(double x);	cmath	Returns smallest integer value not less than **x**
double cos(double x);	cmath	Returns cosine of **x** for **x** in radians
double cosh(double x);	cmath	Returns cosh(**x**)
double exp(double x);	cmath	Returns exp(**x**)
double fabs(double x);	cmath	Returns absolute value of **x**
double floor(double x);	cmath	Returns largest integer not greater than **x**
double log(double x);	cmath	Returns natural log of **x**
double log10(double x);	cmath	Returns base-10 log of **x**
double pow(double x, double y);	cmath	Returns **x** raised to the power of **y**
double sin(double x);	cmath	Returns sine of **x** for **x** in radians
double sinh(double x);	cmath	Returns sinh(**x**)
double sqrt(double x);	cmath	Returns square root of **x**
double tan(double x);	cmath	Returns tangent of **x** for **x** in radians
double tanh(double x);	cmath	Returns tanh(**x**)
int abs(int i);	cstdlib	Returns absolute value of **i**
double atof(char *s);	cstdlib	Converts the string **s** to a **double** value
int atoi(char *s);	cstdlib	Converts the string **s** to an integer value
void *bsearch(void *key, void *base, size_t nelem, size_t size, int (*cmp) (void *ck, void *ce));	cstdlib	Binary search function. Searches ordered array of values and returns address of an element that equals the search key **key** if one exists. Otherwise returns **NULL**. **base** is the start address of the array of **nelem** elements, each of size **size**. **bsearch** calls comparison function at address **cmp**, which must return a negative value if **ck** is less than **ce**, zero if they are equal and a positive value if **ck** is greater than **ce**
void *calloc(size_t nelem, size_t size);	cstdlib	Allocates an array of **nelem** elements, each of size **size**, and all initialized to zero. Returns start address of the array, or **NULL** if allocation unsuccessful. In C++ we normally use **new**. This function is included here to aid reading C code
void exit(int status);	cstdlib	Exits program and returns status to the operating system
void free(void *ptr);	cstdlib	Frees data previously allocated by **calloc** or **malloc** (or **realloc**, not included here). **ptr** is the address of the data returned from the original **calloc** or **malloc** call. In C++ we normally use **delete**. This function is included here to aid reading C code
void *malloc(size_t size);	cstdlib	Allocates an array of size **size**, and returns the start address, or **NULL** if the allocation was unsuccessful. In C++ we

(Continued)

Table 4.5 (*Continued*)

Function prototype	Header file	Operation
		normally use **new**. This function is included here to aid reading C code
void qsort(void *base, size_t nelem, size_t size, int (*cmp)(void *e1, void *e2));	cstdlib	Quicksort function. Sorts array of **nelem** elements, each of size **size**, with start address **base**. **qsort** calls comparison function at address **cmp**, which must return a negative value if **e1** is less than **e2**, zero if they are equal and a positive value if **e1** is greater than **e2**
int rand(void);	cstdlib	Returns a pseudorandom number in the interval [0,**RAND_MAX**]
void srand(unsigned seed);	cstdlib	Stores seed for random number generator. For a given seed, **rand** generates the same sequence of return values
int system(char *s);	cstdlib	Passes the string **s** to the command processor (operating system) to be executed. After execution, the status reported by the command processor is returned
int memcmp(void *s1, void *s2, size_t n);	cstring	Compares two arrays of data, both of size **n**, beginning at **s1** and **s2**, until it finds elements that are not equal. Returns 0 if the arrays were equal, a negative number if the differing element from **s1** were the smaller and a positive number if the differing element from **s1** were the larger
void *memcpy(void *dest, void *source, size_t n);	cstring	Copies **n** bytes from array beginning at **source** to array beginning at **dest**
void *memmove(void *s1, void *s2, size_t n);	cstring	As **memcpy**, but ensures that if the arrays overlap, the copy is not corrupted
void *memset(void *s, int c, size_t n);	cstring	Sets **n** elements beginning at **s** with the value (**unsigned char**) **c**
char *strcat(char *s1, char *s2);	cstring	Copies **s2** (including its terminating null character) onto the end of **s1** (beginning with the element that stores the terminating null character of **s1**)
char *strchr(char *s, int c);	cstring	Searches for first element of **s** that equals (**char**) **c**. If one is found, returns its address, otherwise returns **NULL**
int strcmp(char *s1, char *s2);	cstring	Compares successive elements from strings **s1** and **s2**, until it finds elements that are not equal. Returns 0 if the strings were equal, a negative number if the differing element from **s1** were the smaller and a positive number if the differing element from **s1** were the larger
char *strcpy(char *dest, char *source);	cstring	Copies the string **source** (including its null termination) to the array beginning at address **dest**. Returns **dest**
size_t strlen(char *s);	cstring	Returns number of characters in string **s**, not including its terminating null character
char *strncpy(char *dest, char *source, size_t len);	cstring	Copies the string **source** (NOT including its null termination) to the array beginning at address **dest**. Copies no more than **len** characters from **source**, then stores zero or more null characters in the rest of **dest** until a total of **len** characters have been stored. Returns **dest**
char *ctime(time_t *cal);	ctime	Converts calendar time, returned from a call to **time**, into text representation of local time in the format **Tue Sep 22 10:17:19 1992\n\0**
time_t time(time_t *tod);	ctime	If **tod** is not a null pointer, stores current calendar time in ***tod**. Returns current calendar time

4.25 Chapter end material

Bibliography

The major reference on C++ is

Stroustrup, B. (1997). *The C++ Programming Language*, 3rd edn, Addison-Wesley. Bjarne Stroustrup invented the language and was heavily involved in every stage of its development towards a standard.

An excellent, comprehensive introduction to the language is

Deitel, H. M. and Deitel, P. J. (2003). *C++ How to Program*, 4th edn, Prentice-Hall.

The standard library (which is given little coverage in this book) is well covered by

Josuttis, N. M. (1999). *The C++ Standard Library*, Addison-Wesley.

and

De Lillo, N. J. (2002). *Object-Oriented Design in C++ Using the Standard Template Library*, Brookes/Cole.

Other useful books on C++ are the three by Scott Meyes, all published by Addison-Wesley, *Effective C++*, *More Effective C++* and *Effective STL*.

An authoritative and funny book of answers to frequently asked questions is

Cline, M., Lomow, G. and Girou, M. (1999). *C++ FAQs*, 2nd edn, Addison-Wesley. An online mini-FAQ is available at www.parashift.com/c++-faq-lite/.

The paper referred to in the 'Idiom' sidebar is:

Stroustrup, B. (2002). 'Sibling rivalry', *The C/C++ Users Journal*, July, August, September edition. Available at www.research.att.com/~bs/sibling_rivalry.pdf.

Important writers on C++, notably Stroustrup and Cline *et al.*, can be dismissive of C++ usage that too closely mirrors C (as perhaps this chapter does). Cline *et al.* point out that 'Arrays are evil', and warn of doom if you use them in place of vectors. See page 52 for a response.

5 Object-oriented programming in C++

5.1 The motivation for object-oriented programming

Object-oriented programming is a *paradigm* or organizing principle for software, where the program is structured according to the information it manipulates, rather than the operations it does. Data, not procedures, shape the program. Some programming languages such as Smalltalk and C++ explicitly support object-oriented programming. They provide the tools for programmers to fulfil four rules:

Rule 6

Localize information.

Rule 8

Hide information in public; reveal it in private.

Rule 11

Exploit existing solutions.

Rule 15

Design your program around these six sufficient concepts: states, events, conditions, actions, objects and types.

The localization of information is supported by *encapsulation* – the ability to bundle data and functions into self-contained objects. Exploitation of existing solutions is supported by the language being *extensible* in several powerful ways, meaning that reuse and modification are easy. The following two sections explain how object-oriented programming fulfils encapsulation and extendability.

Objects localize information

In Chapter 4, C++ was used as a *procedural* language. That is, the program is written as a list of instructions, telling the computer, step by step, what to do: open a file, read a number, multiply by 4, display something. Most traditional computer languages like Pascal, C and FORTRAN are procedural.

Procedural programming is fine for small projects. It is the most natural way to tell a computer what to do, and the computer processor's own language, machine code, is procedural, so the translation of the procedural high-level language into machine code is (relatively) straightforward and efficient. What is more, procedural programming has a built-in way of splitting big lists of instructions into smaller lists: the *function*. Functions make a program comprehensible, and if well designed, localize information about how particular things are done. Functions can be grouped into files, which allows further structuring of the program.

However, if the principle of localizing information is going to be really effective, a mechanism for localizing data as well as functionality is required. Pascal, C, etc. do provide some support for data localization. Variables declared within a function are *local* – they are only available for use inside the function, and the function therefore hides them. But the important data in a program usually needs to be available to many different functions. A typical example is a database, where different parts of the program need to read, write and modify items. In a procedural language this is usually achieved by making the important data global. It becomes a shared pool, which many functions can access.

The problems with keeping important data in a shared pool are:

- All programmers must know the format of the shared data structures.
- A function that accidently corrupts the data causes trouble for the whole program.
- A change in the way data are stored means changing many functions.

The solution to these problems is to build a wall around the important data structures, by allowing only a very few functions access to them. Other parts of the program cannot get at the data directly; instead they must use one of the access functions. Chapter 11 discusses modularization based on information hiding, which is the application of this idea at all levels of design.

At the coding level, object-oriented programming provides the structure and syntax for bundling data and access functions together into objects. Usually, the data structures are hidden, which means that access is only via the functions. In this way object-oriented programming supports the localization of information. Building a module that hides data and provides a public access–function interface is accomplished as follows:

1. The programmer defines a data type with the required format. In C++, the keyword class is used to define something which, syntactically, looks like a struct, but includes functions as well as data. Labels within the class show which members are accessible publicly, and which can only be accessed by other members of the class.
2. The programmer declares a variable of the new type. In object-oriented terminology, this is called instantiation of an object, or declaring an instance of the particular type.

Because encapsulation is by achieved by creating types with the required format, any number of objects of a particular type can be

classes and structs

In fact, classes and structs are very similar. Both can contain data *and* functions. Strictly the only difference is that in **structs** all the members are, by default, **public**, and in **classes** all the members are, by default, **private**. Common practice – followed in this book – is to use **structs** just for data and **classes** for data types (i.e. data plus access functions).

instantiated (just as any number of ints can be declared in a program). In each case, the data are hidden within the particular object instance.

In an object-oriented language, existing solutions can be extended powerfully

Object-oriented languages are extensible. C++ provides a way to derive new data types incrementally from existing ones by *type extension* and *inheritance*. It also allows extension of the language itself by *overloading*. These two facilities allow code reuse, and easy application of existing solutions to new problems.

Type extension is the derivation of a new data type from an existing one. The new type inherits a specified existing type's behaviour, but then augments or replaces some of it. The initial type is called a *base type*; the new type is called a *derived type*. Now it is obvious that objects of type A are As. For example, objects of type int are ints. The important thing to remember about objects of type B derived from type A is that they are As as well as being Bs. We say an object of the derived type B *isa* object of the base type A, where *isa* means (literally) that every object of type B is also an object of type A. It is *not* the case that every object of A *isa* B. So A is a more general concept than B; B is more particular than A.

A different way of looking at type extension is first to consider what types of objects a program will handle, then look for common features of those types. Two object types that differ in only one kind of behaviour can share a lot of code. If A *isa* B or B *isa* A, then the more general is the base type and the other can be derived from it. But if there is no *isa* relationship, both A and B can be considered as subtypes of some, more general, *abstract* base type. The base type, C, doesn't really exist, but the common parts of the two concrete types A and B can be abstracted into it, then A and B independently derived from it.

The derivation of types provides a way for old code to call new code. This is in contrast to traditional programming where new code calls old code (by linking in functions from libraries) but addition of a new piece to an old program requires full recompilation. Suppose, for example, you have a graphics system where a type **shape** is defined, then more specialized types like **circle**, **rectangle** and **triangle** are derived from it. Within the base type, there may be a function to add a label, and since all shapes do this in the same way, none of the derived types alter that behaviour. On the other hand, each type needs to be able to draw itself in a particular way, so the base type will have a placeholder for **draw()** behaviour but no implementation. That is, supplied independently by each of the particular derived types. In C++, shape will have a *pure virtual* **draw()** function, meaning that there are no objects actually of the base type (i.e. it is an *abstract class*), only objects derived from it, all of which implement **draw()** in their own way. (In C++, **draw()** might instead have a default implementation in the base class, in which case it will be *virtual* rather than *pure virtual*.) Polymorphism allows a reference or a pointer to a base class also to refer to any derived classes, so you could have

shape *p = &mytriangle, or **shape *p = &mycircle**, for example, then the statement **p->draw()** would determine, at runtime, which of the types derived from **shape p** really pointed to and call the appropriate **draw()** function. Suppose now you wanted to create a **pentagon** class. You would derive **pentagon** from **shape**, thereby inheriting the ability to add labels for free, and you would write a function **draw()** that enables a pentagon to draw itself. Now even if the old, main, program is actually a complicated application, which draws shapes in many varied ways, so long as it does it via **pointer_to_shape->draw()** it will be able to draw pentagons using your new class, without recompilation. Things are not quite this simple in practice, because the addition of pentagons to a drawing package has ramifications for the user interface and other modules too.

In almost all programming languages, there are some operators which are *overloaded*. That is, they can operate on different data types, and produce a result appropriate to the data type.

In the following code segment, the operator + is being used first to add two **int**s, then to add two **float**s.

```
int a = 3;
int b = 4;
int c = a + b;     // + operating on two ints to produce an int
float d = 5.0;
float e = 6.0;
float f = d + e;   // + operating on two floats to produce a float
```

The operator = is also overloaded because it is used to initialize and assign to both **int**s and **float**s.

The language hides from the user the fact that it has to do a lot more work to add two floating point numbers than to add two integers. The operations are very different.

In an object-oriented language, the programmer is allowed to overload operators to work with new data types. For example, suppose a C++ class called **String** was created to store and manipulate strings of characters. Then, through appropriate programming, the = operator could be overloaded to allow one **String** to be made equal to another, and the + operator could be overloaded to allow concatenation of two **String**s. See page 138 for the beginnings of such a **String** class.

The meaning of the overloaded operator for a new type does not have to be the same as its original meaning. For example, consider the << operator. What << does is by default defined only for one kind of object – an integer. An integer followed by << followed by another integer means shift the first integer left the number of positions given by the second. Since no meaning is defined for << for other sorts of objects, programmers can choose their own meanings and program in the appropriate behaviour. The C++ streams library leads the way by defining what << means if the object on its left is an **ostream** (a type for dealing with output from the program). Several different overloadings of << are included in the streams library. The reason the statements

```
cout << "Hello\n";
cout << 3 + 4;
```

work is that in both cases, **cout** is an object of type **ostream**, and there are overloadings of << defined for both of:

Left-hand side type: **ostream**   Right-hand side type: **char** *
Left-hand side type: **ostream**   Right-hand side type: **int**

In each case the library includes code to implement the output of the appropriate data type.

Strictly speaking, operator overloading is merely 'syntactic sugar': the same effects could be achieved with function calls. But overloading makes manipulation of objects concise by exploiting analogies between different types. For example, addition of complex numbers has a clear relationship to addition of integers and floating point numbers, so it is natural to overload '+' for a complex type. On the other hand, overloading allows programmers to do things that might not be expected by users of their code. There's nothing to stop you overloading '+' for **location**s, for example. But what would *that* mean? Therefore, your operator overloading should be done judiciously, to make sure operator behaviour doesn't conceal surprises for other programmers.

5.2 Glossary of terms in object-oriented programming

Several object-oriented programming terms have already been introduced, and more will be needed as C++'s object-oriented features are described. Here some of the most important are briefly defined and illustrated.

Data structure

A data structure is

- one of the base data types in C++: **int**, **char**, **float**, etc.
- anything defined using **struct** (or **class**) that has data structures but no functions
- a collection of data structures such as an array.

---

isa circle an ellipse?

Inheritance means *isa*. A derived type does not take anything away from its base type. This means you have to think carefully about the inheritance hierarchy you want to build. The classic example is that if your base type **bird** has a method (member function) **fly()**, you may not be able to derive **penguin** from it. That problem is fixable by taking **fly()** out of **bird**, and interpolating a level of derived types which might be **flyingbird** (which has **fly()**) and **flightlessbird** (which doesn't). But there are other more subtle traps. You might decide to derive **Real** from **Complex** and **Integer** from **Real**, but do you really want to carry around a zero imaginary value in an inappropriate numerical format in every **Integer** object? It isn't always easy to decide. Is a circle an ellipse? Mathematically it is. What's more, if you have an ellipse type, you could use all the code you've painstakingly developed for drawing, finding areas, etc., by saying that a circle *is* an ellipse with equal major and minor axes and orientation 0. But then again, there are simpler formulae for drawing circles and finding their area than for ellipses, so wouldn't you want to replace everything anyhow? The key to dilemmas like these is whether we want an object of the derived type to function as an object of the base type. In the case of a circle, will we ever want to split its centre into two foci? Almost always the answer is no – whatever we do to a circle, we want to preserve its single centre. It doesn't and never will have a major axis, minor axis or orientation. So the answer (in terms of type derivation) is no, a circle is not an ellipse.

Abstract Data Type (ADT)

An Abstract Data Type (or ADT) is a data structure with a particular behaviour and access functions. The type gives access to the data via the functions but hides the data, revealing nothing about its internal data structures or how the functions work.

The basic data types, **int**, **char**, **float**, **double**, etc. are both data structures and data types. **double**, for example, hides all the information about how floating point numbers are represented, making them accessible only through operators like $+$, $-$, $*$ and $/$.

An example of a useful abstract data type is a stack. A stack stores and retrieves items in a last-in-first-out fashion. Analogous to a stack of trays in a cafeteria, each new item is added to the top and pushes the rest of the stack down; when an item is to be retrieved, the one on top is taken and the rest of the stack pops up one place. The most recently added item still on the stack is always the next retrieved.

Here is procedural-type C++ code for a stack of integers:

```cpp
int stack_array[256];     // Not yet object oriented – This array
                          // is global, not hidden
int stack_pointer = 0;    // Not yet object oriented – This int is
                          // global, not hidden.
int push(const int value)
  {
  if (stack_pointer == 256)
     return(-1);    // Stack full
  stack_array[stack_pointer++] = value;
  return(0);
  }
int pop(int& value)
  {
  if (stack_pointer == 0)
     return(-1);   // Stack empty
  value = stack_array[--stack_pointer];
  return(0);
  }
```

The rest of the program should interface to the stack only through the functions **push** and **pop**. The stack embodies some constraints, such as a limit of 256 on the number of items, so the access functions allow error and success returns. An array is not the only way to implement a stack like this, but the definition of the ADT (its external interface) will not change if the implementation does.

C++ now allows the *encapsulation* of the stack example code into a true data type so that the stack array and stack pointer can be hidden and so the type can be instantiated multiple times. A straightforward translation of the above stack code into a true object-oriented C++ type looks like this:

```cpp
class stack     // class is a keyword to flag the start of an ADT
                // definition. It is analogous to struct.
```

```
      {
      int stack_array[256];
      int stack_pointer;
public:   // public is a keyword: everything that comes after is
          // visible externally
      int push(const int value)
        {
        if (stack_pointer == 256)
           return(-1);          // Stack full
        stack_array[stack_pointer++] = value;
        return(0);
        }
      int pop(int& value)
        {
        if (stack_pointer == 0)
           return(-1);          // Stack empty
        value = stack_array[--stack_pointer];
        return(0);
        }
   };
```

This is not the whole story. The complete class definition for such a
stack will be given in the next section. This partial example shows that
a **class** is syntactically like a **struct**, and contains both data structures
and access functions. The access functions are preceded by the word
public, which ensures that only they and not the data are accessible
from outside. This class definition is like a mould for objects, and now
it is possible to use **stack** in declarations to create any number of variables with this structure *and* this behaviour.

Class

In object-oriented programming, class is another word for Abstract
Data Type (ADT). This book uses 'ADT', or more simply 'type', rather
than 'class' to avoid confusion with the (language-specific) C++ keyword **class** which is the way of setting up ADTs in this particular
language. Most object-oriented programming books, and most C++
programmers, use the word class with both meanings.

Object

An object is an instantiation of an ADT. That is, when a particular data
item is declared, e.g.

stack mystack;

mystack is an object of type **stack**.

Method

A method is an access function for an ADT. So in the stack example **push** and **pop** are methods of the ADT **stack**.

Member function

A member function is the C++ term for a method. In this book the terms 'access function', 'method' and 'member function' are synonymous.

Message

In contrast to conventional programming, where functions call other functions to do things, object-oriented programming encourages the thinking that objects send messages to other objects. In response to received messages, objects execute methods. Now in C++, the separation between methods and messages does not really exist. The only message that can be sent to an object in C++ is one for which it has a method, so what really happens is that the sending object specifies which method the receiving object is to run. This does not necessarily mean that the particular function call is 'compiled in' to the code. Different objects, indeed different ADTs, can have methods of the same name. So in some circumstances, when an object is required to run a method, the system may have to figure out at runtime what type the object is, and therefore which particular method has to be executed.

Base types and derived types

An object-oriented language allows the derivation of ADTs from other ADTs. The derived type is called a derived type (!) and the original type is called the base type. Type derivation is really type *extension* – an incremental modification mechanism. The derived type is partly a copy of the base type, but has other properties or does other things as well. Every object of the derived type is also an object of the base type.

Inheritance

Inheritance is the mechanism whereby a derived type inherits characteristics from its base type. The programmer designs an inheritance hierarchy that defines how types are derived from other types. The purpose of having such a hierarchy is not to put all the program objects into a taxonomy, but to allow code reuse and good modularity. Inheritance also enables the use of polymorphism (coming up next).

Polymorphism

A derived ADT can replace one of the methods from its base ADT. What is more, a pointer to a base type can also point to the derived

---

Polymorphic definitions

Some authors use a different definition of polymorphism. When an ADT can be parameterized over a data type, it is said to be polymorphic. For example, if we had an ADT queue that could directly handle queues of integers, queues of strings, queues of arbitrary structures, that would be polymorphism in this second sense. ADTs with this property are called parameterized types – these are not supported directly in C++, though they can be simulated using templates.

type. Now if that pointer sometimes points to an object of the base type, and sometimes to a derived type object, the system will figure out at runtime which of the methods to run. The ability to have many versions of the same method throughout the class hierarchy is called polymorphism, and the fact that the choice of method is made at runtime is called late binding (as opposed to early binding where function calls are hardwired by the compiler).

5.3 C++ type definition, instantiation and using objects

The general format of a C++ ADT definition is as follows:

```
class name
   {
   // Description of private data items, only usable by member
   // functions
protected:
   // Description of protected data items, if any.
   // Only usable by classes that inherit this class.
public:
   // Description of variables and member functions accessible
   // from outside the class
     name()
        {
        // This is a special member function called a constructor.
        // It is automatically called when the class type is created.
        };
     ~name()
        {
        // This is another special member function called a
        // destructor.
        // It is executed when the program is finished, goes out
        // of scope,
        // or the delete operator is applied on the object name
        };
     // Other member functions defined
     // Declarations of "friend" functions
   };

name::member function name()
   {
   // Large member functions are declared in the class
   // but can be defined outside the class in this way
   }
```

Stack ADT example

Consider the ADT stack for which a partial definition was given earlier. Here is the complete C++ ADT definition:

```
1:   #include <iostream>
```

```
 2:   using namespace std;
 3:   class stack
 4:     {
 5:     // First come the private data items
 6:     int stack_length;
          // To allow the caller to set max length
 7:     int *stack_base;
          // Points to where stack is in memory
 8:     int stack_pointer;
 9:   public:
10:     stack(const int size)        // This is the constructor
11:       {
12:       stack_base = new int[stack_length = size];
13:       stack_pointer = 0;
14:       };
15:     ~stack()   // This is the destructor
16:       {
17:       delete [] stack_base; // NOTE USE OF SQUARE
                                // BRACKETS
18:       }
19:     int push(const int value);   // push and pop are declared
                                     // inside the class definition
20:                                  // but defined below
21:     int pop(int& value);
22:     };
23:
24:   int stack::push(const int value)
25:   // A short function like this could have
26:   // been included in the class definition
27:     {
28:     if (stack_pointer == stack_length)
29:        return(-1); // Stack full
30:     stack_base[stack_pointer++] = value;
31:     return(0);
32:     }
33:   int stack::pop(int& value)
34:     {
35:     if (stack_pointer == 0)
36:        return(-1); // Stack empty
37:     value = stack_base[--stack_pointer];
38:     return(0);
39:     }
```

Here is how the ADT might be used:

```
41:   int main()
42:     {
43:     stack my_stack(10);
44:     // Saying that I won't be putting more
45:     // than 10 items on the stack.
```

```
46:     my_stack.push(345);
47:     my_stack.push(678);
48:     int test;
49:     my_stack.pop(test);
50:     cout << test << " was just popped off the stack\n";
51:     return 0;
52:     }
53:
```

This program introduces most of the C++ ADT syntax. The keyword **class** (line 3) introduces an ADT definition. Private members of the ADT come first (lines 5 to 8), then publicly accessible members, labelled with **public:** (line 9). In this implementation, the function that instantiates a stack (starting line 10) can specify how big it is to be – that is why the pointer **stack_base** is used instead of a fixed-size array.

Methods of an ADT (in C++ these are member functions of a class), can be defined inside the **class** definition or outside – but all have to be declared (prototyped) inside. The ADT stack has methods **stack**, **~stack**, **push**, and **pop**. In this case the first two have been defined inside the **class** definition (lines 10 to 14 and lines 15 to 18); the others are declared inside (line 19 and line 21) but defined outside and written as **stack::push** (lines 24 to 32) and **stack::pop** (lines 33 to 39). The usual practice is to define small member functions inside the **class** definition, and larger ones outside. This makes things readable (i.e. compact where they can be, and with visible structure where they cannot be compact). As well, functions defined inside a **class** definition are automatically *inline*. An inline function is a short function that the compiler deals with in a special way: if it can, when converting the code to assembler, it inserts all the code for the inline function every place where the function is called, instead of representing the function as a regular subroutine, and doing calls to it. This makes for faster execution at the cost of a little extra memory space usage in the program.

The construct **stack::push** (line 24) illustrates the scope resolution operator **::**, in this case telling the compiler that the function being defined is a member function for the **stack** ADT. As already discussed, different types can have member functions of the same name, so **::** is needed to make it clear which type this one goes with.

The member functions **stack** and **~stack** are special. In general, an ADT will have one member function which is the name of the ADT itself, and one which is the name preceded by the ~ symbol. The first of these is called a constructor, and the second is called a destructor. Whenever an object of a type is declared (or *instantiated*), C++ calls the constructor to set up the object. The constructor is therefore a kind of initializer. It can take a parameter to affect how the particular instance is set up. In this case, when an object of type **stack** is declared, the constructor allocates some memory to be the stack and initializes its internal data appropriately. The size of the stack required is given in the declaring function and passed as a parameter to the constructor. The destructor is called automatically when an object goes out of scope – that is, when the function that declared it returns, or when it is explicitly **deleted**.

Location ADT example

Some of the **structs** introduced in Chapter 4 can now be recast in the form of ADTs. Here, for example, is a program that prints out the zones of locations (from a locations database), but now using a **location** class.

```
 1:  //
 2:  // Prints out zones of locations in a location database.
      // Format of the database is:
 3:  // First line: Integer number of records
 4:  // Remaining lines: One record per line: Loc name
      // (20 chars max), latitude,
 5:  // ns character modifier for latitude,
 6:  // longitude, ew character modifier for longitude
 7:
 8:  // fstream required because file input/output streams are
      // used
 9:  #include <fstream>
10:  #include <iostream>
11:  using namespace std;
12:
13:  // An enumerated type is created to associate meaningful
14:  // names with the index values of the zones array
      // immediately below:
15:  enum { nfrigid, ntemperate, torrid, stemperate, sfrigid,
            error };
16:
17:  // The zones array is simply static array of pointers to
18:  // constant strings that specify names of the Earth's
      // zones.
19:  const char *zones[6] = {
20:    "North frigid",
21:    "North temperate",
22:    "Torrid",
23:    "South temperate",
24:    "South frigid",
25:    "None"    // Flag to indicate error return
26:    };
27:
28:  // Now follows the definition of the location class. Most of
29:  // the functions are defined inside because they are short.
30:  // Only member function zone is long enough to have its
      // definition outside the class definition
31:
32:  class location {
33:      char name[20];
```

```
34:    float latitude; char ns;
35:    float longitude; char ew;
36:  public:
37:    location() {};      // Null constructor - nothing to initialize
38:    ~location() {};     // Null destructor - nothing to tidy up
                           // on destruction
39:    void load(istream& in)      // Load data items from an
                                   // input istream object.
40:                       // This will usually be a file (and is in
41:                       // the main program below), but it
                          // could be cin
42:      {
43:      in >> name >> latitude >> ns >> longitude >> ew;
44:      }
45:    void save(ostream& out) const    // Save data items to
                                        // output stream object
46:                                     // (usually a file)
47:                                     // Not used in this example
48:      {
49:      out << name << ' ' << latitude << ' ' << ns << ' '
50:          << longitude << ' ' << ew << '\n';
51:      }
52:    const char *getname() const
53:      {
54:      return(name);
55:      }
56:    const char *zone() const;   // Large member function -
                                   // defined outside class defn.
57:    };
58:
59:  // The member function *zone() was declared inside the
60:  // location class definition, but not defined. Here is its
     // definition:
61:  // The zone member function returns a pointer to a string
     // stating which zone of the earth this particular place is
62:  // located
63:  const char *location::zone() const
64:    {
65:    if (!name[0])            // No location loaded in this object
66:       return zones[error];      // Error return
67:    if (latitude < 23.5)
68:       return zones[torrid];     // Using the values of enum to
                                    // index the zones array
69:    if (latitude < 66.5)
70:      {
71:      if (ns == 'N')
72:         return zones[ntemperate];
73:      return zones[stemperate];
74:      }
75:    if (ns == 'N')
76:       return zones[nfrigid];
77:    return zones[sfrigid];
```

```
78:   }
79:
80:   //
81:   // Here is the main program which uses a location object
      // to read in locations from a database one by one,
82:   // and print out their zones.
83:   //
84:   int main(int argc, char *argv[])
85:     {
86:     if (argc != 2) {
87:       cout << "Usage: tellzone locationsfile\n";
88:       return -1;
89:       }
90:     ifstream infile(argv[1]);  // Opening the file specified
91:     // on the command line as an input file stream object
92:     if (!infile) {
93:       cerr << "Can't open " << argv[1] <<
              " for reading\n";
94:       return -1;
95:       }
96:     int numrecords;
97:     infile >> numrecords;
      // The first line of the database contains a single
      // number
98:     // specifying the number of lines or "records" in the
      // file
99:     location current_location;
100:    for (int i = 0; i < numrecords; i++)
101:      {
102:      current_location.load(infile);   // Load the next line
103:      // of infile into the current_location object
104:      cout << "Location " << current_location.getname()
105:        << " is in the " << current_location.zone() <<
              " zone\n";
106:      }
107:    infile.close();    // Close the file input stream when all
                          // lines are read
108:    return 1;
109:    }
```

The only new syntactic construct in the program above is the addition
of the word **const** after the declaration or definition of an access function. For example, on line 56,

const char *zone() const;

const used in this context is a guarantee that the access function will not
change the object itself. **zone** returns a pointer to a string indicating
the zone of the object, but it does not alter any of the internal values
of the object.

   Observe how in this program the ADT **location** encapsulates the
way it reads and writes records from and to database files (lines 39 to

44 for reading, 45 to 51 for writing). It is a good strategy to make types responsible for loading and saving themselves: the format of the database then becomes localized in the type.

The **current_location** variable or object in the main program (declared line 99) is repeatedly loaded with new records from the database (line 102). The **zone()** member function is then called (line 105) to calculate the zone based on those current values. So although only a single object of type **location** is used in the main function, it is a true variable.

> The recurring latent bug (see pages 95 and 105) is in this program too! Part of the problem is in the data items of location and part in the way they are loaded. The bug will happen, for example, when a location like Sutton-under-Whitestonecliffe is read from the database. Do you see the problem? More on page 144.

Vector ADT example

Here is an ADT definition for dealing with vectors of up to eight real-valued components:

```cpp
const int max_dimensions = 8;
class real_vector
    {
    int dimensions;
    double vector_base[max_dimensions];
        // The vector itself is stored in this array
public:
    real_vector(const int ndim)   // The constructor takes a
                                  // parameter which specifies
                                  // how many dimensions this
                                  // vector is to have
      {
      if (ndim > max_dimensions)
         dimensions = max_dimensions;
      else
         dimensions = ndim;
      }
    ~real_vector()     // The destructor does not have to do any
                       // tidying up, so it is empty.
      {
      }
    int set_component(const int comp, const double value);
    int get_component(const int comp, double& value) const;
    int tell_num_dimensions() const
      {
      return(dimensions);
      }
```

```cpp
    real_vector sum_with_me(const real_vector& u);
        // Returns a vector sum
        double dot_with_me(const real_vector& u);
        // Returns dot product
};

int real_vector::set_component(const int comp, const double
    value)
{
if ((comp < 0) || (comp > dimensions-1))
    return(-1);
vector_base[comp] = value;
return(0);
}
int real_vector::get_component(const int comp, double& value)
    const
{
if ((comp < 0) || (comp > dimensions-1))
    return(-1);
value = vector_base[comp];
return(0);
}
real_vector real_vector::sum_with_me(const real_vector& u)
// Returns a vector that is the sum of this object and the
// vector given as a parameter
{
real_vector sum(dimensions);
// Set up a vector with the same number
// of dimensions as me.
if (u.tell_num_dimensions() != dimensions)      // ***
// If the vector I am being summed with has a different
// number of dimensions
    return(sum); // Essentially an error return -- zero vector
for (int i = 0; i < dimensions; i++)
    sum.vector_base[i] = u.vector_base[i] + vector_base[i];
        //+++
return(sum);
}
double real_vector::dot_with_me(const real_vector& u)
// Returns dot product of this object and the vector given as a
// parameter.
// (The dot product multiplies each pair of components, then
// sums the results)
{
double dot_product = 0;
if (u.dimensions != dimensions)
    return(dot_product);
for (int i = 0; i < dimensions; i++)
    dot_product += u.vector_base[i] * vector_base[i];
        //+++
return(dot_product);
}
```

This class definition does not introduce any new syntax, but will repay close study. Note in particular how the line marked *** treats information hiding in a different way to the lines before +++. In the *** line, the second vector (**u**) is treated as a 'distant' object, which must be accessed through its public access functions; in the +++ lines, its private data items are accessed directly. Because **u** is of the same type as the object whose member function is being called (**real_vector**), its **private** data items are available to the method.

Here is a program using real vectors.

```cpp
#include <iostream>
using namespace std;
#include "vec_def.h"        // Including the above vector class
                            // definition file
int main()
  {
  real_vector a(3), b(3);   // These will be set up as three
                            // dimensional vectors
  a.set_component(0,0.0);
  a.set_component(1,1.0);
  a.set_component(2,2.0);
  b.set_component(0,3.0);
  b.set_component(1,4.0);
  b.set_component(2,5.0);
  real_vector c = a.sum_with_me(b);
  double x, y, z;
  c.get_component(0,x);
  c.get_component(1,y);
  c.get_component(2,z);
  cout << x << "," << y << "," << z << "\n";
  return 0;
  }
```

The summation of two **real_vector**s is done by having **a** add **b** to itself and return the result. This is the standard object-oriented way of doing operations on objects: send a message (in C++, call a member function) and get the object to do the operation to itself. An alternative, more conventional, way of summing two vector objects would be to use an ordinary function (not an ADT method) that takes the vectors to be summed as arguments. Here is such a function, not part of the **real_vector** ADT:

```cpp
real_vector sum(const real_vector& a, const real_vector& b)
  {
  int dim;
  dim = a.tell_num_dimensions();
  real_vector the_sum(dim);
  if (dim != b.tell_num_dimensions())
    return(the_sum);       // Error return – a zero vector
  int i;
  for (i = 0; i < dim; i++) // For each component in turn,
```

```
// get the values of the two input vectors, add them, and put
// them in the sum.
  {
  double acomp, bcomp;
  a.get_component(i, acomp);
  b.get_component(i, bcomp);
  the_sum.set_component(i,acomp+bcomp);
  }
return(the_sum);
}
```

What a lot of overhead the **sum** function suffers to do something so simple! All those function calls are necessary because **sum** has to go through **real_vector**'s member functions to get what it wants. There is a way to make a function like **sum** more efficient, by allowing it access to **real_vector**'s private data, even though it is not itself part of **real_vector**. The way is to make **sum** a *friend* of **real_vector**.

Friend functions are C++'s nod to conventional procedurally oriented programming. They break the walls of an ADT, so that a privileged, named, friend can get at the private data. They are syntactically easy to deal with: all that is required is to put a line in the **class** definition (usually at the end) of the form:

friend *prototype_of_function_that_is_to_be_a_friend_of_this_class*

Then the named function has full access to everything in the class.
In the **sum** example, the following line would be added within the **class** definition for **real_vector**:

friend real_vector sum(const real_vector& a, const
    real_vector& b);

then the function **sum** could make use of its privileged access to the internals of **real_vector** ADTs. It could be rewritten:

```
real_vector sum(const real_vector& a, const real_vector& b)
  {
  real_vector sum(a.dimensions);
  if (a.dimensions != b.dimensions)
    return(sum);
  for (int i = 0; i < a.dimensions; i++)
    sum.vector_base[i] = a.vector_base[i] + b.vector_base[i];
  return(sum);
  }
```

Friend functions are somewhat contrary to the spirit of object-oriented programming – they break the wall of the ADT. Used with discretion they help closely related ADTs communicate concisely.

5.4 Overloading

C++ allows several functions to be defined with the same name, even if those functions have the same scope. That is, they could be ordinary functions (not methods) that each have the scope of the whole program, or they could be member functions of a single ADT. The constraint is that each such function should be unique in terms of its argument list and return type. It must not have the same number of arguments of the same types, as another version of the function. The compiler works out which particular function is needed in which situation, according to the types of the arguments for a particular call. So if a program needs a function for sorting integer arrays and a function for sorting arrays of **doubles**, it is legitimate to define:

```
int sort(int *start, const int num_elements)
   {
   ...
   }
int sort(double *start, const int num_elements)
   {
   ...
   }
```

and then have code like:

```
double array1[5] = {0.6, 0.4, 3, 21, 0};
int   array2[4] = {9, 8, 5, 6};
sort(array1, 5);
sort(array2, 4);
```

The appropriate function will be called in each case. **sort** is said to be 'overloaded'.

Within class definitions, overloading is often used for constructors. For example, the vector type defined above allowed the creation of a **dim**-dimensional vector. Given that 2D and 3D vectors are most commonly used, the program might include three definitions for the constructor in the **class** definition:

```
const int max_dimensions = 8;
class real_vector
   {
   int dimensions;
   double vector_base[max_dimensions];
public:
   real_vector(const int ndim)
      {
      if (ndim > max_dimensions)
         dimensions = max_dimensions;
      else
         dimensions = ndim;
      }
   real_vector(const double x, const double y)
      {
      dimensions = 2;
      vector_base[0] = x;
      vector_base[1] = y;
      }
```

```
        real_vector(const double x, const double y, const double z)
          {
          dimensions = 3;
          vector_base[0] = x;
          vector_base[1] = y;
          vector_base[2] = z;
          }
    ...
```

The main program (page 131) then becomes much simpler:

```
#include <iostream>
using namespace std;
#include "vec_def.h"      // Including the vector class definition
                          // file
int main()
  {
  real_vector a(0.0, 1.0, 2.0);
  real_vector b(3.0, 4.0, 5.0);
  real_vector c = a.sum_with_me(b);
  // or perhaps with the friend function sum:
  // real_vector c = sum(a,b);
  double x, y, z;
  c.get_component(0,x);
  c.get_component(1,y);
  c.get_component(2,z);
  cout << x << "," << y << "," << z << "\n";
  return 1;
  }
```

Operator overloading

In Chapter 4 we saw the way that **ostream** objects respond to the operator << on the basis of parameter data type. For example, when **cout** sees the parameter "Hello\n", it also sees and responds to the particular data type – a string. This is an example of operator overloading.
   Consider the following program:

```
#include <iostream>
using namespace std;
int main()
  {
  int   a = 123;
  double  b = 12.34;
  cout << "a = " << a << "; b = " << b << "; their product is "
      << a*b << '\n';
  return 0;
  }
```

The output of the program is:

a = 123; b = 12.34; their product is 1517.82

How does this work? Not only has the object **cout** correctly dealt with objects of type **char *** (i.e. null-terminated string), **int**, **double**, and **char**, but also several << operators have been chained together.

<< operators are always evaluated left to right (see Table 4.4 on page 83), and for an **ostream** object, the return value of the operation is a reference to the object itself. So the above program does **cout** << "a = " first, returns **cout** from this operation, then goes on to the next operator to do **cout** << a. This again returns **cout** ready to do **cout** << "; b = ", and so on through the chain of <<.

Now it would be perfectly possible to define new meanings for << when the type on its left is something different from **ostream**. More commonly, programs can define appropriate ways for **ostream** to respond to << for new types of object on the right-hand side.

Operators can be overloaded in two ways. The first, and least object-oriented, way is with the following syntax:

type_of_result **operator** *operator_symbol (type_of_arg1*
    arg1_name, type_of_arg2 arg2_name)
    {
    Statements written just as in a conventional function
    }

So the syntax is very like that of a conventional function definition, except for the keyword **operator**. This function is not 'called' in a conventional sense; rather it is applied when the operator appears, operating on two objects of the appropriate types. The syntax above is for a binary operator where the arguments/operands are written on either side of the operator; it is also possible to overload unary operators (one operand), in which case there is only one argument in the definition.

Here is an example of operator overloading. Consider the data structure **student**.

```
struct student
    {
    char name[30];
    unsigned long id;
    float mark;
    };
```

and assume that a variable **alice** of this type stores the information for Alice Nonesuch.

A mechanism is required for printing out **student** objects, formatted as: ID number, tab, name, tab, mark; for example:

98765432 Alice Nonesuch 88.2

The natural way to do this would be to overload the << operator for the case where the object on its left is an **ostream** (like **cout**), and the object on its right is a **student**. Then the simple line,

cout << **alice**;

would do the job.

Here is the overloaded operator definition which accomplishes this:

```
ostream&operator << (ostream& stream_name, student& st)
  {
  stream_name << st.id  << '\t' << st.name << '\t'
      << st.mark << '\n';
  return(stream_name);
  }
```

The reason for returning the **stream_name** reference parameter here is so that << operators can be chained, just as discussed for the example at the start of this section. Usually operators return a reference to their first argument, so the same object is used for the next operation in the chain; stream objects like **ostreams** should *always* be passed by reference.

The second, and more object-oriented, syntax for operator overloading is used within a class definition. Rather than a conventional member function, an operator function can be defined:

```
type_of_result operator operator_symbol (type_of_arg
    arg_name)
  {
  Statements written just as in a conventional function
  }
```

Again, this is syntax for a binary operator. The reason is that the first argument (the operand appearing on the left of the operator) is implicitly the object for which this is a member function. So the argument in the definition is the object which will appear on the right of the operator.

The **real_vector**s introduced earlier included member functions for vector addition and dot products. The first of these might be replaced with an overloaded operator +. Then instead of

```
real_vector c = a.sum_with_me(b);
```

we could write:

```
real_vector c = a + b;
```

To make this happen, + can be overloaded for **real_vector** objects by adapting the **sum_with_me** method. Here is what it looked like on page 130:

```
real_vector real_vector::sum_with_me(const real_vector& u)
  {
  real_vector sum(dimensions);
  if (u.tell_num_dimensions() != dimensions)
    return(sum);
  for (int i = 0; i < dimensions; i++)
    sum.vector_base[i] = u.vector_base[i] + vector_base[i];
  return(sum);
  }
```

Conversion to using an overloaded operator is easy: all that is needed is the keyword **operator** followed by the appropriate symbol. Here is the new version:

```
real_vector real_vector::operator+ (const real_vector& u)
    {
    real_vector sum(dimensions);

    if (u.dimensions != dimensions)  // Now directly accessing
        return(sum);                 // u's private data
    for (int i = 0; i < dimensions; i++)
        sum.vector_base[i] = u.vector_base[i] + vector_base[i];
    return(sum);
    }
```

Alternatively, instead of being a member function (using the second syntax for operator overloading), the operator function can be an ordinary function, as **sum** was (using the first syntax). Recall that with **sum** a friend of **real_vector**, its definition was:

```
real_vector sum(const real_vector& a, const real_vector& b)
    {
    real_vector sum(a.dimensions);
    if (a.dimensions != b.dimensions)
        return(sum);
    for (int i = 0; i < a.dimensions; i++)
        sum.vector_base[i] = a.vector_base[i] + b.vector_base[i];
    return(sum);
    }
```

Converting to the overloaded + operator gives:

```
real_vector operator+ (const real_vector& a, const real_vector& b)
    {
    real_vector sum(a.dimensions);
    if (a.dimensions != b.dimensions)
        return(sum);
    for (int i = 0; i < a.dimensions; i++)
        sum.vector_base[i] = a.vector_base[i] + b.vector_base[i];
    return(sum);
    }
```

The line

```
friend real_vector operator+ (real_vector&, real_vector&);
```

would also have to be added to the **real_vector** definition.

Whether + is overloaded within the ADT or not, the summation can now be written

```
c = a + b.
```

The dot product could be written in operator form as well. However, the period symbol . is one of the few operators that cannot be overloaded, so another symbol would have to be used. The other operators that *cannot* be overloaded are .* :: and ?:. All others (see page 83) can. Note too that the symbol # is a preprocessor symbol, not a language operator, so cannot be overloaded.

Finally, returning to overloading of << for **ostreams**, here is a function to print out **vector**s in a reasonable way:

```
ostream&operator<<     (ostream& str, real_vector& a)
  {
  int dim;
  double comp;
  dim = a.tell_num_dimensions();
  str << "( ";
  for (int i = 0; i < dim-1; i++) // Stops at dim-1 so as not to
                                  // print a terminal comma
    {
    a.get_component(i,comp);
    str << comp << ", ";
    }
  a.get_component(i,comp);
  str << comp << " )";
  return(str);
}
```

This defines how the operator << is to function when the object on its left is of type **ostream** and the object on its right is **real_vector**. The function itself steps through the components of the vector, sending them to the **ostream** (which already knows how to handle << on all the standard types).

5.5 Building a String class

To consolidate the ideas so far, and introduce a few new syntactical details, we now step through a complete program line by line. The program embodies a simple String abstract data type, and includes a main function which exercises the type. String types are a standard part of many C++ class libraries. They allow programmers to manipulate text strings in powerful and compact ways; for example, one string may be copied to another simply using the = sign. The String type illustrated here is relatively straightforward, but shows how the ideas introduced so far lead to the construction of very powerful objects. Commentary on the code is interleaved with it. The name **String**, with a leading capital S, is used throughout to differentiate the type from C++ standard library **string** type and the generic use of the word string, which has been used so far to describe a null-terminated array of characters.

```
1   #include <iostream>
2   #include <cstring>
```

iostream provides prototypes for the input/output, and **cstring** for the C-style character string handling routines like **strcpy()**.

```
3   using namespace std;
4   int error(char *p)
```

```
5    {
6    cerr << p << "\n";
7    return 0;
8    }
```

The **error()** function given in lines 4 to 8 is a minimal implementation
of an error reporting system. **cerr** is the name of the 'standard error'
output – usually things put to this stream just go to the computer monitor or display in the same way as standard output.

```
9
10   class String
11   {
12     char *buf;
13     int len;    // Length does not include terminating '\0'
```

buf will point to the null-terminated text stored in the **String** object.
The length of that text (not including the terminating '\0') will be
maintained in **len**. These two items are private to the class and can
only be manipulated by its member functions (methods) or its friends.
They are the hidden part of the ADT, and, in principle, the way that
Strings are represented is a secret of **String**. For example, **buf** could
point to **len** characters without a null termination – the question of
what constitutes a String within the ADT is strictly the ADT's own
business. As it turns out, the program does use a null-terminated
string.

```
14   public:
15   // Constructors
16     String (const char *p)      // String x("abc");
17       {
18       len = strlen(p);
19       buf = new char[len + 1];
20       strcpy(buf, p);
21       }
22     String () { buf = new char; buf[0] = 0; len = 0;}
           // String x;
23     // Note that String of length 0+1 created, so that delete
24     // buf will work correctly when the destructor is called.
```

Two constructors are given for this class; the first initializes the **String**
object with a particular string of text; the second is used when the text
the object is to store is not known at the time of creation. Note that the
comments on lines 16 and 22 show the context for use of that particular constructor. The line 16 constructor does not affect the initialization text, and so it is declared as **const**. The line 22 constructor sets up
a minimal text array: 0 characters of text followed by a null terminator – 1 character. Then, when the String is destroyed and **delete** is
applied to **buf** (see the destructor), there will be something to delete.
Not included yet is a copy constructor, **String(const String& str);**.
See the exercise on page 144.

```
25
26      // Destructor
27      ~String() { delete [] buf;}
```

Note the use of square brackets in the **delete** statement to ensure that the whole buffer array is deleted.

```
28
29      // Assignment to and from Strings
30      String& operator=(const char *p); // x = "abc";
```

The assignment operator (the single equals sign) for the case when a character pointer is assigned to a **String** object leaves that character pointer unchanged (note the use of **const**), and returns a reference to a **String** object. Assignment operators return references to allow things like **a** = **b** = **c**;. This is evaluated right to left, which means that **b** is made equal to **c**, then the result of that operation (i.e. what the equality operator returns) is available, and **a** is made equal to it. So an equality operator should return a reference to whatever is on its left-hand side.

```
31      String& operator=(const String &str); // y = x;
```

The assignment operator for the case when a **String** object is assigned to another **String** object returns a reference to the object that it has just been assigned to (i.e. the left-hand side of the equality sign).

```
32      char& operator[](int i);    // c = x[4]; x[5] = 'a';
```

The subscripting operator [] can be used to index any ADTs for which indexing has some meaning. In the case of a **String**, the index is used to get at the ith character of the String. So although the ADT is not an array, you can subscript it as if it were. In line 32, the subscripting is done with integers – this is the meaning of the (**int i**), but C++ is powerful enough that the [] operator can be used with other things than integers between the brackets. The declaration returns a reference to a character. Observe the two examples given in the comment to this line. The first example **c** = **x**[4] is what we would expect: **x** is a String, **4** is an integer, so **x**[4] is using the subscripting operator to get at something – the fourth character of the text stored in the **String** object. This character is then copied into **c**. But the second example is slightly more surprising: the character 'a' is being copied into the fifth location of the text stored in the object – the [] operator is on the left-hand side of the expression. This works because a reference can appear on the left-hand side of an assignment.

```
33
34      friend ostream& operator<<(ostream& out, const String
            &x);
35      friend istream& operator>>(istream& in, String &x);
```

Lines 34 and 35 augment the standard stream's library so that objects of type **String** can be input and output using the put to and get from operators.

```
36
37     // Tests
38     friend int operator == (const String &x, const char *s)
39        {
40          return (strcmp(x.buf, s) == 0);
41        }
42     friend int operator == (const String &x, const String &y)
43        {
44          return (strcmp(x.buf, y.buf) == 0);
45        }
46     friend int operator != (const String &x, const char *s)
47        {
48          return (strcmp(x.buf, s) != 0);
49        }
50     friend int operator != (const String &x, const String &y)
51        {
52          return (strcmp(x.buf, y.buf) == 0);
53        }
```

Lines 38 to 53 overload the standard C++ relational operators. These should be compatible with their usual function, so a non-zero return value means true, or success, and a zero return value means false, or failure. Remember that **strcmp()** returns zero when the two Strings it compares are equal. The above four inline functions allow **String**s to be compared with each other, and with conventional null-terminated strings.

```
54     friend int strlen(const String &x)
55        {
56          return x.len;
57        }
```

The standard library **strlen()** function is overloaded to handle String objects. Because the **String** ADT contains a data item for the String length, this is a more efficient **strlen()** than the conventional one which must count the number of characters from its start position to a terminating '\0'.

```
58     };
59
```

Now comes the definition of the member functions that were too big to be included in the class definition.

```
60   String& String::operator=(const char *p)
61      {
62      // Delete old String
63      delete [] buf;
```

When a String is assigned to, whatever it contained before is over-written. Thus the old buffer must be deleted.

```
64      len = strlen(p);
65      buf = new char[len + 1];
```

```
66      strcpy(buf, p);
67      return *this;
```

Here is a new piece of syntax. The keyword **this** is used to refer to the current object – it is its address. So although there may be many objects of a given ADT instantiated at any moment in a program, the one executing this particular method at this particular time can always get at its own address through the **this** pointer. In the case of line 67, the object is returning a reference to itself. As mentioned above, the assignment operator should return a reference to the thing just assigned to, and that is what line 67 achieves.

```
68      }
69
70   String& String::operator= (const String &str)
71      {
72      if (str.buf != buf) // Protect against x = x;
```

When writing assignment methods, it is always good to protect against the possibility that a thing is being assigned to itself. In this case, if the test had not been included, **buf** would get deleted before the things it contained could get copied (to a new version of itself).

```
73         {
74         delete [] buf;
75         len = str.len;
76         buf = new char[len + 1];
77         strcpy(buf, str.buf);
78         }
79      return *this;
80      }
81   char& String::operator[](int i)
82      {
83      if (i < 0 || i >= len) {
84         error("index out of range");   i = 0;
85         }
86      return buf[i];
87      }
88   ostream& operator<<(ostream& out, const String &x)
89      {
90      return out << x.buf;
91      }
92   istream& operator>>(istream&in, String &x)
93      {
94      char tempbuf[256];
95      // Set maximum number of chars to read
96      in.width(256);
97      // And get a whitespace-separated word
98      in >> tempbuf;
```

The **iostream** 'get from' function >> is overloaded to read a whitespace-separated word into the **String**. First, though, the **iostream**

width() member function is used to limit the maximum number of characters that **>>** will read to 255. In general, **width()** limits to one less than its argument so there's room for a '\0'. This means that words longer than 255 letters will be truncated, but that's much better than them overflowing the buffer. So the get from operator for **Strings** does the same thing as the get from operator for **char *** type strings, but in a safer way.

```
 99
100   // Copy string over
101   x = tempbuf;
```

The assignment in line 101 uses the assign-**char ***-to-**String** function from lines 60 to 68.

```
102   return in;
103   }
104
105   main()
106   {
107   String st[100];
```

Array of uninitialized **Strings** constructed according to line 22.

```
108   int n;
109   cout << "Enter several lines of text:\n";
110   for (n = 0; n < 100; n++)
111     {
112     cin >> st[n];
```

Line 112 shows standard input into **String** objects; uses function on lines 92 to 103.

```
113   String test;
114     cout << (test = st[n]) << '\n';
```

Line 114 uses assignment of **Strings** to **Strings** (lines 70 to 80) and standard output (lines 88 to 91).

```
115   for (int j = 0; j < n; j++)     // Search for a match
116     {
117       if (st[j] == test)
```

Test for equality of Strings using function at lines 42 to 45.

```
118         cout << "Seen before at line " << j << '\n';
119     }
120   if (test == "done")
```

Test for equality of a **String** and a conventional null-terminated string uses function at lines 38 to 41.

```
121       break;
122     }
```

```
123   cout << "Here are the input lines in reverse order, with\
          modifications\n";
124   for (int i=n-1; i >= 0; i--)
125     {
126     st[i][i] = 'X';
```

Line 126 uses the subscripting operator to alter the ith character of the ith **String** (lines 82 to 87). Note that the test on line 84 may succeed depending on the lengths of the input lines.

```
127     cout << st[i] << '\n';
128     }
129   return 0;
130   }
```

Enter, compile and test the above program. Enhance the main function to do more rigorous testing of the String ADT. Extend the **String** type in the following ways:

- Add a constructor that takes a reference to a **String** as its argument. This will differ from the assignment operator. How?
- Now you have both an assignment operator and a copy constructor, seed both with **cout** statements that indicate which has been called and report the state of **buf** before returning. Then embed the following code in **main**:

```
String a("Hello");
String b = "Goodbye";
String c(b);
b = a;
```

Explain the roles of the copy constructor and assignment operator in each of these statements by examining the output of the program.

- Overload the operator + for **String**s to allow the concatenation of two **String**s.
- Overload the operator += for **String**s to allow addition to the end of a **String**.
- Overload the operator -- for **String**s to allow a **String** to be shortened by one letter.
- Add a member function which removes all white space from a **String**.
- Replace the **char[]** buffers in the **location** and **student** data structures with **String**s. How does this remedy the bugs identified in exercises on pages 95, 105 and 129? How does this confirm the value of the rule:

Rule 16

Remove magic numbers and threshold.
Reduce the number of parameters.

String classes and reference counting

This simple **String** class copies string objects by copying the array of characters. An alternative is to have a helper class (e.g. **string_helper**) that contains the array plus a reference count, indicating how many string objects are referring to the same **char** array. Within the string object will be a pointer to the helper object (e.g. **string_helper *help**). Then, if A and B are strings, a statement **A = B**; will make A's **help** pointer equal to B's **help** pointer and increment the reference count in the helper object. This is fast: there is no copying of what might be a very long array of characters. On the other hand, the array *will* have to be explicitly copied as soon as one of the objects referring to it changes it. This method is known as *copy-on-write* and is used in many practical classes that manipulate large arrays.

5.6 Derived types, inheritance and polymorphism

Locations and mountains example

Any ADT can be used as a base type for deriving new ADTs. This means that existing code can be reused in the creation of similar but different objects. For example, suppose we wanted to extend the programs developed earlier for dealing with locations to handle mountains and cities as well. Mountains are locations – they may be specified by a name, a latitude and a longitude, but they also have a height. Cities are locations too, but they also have a population size. Given the definition of a **location** ADT (page 126, perhaps with the substitution of **char[]** buffers by **String**s), a derived type **mountain** can be defined as follows:

```cpp
class mountain : public location {
  float height;
public:
  mountain() {};
  ~mountain() {};
  void load(istream& in) {
    in >> name >> latitude >> ns >> longitude >> ew >>
        height;
  }
  void save(ostream& out) const {
    out << name << '' << latitude << '' << ns << ''
      << longitude << '' << ew << '' << height << '\n';
  }
  float getheight() const
    {
    return(height);
    }
};
```

mountain is defined as a class derived from a particular 'parent', and its definition is really about the way that parent type has to be modified to make the **mountain** type. The first line of the definition gives the parent as **location**. The symbol : means 'derived from', and the **public** in the first line means that public members of the base class **location** will also be public members of **mountain**. Everything in the parent that the derived type definition is silent about is *inherited*, unchanged, from the parent. So **mountain** *inherits* the **name**, **latitude**, **ns**, **longitude**, and **ew** private data members from **location**, as well as the method **zone** that calculates the zone of the current object. **height** is a new private member of **mountain** and **getheight** is a new public member. And because **mountain** has explicit methods of its own called **load** and **save**, these replace the ones it would have otherwise inherited. Observe that these new versions assumed that the format of a **mountain** record in a database file would be like a simple **location**, but with a floating point number appended giving the height of the mountain.

A **city** type can be derived similarly from **location**:

```
class city : public location {
  long population;
public:
  city() {};
  ~city() {};
  void load(istream& in) {
    in >> name >> latitude >> ns >> longitude >> ew >>
        population;
  }
  void save(ostream& out) const {
    out << name << '' << latitude << '' << ns << ''
      << longitude << '' << ew << '' << population << '\n';
  }
  float getpopulation() const
    {
    return(population);
    }
};
```

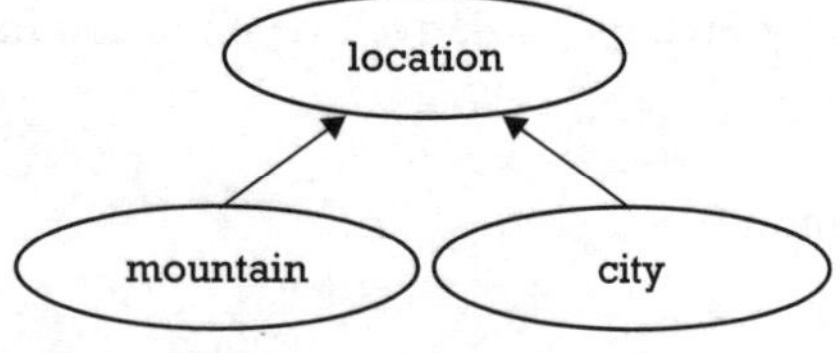

Figure 5.1 *Class diagram showing the derivation of* **mountain** *and* **city** *from* location

Now we have our first inheritance hierarchy. We can show this with the class diagram in Figure 5.1.

Note that arrows point *up* the hierarchy, is the direction of *isa*.

Before moving on to a modified zone-printing program that works for both locations and mountains, it is appropriate to consider how the program should be organized into files. Chapter 6 includes general rules for dividing programs into files, but in C++ the basic principle is this: put ADT (i.e. **class**) definitions in 'header' files, with a suffix of **.h**, and functions (including member functions not included in the ADT definitions) in **.cpp** files, which **#include** the headers they need. So the first file, called **loc.h**, is this:

```
#include <iostream>
using namespace std;
class location {
protected:
  char name[20];          // String name; would be safer.
  float latitude; char ns;
```

```
    float longitude; char ew;
public:
  location() {};
  ~location() {};
  virtual void load(istream& in)
    {
    in >> name >> latitude >> ns >> longitude >> ew;
    }
  virtual void save(ostream& out) const
    {
    out << name << ' ' << latitude << ' ' << ns << ' '
      << longitude << ' ' << ew << '\n';
    }
  const char *getname() const
    {
    return(name);
    }
  const char *zone() const;
};
```

loc.h contains the definition of the **location** type. There are two changes from the earlier definition. The first is the label **protected** in front of the previously private data items. The keyword **protected** makes the data items that follow it accessible to ADTs derived from this one. In other words, objects of derived types can, in their own methods, have direct access to the data, but other objects cannot. **protected** therefore looks like **public** to derived types, and like **private** to other types. In this case, **mountain** objects must have access to **name**, **latitude**, etc., because they load and save values in these data items.

The second change to the earlier definition is the addition of the keyword **virtual** before the methods **load** and **save**. Though not strictly necessary for this example, we are establishing here a practice that is worth adopting habitually: *Every member function of a base type which is superseded in any derived type is labelled* **virtual** *in the base type*. The reason for doing this is polymorphism, explained and illustrated soon. Usually there is little penalty in using the **virtual** label even when polymorphism is not going to be used, and it serves as a flag that other types are derived from this one, and at least one of those derived types replaces the method labelled **virtual**.

Associated with **loc.h** is **loc.cpp**, the file that contains the methods for objects of this type:

```
#include "loc.h"
enum { nfrigid, ntemperate, torrid, stemperate, sfrigid, error };
const char *zones[6] = {
  "North frigid",
  "North temperate",
  "Torrid",
  "South temperate",
  "South frigid",
  "None"     // Flag to indicate error return
};
```

```cpp
const char *location::zone() const
  {
  if (!name[0])      // No location loaded in this object
    return zones[error]; // Error return
  if (latitude < 23.5)
    return zones[torrid];
  if (latitude < 66.5)
    {
    if (ns == 'N')
      return zones[ntemperate];
    return zones[stemperate];
    }
  if (ns == 'N')
    return zones[nfrigid];
  return zones[sfrigid];
  }
```

loc.cpp includes **loc.h**, and it contains the code for the **zone** method.
loc.cpp nicely localizes all the information about the names of zones
and how they are stored in the program.

   mountain.h will contain the **mountain** class definition we have
already seen:

```cpp
class mountain : public location {
  float height;
public:
  mountain() {};
  ~mountain() {};
  void load(istream& in)
    {
    in >> name >> latitude >> ns >> longitude >> ew >>
        height;
    }
  void save(ostream& out) const
    {
    out << name << ' ' << latitude << ' ' << ns << ' '
      << longitude << ' ' << ew << ' ' << height << '\n';
    }
  const float getheight() const
    {
    return(height);
    }
};
```

Finally, the thing that this particular program is concerned with, printing out a list of places and their zones, is localized with a main program.

```cpp
//
// Prints out zones of locations and mountains in a database.
// Format of database is:
// First line: Integer number of _location_ records (put into
// numrecords)
```

```cpp
// Next numrecords lines: One record per line:
// Loc name (20 chars max), latitude,
//        ns character modifier for latitude,
//        longitude, ew char modifier for longitude
// Next line: Integer number of _mountain_ records (put into
// numrecords)
// Next numrecords lines (= number of mountains): One record
// per line: As location with height appended.
#include <fstream>
#include "loc.h"
#include "mountain.h"

int main(int argc, char **argv)
  {
  if (argc != 2)
    {
    cerr << "Usage: tellzone locationsfile\n";
    return -1;
    }
  ifstream infile(argv[1]);
  if (!infile)
    {
    cerr << "Can't open " << argv[1] << " for reading\n";
    return -1;
    }
  int numrecords;
  infile >> numrecords;
  location current_location;
  for (int i = 0; i < numrecords; i++)
    {
    current_location.load(infile);
    cout << "Location " << current_location.getname()
       << " is in the " << current_location.zone() << " zone\n";
    }
  infile >> numrecords;
  mountain current_mountain;
  for (int i = 0; i < numrecords; i++)
    {
    current_mountain.load(infile);
    cout << "Mountain " << current_mountain.getname()
       << " is in the " << current_mountain.zone() << " zone\n";
    }
  infile.close();
  return 1;
  }
```

This main program contains two closely equivalent loops, the first
dealing with the pure locations, and the second dealing with
mountains.

To get this program to run, the two **.cpp** files have to be compiled
and then linked. In an integrated environment like Visual C++ or

Borland C++, this is usually achieved by creating a 'project' (for example, called **tellzone.prj**), and adding the two files to it. 'Making' the program using the integrated environment will ensure that all files in the project are compiled and linked into an executable.

The complete program will read a database such as:

```
5
Toronto    43.66 N 79.33 W
London     51.5 N 0.17 W
Delhi      28.63 N 77.23 E
Sydney     33.87 S 151.22 E
Lima       12.05 S 77.05 W
1
Everest    30 N 85 E 29028
```

and print out:

```
Location Toronto is in the North temperate zone
Location London is in the North temperate zone
Location Delhi is in the North temperate zone
Location Sydney is in the South temperate zone
Location Lima is in the Torrid Zone
Mountain Everest is in the North temperate zone
```

The two loops in the main program look very similar. It would be nice to collapse them into one to make the program more compact. That is, because this program only deals with mountains in terms of their locations, it should be able to talk about them in terms of the base type **location**. This is possible in C++, and the mechanism to achieve it is as follows: a pointer to a base class may also point to an object of any class derived from the base class. Here is a rewrite of **tellzone.cpp** which illustrates this. There are still two separate loops for **location**s and **mountain**s, so not a great deal has been gained by using a base class pointer. But we shall soon seen how this mechanism provides considerable power.

```
#include <fstream>
#include "loc.h"
#include "mountain.h"

int main(int argc, char **argv)
    {
    location *ploc;
    if (argc != 2)
        {
        cerr << "Usage: tellzone locationsfile\n";
        return -1;
        }
    ifstream infile(argv[1]);
    if (!infile)
        {
        cerr << "Can't open " << argv[1] << " for reading\n";
```

```
      return -1;
    }
  int numrecords;
  infile >> numrecords;
  location current_location;
  ploc = &current_location;        // ploc, a pointer to a location,
                                   // now points at a location

  for (int i = 0; i < numrecords; i++)
    {
    ploc->load(infile);
    cout << "Location " << ploc->getname()
       <<" is in the " << ploc->zone() << " zone\n";
    }
  infile >> numrecords;
  mountain current_mountain;
  ploc = &current_mountain;
  // A pointer to a base class can point to an object of a derived
  // class. So ploc, a pointer to a location now points at a
  // mountain
  for (i = 0; i < numrecords; i++)
    {
    ploc->load(infile);   // Calling mountain's load method
    cout << "Mountain " << ploc->getname()
       << " is in the " << ploc->zone() << " zone\n";
    //The previous two lines call location's getname and zone
    // methods, which mountain inherits
    }
  infile.close();
  return 1;
  }
```

Now the two loops look very similar indeed.

Something profound is happening in the above code. A pointer to a base type is being moved between objects of that base type and objects of a derived type. Not until the program runs does a pointer 'know' for sure whether it is pointing at a base object or a derived object. So, at runtime, the program has to figure out which methods to call based on what is currently being pointed to. This is called *polymorphism* and is a very powerful facility associated with inheritance.

Student marks example

To illustrate the power of polymorphism, we consider again the students database from Chapter 4. In this section, the database will be extended to deal with different types of students.

Omitting the **load_data** function, the program was given on page 106 as follows:

```
#include <iostream>
#include <fstream>
```

```cpp
#include <cstdlib>
using namespace std;
#define  MAXSTUDENTS 80
#define  MAXNAMELENGTH 30

struct student
   {
   char name[MAXNAMELENGTH]; // Maybe we should use
                             // Strings!
   long id;
   float mark;
   };
float average_mark(int, student *);
int load_data(char *, student **);

int main(int argc, char **argv)
   {
   student *syde321;
   int num_students;
   if (argc != 2)
      {
      cerr << "Usage: average filename\n";
      return(-1);
      }
   if ((num_students = load_data(*(argv+1),&syde321)) < 0)
      return(-1);

   cout << "Average was " << average_mark(num_students,
         syde321) << "\n";
   return(0);
   }

float average_mark(int num_students, student *p)
   {
   int i;
   float sum_marks;
   sum_marks = 0.0;
   for (i = 0; i < num_students; i++)
      sum_marks += p[i].mark;
   return(sum_marks/num_students);
   }
```

First, the **student** structure must be turned into a C++ class – an abstract data type. As discussed in Chapter 7, choice of data structure is vital to good design, and the facilities of the language do have an impact on that choice. The C++ version presented below has a fundamental change in data structure. Instead of **load_data** setting up an array of **student** structures of the appropriate length, the C++ version maintains a fixed length array of pointers to **student** objects. This, of course, means a change to **load_data** as well as to the functions given. The virtual function

virtual void calc_mark()

introduced in the class definition is the mechanism by which polymorphism will happen. In the base **student** type, **calc_mark** does nothing.

```cpp
#include <iostream>
using namespace std;
#define    MAXSTUDENTS 80
#define    MAXNAMELENGTH 30

class student {
protected:
  char name[MAXNAMELENGTH];
  long id;
  float mark;
public:
  student(char *n, long i, float m)      // load_data uses and
                                         // sets up address in
                                         // student_list

    {
    strncpy(name, n, MAXNAMELENGTH);
    name[MAXNAMELENGTH-1] = '\0'; // Ensure null-
                                        // terminated
    id = i;
    mark = m;
    }
  float get_mark()
    {
    return(mark);
    }
  virtual void calc_mark()
    {
    }
  };

class student_list {
protected:
  student *st[MAXSTUDENTS];
  int   num_students;
public:
  student_list()
    {
    num_students = 0;     // Will be set by load_data
    }
  int   load_data(char *);
  float  average_mark();
  };

float student_list::average_mark()
  {
  float sum_marks = 0.0;
  for (int i = 0; i < num_students; i++)
    sum_marks += st[i]->get_mark();
```

```
    return(sum_marks/num_students);
    }

int main(int argc, char **argv)
  {
  student_list software_design;
  if (argc != 2)
    {
    cout << "Usage: average filename\n";
    return(-1);
    }
  if (software_design.load_data(*(argv+1)) < 0)
    {
    cout << "Error while loading data\n";
    return(-1);
    }
  cout << "Average was " << software_design.average_mark()
       << '\n';
  return(0);
  }
```

Now suppose that students for a software design course could choose their method of assessment between two alternatives: stream A marked on the basis of a midterm exam (50%) and a final exam (50%), and stream B marked on the basis of a project (40%) and a final (60%). Here is the first new (derived) ADT for dealing with this situation.

```
class a_stream_student : public student {
  float midterm;
  float final;
public:
  a_stream_student(char *n, long i, float mid, float fin);
  void calc_mark();
  };
a_stream_student::a_stream_student(char *n, long i, float mid,
    float fin) : student(n, i, 0.0)
  // The above syntax means that this derived type's constructor
  // calls the base type's
  // constructor first, with the arguments n, i, 0.0.
  {
  midterm = mid;
  final = fin;
  }
void a_stream_student::calc_mark()
  {
  mark = midterm*0.5 + final*0.5;
  }
```

This ADT is derived from **student**, and inherits the data and methods of this parent. **midterm** and **final** are new private members of the ADT, and the method **calc_mark** is a new public member. The constructor for **a_stream_student** begins by calling the base class constructor (note the syntax), then adds its own code. Now, a base class constructor will be automatically called with no arguments if it is not explicitly named in the derived class constructor; it is good practice always to show an explicit call of the base class constructor to avoid ambiguity.

The method **calc_mark** has the weightings hard-coded for illustrative purposes. In practice, they would be variables.

It would now be possible to write code like the following:

```
a_stream_student    the_ideal_student("Alice
                        Nonesuch",12345678,100.0,100.0);
the_ideal_student.calc_mark();
cout << "The ideal student scored "
    << the_ideal_student.get_mark() << '\n';
```

Here the method **calc_mark** is special to the derived ADT, while the method **get_mark** comes from the base ADT. In any case, the object declared, **the_ideal_student**, would be able to do both things without problem.

A **b_stream_student** class may be defined similarly:

```
class b_stream_student : public student {
    float project;
    float final;
public:
    b_stream_student(char *n, long i, float pro, float fin);
    void calc_mark();
    };
b_stream_student::b_stream_student(char *n, long i, float pro,
        float fin) : student(n, i, 0.0)
    {
    project = pro;
    final = fin;
    }
void b_stream_student::calc_mark()
    {
    mark = project*0.4 + final*0.6;
    }
```

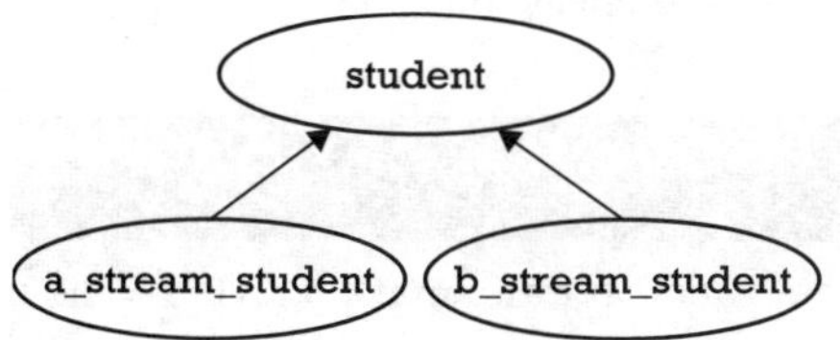

Figure 5.2 *Class diagram showing the derivation of a_stream_student and b_stream_student from student*

So now there are two ADTs derived from the **student** ADT, each inheriting the data and methods of the base class, and each adding data and methods of its own. Figure 5.2 shows the class diagram. Again note that the arrows point from derived types to the base type.

The final thing we need is a new **student_list** ADT. In fact, we derive a type from the existing **student_list**:

```
class software_design : public student_list {
public:
    int work_out_marks();
    };
```

The name of this derived type **software_design** illustrates that it is perfectly legitimate to derive a type (or create a type) that is only ever going to be instantiated once – in this case to store the marks for a particular course. It is not the case that types have to be general-purpose things with lots of parameters, and that instantiations have to specify their particulars by setting all those parameters. There is nothing wrong with making a type for just one instance.

Now, a pointer to an ADT can point to any ADTs derived from it. So when **load_data** sets up the list of students, it creates an **a_stream_student** or a **b_stream_student** as appropriate, then takes the address and stores it in the **st** array. Here is where the magic happens …

```
int software_design::work_out_marks()
  {
  for (int i = 0; i < num_students; i++)
    st[i]->calc_mark();
  return(0);
  }
```

The method **work_out_marks** of the ADT **software_design** (which has been derived from **student_list**), steps through the **st** array (which is a data structure inherited from **student_list**) for **num_students** (also inherited from **student_list**). For each pointer to **student** in the array, the method **calc_mark()** is called. In the base class **student** this method is marked **virtual**. This is a special signal to the compiler which means 'work out at runtime which method is to be run'. So, if **st[i]** is pointing at an object of type **a_stream_student**, then the method executed will be **a_stream_student::calc_mark()**. If **st[i]** is pointing at a **b_stream_student** object, the method executed will be **b_stream_student::calc_mark()**. To summarize, polymorphism's mechanism in C++ is: (a) a pointer to a type can point to any derived type; (b) a function in a base type marked **virtual** means wherever a pointer to this type accesses this method, work out what derived type it is actually pointing at, and call the method for that derived type. Polymorphism will work through references too.

Virtual and pure virtual functions

What happens if an object of type **student** is instantiated and its member function **calc_mark()** called? The answer is nothing, because we had

```
virtual void calc_mark()
{
}
```

in the definition of **student**. We could define a different default behaviour for **calc_mark()**, but first we should consider whether there is actually such a thing as a default student. Aren't all students on one stream or another? If that is the case, then we never want to make an object of type **student**. We just want **a_stream_students**, **b_stream_students**, and, perhaps later, **c_stream_students**. By replacing the above lines in the **student** definition with

```
virtual void calc_mark() = 0;
```

we can turn **calc_mark()** into a *pure* virtual function, and **student** into an *abstract base class*. With this definition of **calc_mark()** the compiler knows that we will never declare an object of the base type, only objects of types derived from it. So it will warn if we ever do try to instantiate a **student**. That's probably what we want for this program.

Because this example includes so many of the fundamentals of C++, the complete program is now given. Note that the structure of the database file has changed from that used in the previous chapter, to accommodate the two marks for each student. Again, the program is divided into files.

First, the header file, **student.h**, which contains all the ADT definitions:

```cpp
#include <cstring>
using namespace std;
#define  MAXSTUDENTS 80
#define  MAXNAMELENGTH 30
class student {
protected:
  char name[MAXNAMELENGTH];
  long id;
  float mark;
public:
  student(const char *n, const long i, const float m)
    {
    strcpy(name, n);
    id = i;
    mark = m;
    }
  float get_mark() const { return(mark); }
  virtual void calc_mark() {}
};

class a_stream_student : public student {
  float midterm;
  float final;
public:
  a_stream_student(const char *n, const long i, const float mid,
      const float fin) : student(n, i, 0.0)
                      // This is called explicitly here to pass
                      // the arguments down
    {
    midterm = mid;
    final = fin;
    }
  void calc_mark()
    {
    mark = midterm*0.5 + final*0.5;
    }
};
class b_stream_student : public student {
  float project;
  float final;
public:
  b_stream_student(const char *n, const long i, const float pro,
      const float fin) : student(n, i, 0.0)
    {
```

```cpp
        project = pro;
        final = fin;
      }
    void calc_mark()
      {
      mark = project*0.4 + final*0.6;
      }
  };

class student_list {
protected:
  student *st[MAXSTUDENTS];
  int num_students;
public:
  student_list() { num_students = 0; }      // Will be set by
                                            // load_data

  ~student_list()
    {
    for (int i = 0; i < num_students; i++)
      delete st[i];
    }
  int load_data(const char *);
  float average_mark() const;
  };

class software_design : public student_list {
public:
  software_design() : student_list() {};
  int work_out_marks();
  };
```

Next comes the file containing the methods for the classes not specified in the header. This file is **student.cpp**. Note how it hides the
information relating to database structure.

```cpp
#include <fstream>
using namespace std;
#include "student.h"
float student_list::average_mark() const
  {
  float sum_marks = 0.0;
  for (int i = 0; i < num_students; i++)
    sum_marks += st[i]->get_mark();
  return(sum_marks/num_students);
  }

// The load_data method is the only one that needs to know the
// format of the data in the database file.
// This is shown in the following example line:
// a Alice_Brown 12345678 75.2 82.3
// The first character indicates the stream.
int student_list::load_data(const char *fname)
```

```cpp
{
ifstream in(fname);
if (!in)
  return(-1);
in >> num_students;
int i;
  char name[MAXNAMELENGTH];
  long idnumber;
  float mark1, mark2;
  char type;
for (i = 0; i < num_students; i++)
  {
    in >> type >> name >> idnumber >> mark1 >> mark2;
    // Now set up appropriate type of student structure
    // and set pointer in st[]
    if (type == 'b')
      st[i] = new b_stream_student(name,
           idnumber, mark1, mark2);
    else
      st[i] = new a_stream_student(name,
           idnumber, mark1, mark2);
  }
in.close();
return(num_students);
}

int software_design::work_out_marks()
  {
  for (int i = 0; i < num_students; i++)
    st[i]->calc_mark();
  return(0);
  }
```

Finally, the file to do the particular job of this program is average.cpp.

```cpp
// Program to calculate average mark in a course mark database.
// Along the way, derived types, inheritance and
// polymorphism are illustrated.
#include <iostream>
using namespace std;
#include "student.h"

int main(int argc, char **argv)
  {
  software_design software_design;
  if (argc != 2)
    {
    cout << "Usage: average filename\n";
    return(-1);
    }
  if (software_design.load_data(*(argv+1)) < 0)
    {
    cout << "Error while loading data\n";
```

```
return(-1);
   }
software_design.work_out_marks();
cout << "Average was " << software_design.average_mark()
      << '\n';
return(0);
}
```

> Hand execute the program to interpret its use of classes, objects and polymorphism.

> Substitute the **char[]** buffers in this program with **String**s. Test to break both versions. Are **String**s really more error-resilient? (They certainly should be!)

5.7 Exceptions

C++ has a method for dealing with exceptions: times when something unexpected happens and special code has to be used for dealing with it. The basic mechanism is for a function that detects the exceptional situation to 'throw' an exception with a statement of the form

throw
     exception_type(constructor_parameter_for_exception_type)

This might be used in an expression like

```
if (invalid_token(x))
    throw runtime_error("Invalid x token");
```

where **runtime_error** is an exception type defined in the system header file **stdexcept**. You can use any type for the exception, but the types defined in **stdexcept** are usually enough. An object of the specified type gets constructed (in this case with a parameter string "**Invalid x token**") and thrown to the nearest **catch** statement that can deal with it. To be more precise, when an exception is thrown, the current block or function is left, as if there were a **return** statement. All the objects associated with the block or function are destroyed and control returns to the caller, which also immediately returns (destroying all its objects) unless it is equipped to catch the exception. This goes all the way up the call tree (i.e. the stack is unwound) until a suitable exception handler is found. Specification of a suitable exception handler is like this:

```
try {
  // Do many things, one of which causes the exception
  ...
  }
catch (exception_type) {
  // Handle exception
  ...
  }
```

So any exception of type **exception_type** thrown in the **try** block is caught by the **catch** block. A **catch** block can process all exceptions of any type by being specified with an ellipsis:

```
catch (...) {
  }
```

and a **catch** block can rethrow an exception with

```
  throw;
```

to pass the exception up to a higher level handler if it can't clean things up alone.

Here is an example of a program that uses the mechanism.

```
#include <iostream>
using namespace std;
int main() {
  int i(0);
  char *p[4*1024];
  try {
    while(i < 4*1024) {
      p[i++] = new char[1024*1024];
      cout << "Megabyte number " << i <<
          " allocated starting at ";
      cout << (unsigned int) p[i - 1] << endl;
      }
    cout << "Successfully allocated 4 Gigabytes\n";
    }
  catch(bad_alloc) {
    cout << "Out of memory after " << i << " Megabytes\n";
    }
  for (int j = 0; j < i; j++)
    delete [] p[j];
  return 0;
  }
```

As you see, **throw** is missing from this program. That is because we are relying on **new** to throw a **bad_alloc** error, which it does when it cannot allocate any more memory.

You can use this program to find out how much memory is allocatable on your system (up to 4 Gbytes). The amount will depend on how much virtual disk-based memory the system uses in addition to real chip memory. On my laptop running Linux, memory runs out at 2931 Mbytes. (Under Windows, compiled with Visual C++ 98, this program reports that all 4 Gbytes are allocated! But closer inspection reveals that **new** is returning 0 for the later calls: in other words, VC++ is using the old fashioned **new** which returns 0 when it fails to allocate and doesn't throw an exception. See page 106.)

In this book we will mostly forego **try**, **throw** and **catch**, and stick with the return value mechanism for reporting exceptional conditions and errors. C++ exception handling allows neat separation of error handling from other program code, but this is unnecessary in most of the programs presented here.

An undefined exceptional situation

What should happen when you divide a number by zero? In real life (i.e. mathematics), division by zero is undefined. Some people informally say that division of anything positive by zero equals infinity, and division of anything negative by zero equals minus infinity. These are really shorthand statements about limits as the denominator tends to zero, so are forgivable. Division of zero by zero, on the other hand, is not equal to anything, even informally. Not zero, not one and not infinity.

In C++, division by zero isn't defined either. But saying something is undefined in a programming language is different from saying it's undefined in mathematics. In a programming language, you have to consider what happens when someone actually *does* divide something by zero. Although the language doesn't define that, something must happen, and you'd expect it to be an exceptional something.

Here's a program that you can run to find out what happens when you divide 1 by 0 on your computer:

```cpp
#include <iostream>
using namespace std;
int main() {
  typedef double numtype;
  numtype a(0);
  numtype b(1);

  try {
    numtype c = b/a;
    cout << "c = " << c << endl;
  }
  catch(...) {
    cout << "Division by zero was exceptional!\n";
    throw; // rethrow to see what the system makes of it
  }

  try {
    numtype d = a/a;
    cout << "d = " << d << endl;
  }
  catch(...) {
    cout << "Dividing zero by zero was exceptional!\n";
    throw;
  }
  return 0;
}
```

The program is written to explore the possibilities that division worked and generated a value, or that an exception was thrown. The **throw** statements rethrow the exception to any higher-level exception handlers that might catch it. This allows us to see whether the operating system that calls **main()** does anything with an exception.

The **typedef** is used so that we can see what happens for different types – especially to check the difference between floating point types like **double**, and integer types.

Well here's what happens on my machine. With **doubles**, as in the example above, using Visual C++, c prints out as 1.#INF and d prints as −1.#IND; using GNU g++, c prints as inf and d prints as nan. No exceptions are thrown. Both implementations claim to implement standard floating point functionality, so we can infer that 1.#INF for VC++ equals inf for g++ equals an informal infinity, while −1#IND equals nan equals not a number. The g++ results seem unambiguously correct; the VC++ results questionable. However, when we edit the **typedef** to use integers instead of floats, the story is different. VC++ throws exceptions. The first exceptional statement is printed out and the exception rethrown, to be caught by the programming environment, bringing up a dialog box signalling an integer division by zero. Full marks! g++ on the other hand neither sets c to anything nor throws a C++ exception. Because division by zero is undefined by the language, it can't really be blamed. But what it does instead is bizarre: the program terminates and prints 'Floating point exception'. Given the absence of floating point types in the revised program, this is misleading to say the least.

5.8 Templates

If you create a program that works with objects of one type, how could you generalize it so that it works with objects of other types? There are at least three answers to this question. First, suppose you have a conventional program, broken up into several files, where everything works for some data type, for example **double**. Now you want to create a sister program that does exactly the same thing for ints instead of **doubles**. The most straightforward way to proceed is to change every relevant occurrence of **double** in every file to some name of your own, e.g. **itemtype**, then put the line

typedef double itemtype;

at the top of your header file. Generating the sister program is then as simple to changing the **double** to int:

typedef int itemtype;

and recompiling. This is what we did in the divide by zero test program on page 162.

The second possibility for working with multiple types is to derive the types in question from a common base class then use polymorphism to select between them. This is a true object-oriented solution and, if properly implemented, means your program will be able to work with any mixed assortment of the two types. However, it relies on the system working out the type of each object at runtime so can be very inefficient, especially if any one program (or any one use of your collection type) always uses objects of the same type.

The third alternative is to use the C++ template facility. Conceptually this is similar to the first, **typedef**, approach, but it is much more flexible and allows you to cut programs that use *any* type out of a general template.

A template is a specification to the compiler of how to generate a function or a class using a type or types that are not yet known. On its own it can do nothing, but when the template is *specialized* with type names later in the program, the compiler fills everything in. For example

```
template <class itemtype>
const itemtype& max (const itemtype& x, const itemtype& y) {
    return ((x > y) ? x : y);
}
```

So long as **itemtype** has the concept of a greater than > operator, this function will return the maximum of the two objects passed to it. To actually cut a **max** function from this template, we might write a program like:

```
int main()
    {
    cout << max(3.1,4.2) << " was largest\n";
    return 0;
    }
```

which, because the compiler recognizes the parameters to **max** as both floating point constants, will generate a version of **max** that works on floating point values. But relying on the compiler to spot the type of the arguments is just a shortcut that doesn't always work. We can be explicit:

```
int main()
  {
  cout << max<int>(3.1,4.2) << " was largest\n";
  return 0;
  }
```

This time the compiler is told that the version of **max** to create is one that replaces **itemtype** with **int**. This will work: the floating point constants will be type converted to **int**s before being sent to the **int** version of max.

The examples above illustrate template syntax. Any time you are about to use a template, you write

template <class T>

before the parameterized function or class. T can be any label you want. Then within the function or class, you just use T to refer to the as-yet-unknown type. When it comes to specializing the template (i.e. using a function or class with a particular type), you put the appropriate type name in angle brackets next to the function or class name.

So it's easy to convert an existing program specialized for a particular type to a template. Indeed this is the best way to develop a template class or function: develop for a specific type first, then replace instances of that type with the template parameter. Here's how that works for the stack that we developed on page 123. The code here is identical to that before except that the **int**s that refer to items stored on the stack have been converted to **itemtype**s, and the appropriate template syntax has been used before the class definition, the member function definitions, and the actual instantation of a **stack**.

```
#include <iostream>
using namespace std;
template <class itemtype>
class stack
  {
  // First come the private data items
  int stack_length;        // To allow the caller to set max length
  itemtype *stack_base;  // Points to where stack is in memory
  int stack_pointer;
public:
  stack(const int size)     // This is the constructor
    {
    stack_base = new itemtype[stack_length = size];
    stack_pointer = 0;
    };
  ~stack()                          // This is the destructor
    {
    delete [] stack_base;  // NOTE USE OF SQUARE BRACKETS
```

```cpp
      }
    int push(const itemtype value);  // push and pop are declared
                                     // inside the class definition
                                     // but defined below
    int pop(itemtype& value);
    };

template <class itemtype>
int stack<itemtype>::push(const itemtype value)
// A short function like this could have
// been included in the class definition
    {
    if (stack_pointer == stack_length)
        return(-1); // Stack full
    stack_base[stack_pointer++] = value;
    return(0);
    }
template <class itemtype>
int stack<itemtype>::pop(itemtype& value)
    {
    if (stack_pointer == 0)
        return(-1); // Stack empty
    value = stack_base[--stack_pointer];
    return(0);
    }

int main()
    {
    stack<int> my_stack(10);
    // Saying that I won't be putting more
    // than 10 items on the stack.
    my_stack.push(345);
    my_stack.push(678);
    int test;
    my_stack.pop(test);
    cout <<test << " was just popped off the stack\n";
    return 0;
    }
```

With this version of a stack it is easy to make the stack store any kind of objects. For example, to use a stack of **doubles**, we could just change **main()** to:

```cpp
int main()
    {
    stack<double>my_stack(10);
    // Saying that I won't be putting more
    // than 10 items on the stack.
    my_stack.push(345.543);
    my_stack.push(678.876);
    double test;
    my_stack.pop(test);
```

```
    cout << test << " was just popped off the stack\n";
    return 0;
}
```

Templates have a downside. Although the C++ language definition includes an **export** keyword to allow template implementations to be in files distinct from where they are used, this does not seem to have been implemented in any compilers. Consequently all template code has to go in header files for the time being. This neither hides information nor allows efficient maintenance of code.

A further problem with templates is that programmers often use a generic type name of T (or T, S, R, … as multiple types are templated simultaneously). This makes code cryptic.

5.9 Streams

Perhaps the most important class library a user of C++ encounters is the C++ streams library. We have already used its features extensively in dealing with input and output.

A stream is an abstraction that refers to any flow of data from a source (or producer) to a sink (or consumer). By means of a careful sequence of type extensions, this abstraction is applied to different kinds of sources and sinks by different parts of the streams library. At a fairly high level in the inheritance hierarchy the facilities for formatting input and output are implemented. The ADT **ios** provides all such formatting, including the overloaded definitions for the << and >> operators. Further down the hierarchy are the facilities that deal specifically with files, memory buffers, etc.

The relationship between ADTs (classes) in the streams library is complicated by the liberal use of *multiple inheritance* in the class library. Multiple inheritance is not developed in this book, but its principle is simple – to allow types to derive from multiple base classes. In the streams library it is used to provide an efficient set of derived types for all kinds of input and output.

The classes that provide the easiest file input/output are **ifstream** and **ofstream**. The following program shows how these may be instantiated and their methods used to copy one file to another.

```
// Program to copy one file to another
#include <iostream>
#include <fstream>
using namespace std;
int main(int argc, char **argv)
  {
  if (argc != 3)
    {
    cerr << "Usage: copy source dest\n";
    return(-1);
    }
ifstream in(*(argv+1));
  // If working on a PC e.g. Borland C++, have to use
  // ifstream in(*(argv+1), ios::binary);
  if (!in)
```

```
        {
        cerr <<"Can't open " <<*(argv+1) <<
            " for reading\n";
        return(-1);
        }
    ofstream out(*(argv+2));
      // If working on a PC e.g. Borland C++, have to use
      // ofstream out(*(argv+2), ios::binary);
      if (!out)
        {
        cerr <<"Can't open " << *(argv+2) << "for writing\n";
        return(-1);
        }
      char c;
      while(in.get(c))
        out.put(c);
      // The above three lines could be replaced by the more
      // efficient:
      // char buffer[256];
      // while(in.read(buffer,255))
      // out.write(buffer,in.gcount());
      // out.write(buffer,in.gcount());
      // One last one for the last partial read.
      in.close();
      out.close();
      return(0);
      }
```

The files for reading and writing are opened by the **ifstream** and **ofstream** constructors, the second argument being optional and giving the file mode. The **get** and **put** methods are available for moving single characters. In fact **get** is overloaded to be able to fetch several characters. **read** and **write** can be used for larger block read and writes. **gcount** always returns the last number of characters read, so when **read** fails because there were fewer than 256 characters left to read, **gcount** still gives the correct number.

In this program, only the instantiation of **in** and **out** use member functions from **ifstream** and **ofstream** (i.e. constructors). The **get**, **put**, **read**, **write**, and **gcount** methods are all inherited from **istream** or **ostream** as appropriate. **close** is inherited from **fstreambase**. In other words, all the file manipulation facilities, except for opening and closing, are (from the outside) identical to facilities for other kinds of input/output. This is why earlier programs were able to use the put to and get from operators to manipulate formatted text on disk files.

So, the basic set of classes for input and output consists of:

- **istream**   Input stream (**cin** is an object of this type).
- **ostream**   Output stream (**cout** and **cerr** are objects of this type).
- **ifstream**   Input file stream (a type extension of **istream** for dealing with disk files).
- **ofstream**   Output file stream (a type extension of **ostream** for dealing with disk files).

Table 5.1 *File modes – optional extra argument to **ifstream** and **ofstream** instantiations*

File mode	Meaning
ios::app	Append data, writing at the end of the file
ios::ate	Seek to the end of the file upon the original open
ios::in	Open for input (implied with **ifstream**)
ios::out	Open for output (implied with **ofstream**)
ios::binary	Open file in binary mode (on a DOS system: default is text mode)
ios::trunc	Discard existing contents of the file if it exists
ios::nocreate	If the file does not exist, open fails
ios::noreplace	If the file exists, open for output fails unless **ate** or **app** is set

The basic set of member functions for *file* input and output are:

- ifstream(const char *filename)   Constructors for opening a file.
  ifstream(const char *filename,   For input (used in above
  const int mode)   example).
- ofstream(const char *filename)   Constructors for opening a file.
  ofstream(const char *filename,   For output (used above).
  const int mode)
- close()   To close the file.

The mode argument to the constructors may be any of those shown in Table 5.1.

In addition, file input and output have in common with standard input and output the member functions shown in Tables 5.2 and 5.3.

Note that the **get()** and **getline()** member functions are useful for getting strings with embedded white space.

Create a **getline()** function for getting a line of input into a **String**.

Streams have another kind of operator called a *manipulator*. Manipulators are used to control the format of text fields input and output by the get from and put to operators. Here is an example:

```
int i = 123;
cout << setw(6) << setfill('#') << i;
```

which prints out

```
###123
```

setw() and **setfill()** are manipulators inserted into a chain of put to operations. **setw()** sets the field width to its argument, so the output will be printed in a field of 6 characters; **setfill()** sets the field fill character to its argument, so that unused spaces in the output field will be filled with that character. In this case, 123 is a three-digit number to be printed in a field of size 6, so there are three spaces to fill and # is used in each of these. Table 5.4 shows the most commonly used manipulators.

Table 5.2 *istream member functions*

int eof()	Non-zero on end of file
int clear()	Resets any failure or end-of-file state. After an end of file or an error, you must call **clear()** before you can do anything more with the stream
int gcount()	Returns the number of characters last extracted
int get()	Extracts the next character or EOF (end of file)
istream& get(char *buf, int len, char delim = '\n')	Extracts characters into the given **buf** until the delimiter (third parameter) or end of file is reached, or until (len-1) types have been read. A terminating '\0' is appended to the string, but the delimiter is not. Fails only if no characters were extracted
istream& get(char&c)	Extracts a single character into **c**
istream& getline(char *buf, int leng, char delim = '\n')	Same as **get** except the delimiter is also extracted
istream& ignore(int n=1, int delim = EOF)	Causes up to **n** characters in the input stream to be skipped; stops if **delim** is encountered
int peek()	Returns next character without extraction
istream& putback(char c)	Pushes **c** back onto the stream
istream& read(char *buf, int n)	Extracts n characters into buf. Use **gcount()** for the number of charcters actually extracted if an error occurred
istream& seekg(long pos)	Moves to position **pos** in file
istream& seekg(long pos, seek_dir dir)	Moves to a position relative to the current position, following the definition **enum seek_dir {beg, cur, end}**;
long tellg()	Returns the current stream position
>> (Get from)	Overloaded formatted input operator for text input

Table 5.3 *ostream member functions*

ostream& flush()	Flushes the stream
ostream& seekp(long pos)	Moves to position **pos** in file
ostream& seekp(long pos, seek_dir dir)	Moves to a position relative to the current position, following the definition **enum seek_dir {beg, cur, end}**;
ostream& put(char c)	Inserts the character **c**
long tellp()	Returns the current stream position
ostream& write(char *buf, int n)	Inserts **n** characters from **buf** into the stream
<< (Put to)	Overloaded formatted output operator for text output

Table 5.4 *Stream manipulators – alter stream states by embedding manipulators in chains of put to or get from operations*

Manipulator	Action
dec	Output/input integers in decimal
hex	Output/input integers in hexadecimal
oct	Output/input integers in octal
ws	Extract whitespace characters
endl	Insert newline '\n' and flush the stream
ends	Insert terminating '\0' in string
flush	Flush an ostream
setfill(int c)	Set the fill character to **c**
setprecision(int n)	Set the floating-point precision to **n**
setw(int n)	Set the field width to **n**

A useful file-changing program

Many of the programs shown in this book are prefaced by line numbers. I have two programs, **addnumbers.cpp**, and **remnumbers.cpp** that convert between versions of the program text with and without line numbers. Adding a number to the beginning of each line is, in principle, easy, but it's nice to format the line numbers so they all take the same amount of space (so the colons line up). The amount of space allocated is just enough for the longest line number plus a gap for an asterisk or other marker if one is needed later. Here is the program for adding line numbers. You will see that it uses many of the stream manipulation facilities introduced in the main text, including testing for end of file, clearing the eof state, seeking to the beginning of the file (rewinding), and the **setw** manipulator.

```cpp
//
// Add line numbers to a source code listing
// Prepend to each line a string: ' ',linenum in fixed width,": "
// To decide width of field, have to count the number of lines
//
// John Robinson 7 September 2003
#include <iostream>
#include <fstream>
#include <iomanip>
using namespace std;
int main(int argc, char *argv[]) {
   char linebuffer[2048];
  if (argc != 3) {
     cerr << "Usage: addnumbers in_codefile out_numbfile\n";
     return -1;
     }
  ifstream infile(argv[1]);
  if (!infile) {
     cerr << "Can't open " << argv[1] << endl;
     return -1;
     }
  // Count lines in infile
  int numlines = 0;
  while(!infile.eof()) {
     infile.getline(linebuffer, 2048);
     numlines++;
     }
  numlines--;          // Because last one was eof
  infile.clear();       // Clear eof flag
  infile.seekg(0);     // Rewind to start of infile
  // Work out field width for linenumbers
  int fieldwidth = 0;
  while (numlines) {
     fieldwidth++;
     numlines /= 10;
     }
  ofstream outfile(argv[2]);
  if (!outfile) {
     cerr << "Can't open " << argv[2] << endl;
```

```
    return -1;
  }
int linenumber = 0;
while(1) {// Until we break out by finding end of file
  infile.getline(linebuffer, 2048);
  if (infile.eof())
     break;
  linenumber++;
  outfile << ' ' << setw(fieldwidth) << linenumber << ": ";
  outfile <<  linebuffer << endl;
  }
infile.close();
outfile.close();
return 0;
}
```

The program is called from the command line by typing 'addnumbers program.cpp program.numbered'.

5.10 C++ and information localization

This chapter has introduced the main concepts of object-oriented programming in C++. Many things have been omitted, but the essentials are here. If you have worked through the examples, you will have a good grasp of how object-oriented programming works. It is worth asking how well it achieves information localization, since that was the principle cited earlier.

In one sense, C++ does not really localize information. The reason is that **class** definitions must be known in any file that makes use of them. So the user of a particular ADT can always see something of how that ADT is implemented, in particular what its **private** data and functions are. It is standard and good practice in C++ to put class declarations in header files (**iostream** is an example), and to put methods and other procedures in ordinary .**cpp** files. This localizes the information. But it does not hide it completely.

Object-oriented programming gives not just modules with access functions, but ADTs (with methods) that can be repeatedly instantiated. The concept of derived types is also fundamental. This too provides information localization, because the commonalities of two derived types are localized in their base type.

Object-oriented programming is not the holy grail of software design. Indeed, Brooks in his 'Silver bullet' article on software engineering (see page 18) saw it as attacking only the accidents of software. But its power has been shown in real design projects. Faster development, code reusability, and often increased performance result from object-oriented programming.

5.11 Chapter end material

Bibliography

The references for Chapter 4 are equally applicable to this chapter.

6 Program style and structure

Principle 2

**The text of the program itself – the code –
is the canonical expression of the design.**

Rule 13

Program defensively.

The literary view of software encourages care in the writing of programs. Pragmatically, the four most important reasons for adopting good programming style are

- to prevent bugs
- to make bugs easy to discover when they occur
- to make the program clear to its author and therefore increase programming efficiency
- to make the program clear to others and therefore maintainable.

Of these, the first is most important.

Most programmers spend too long on debugging. Even if successive working versions of a program as it grows are not officially labelled as discrete versions, each involves getting over the same three hurdles. The new code is written, made to compile, then debugged for logical errors. Whether it takes minutes or hours to write the initial code for a version, debugging can take significantly longer (see Figure 6.1). And that is before we've even started doing *real* testing (which will identify things like module incompatibilities, efficiency problems and yet more logical errors).

The message of this chapter is to avoid Figure 6.1. Easier said than done? Well, nothing comes for free. What you should aim at instead is Figure 6.2 – taking a little more effort on the coding, wringing more useful help out of the compiler, and so reducing the debug hurdle to a manageable size.

This chapter is about program style with the focus on avoiding bugs. First, we look at the most common types of logical errors and how to avoid them. Then come more general style guidelines for bug avoidance and overall clarity. Finally we consider how to divide your program into files so that its high-level structure is clear. This last section might be considered an extension of the previous two chapters because it includes C++ coding with standard library **strings** and templates to provide development tools. There are sidebars dealing

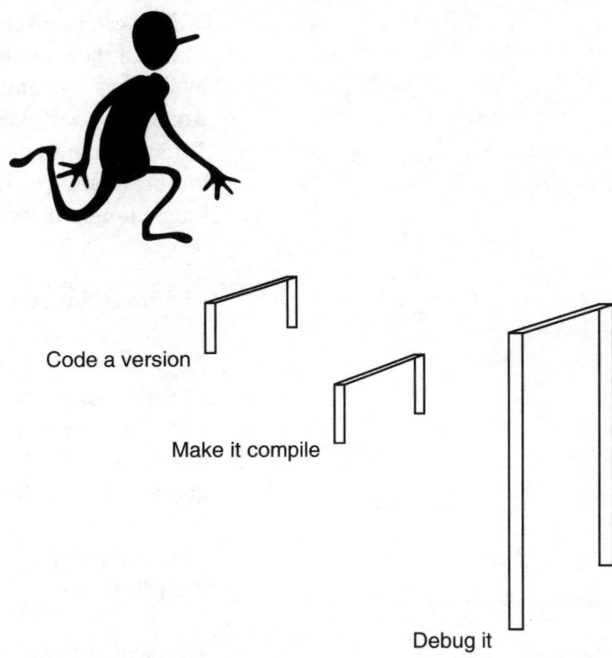

Figure 6.1 *The problem*

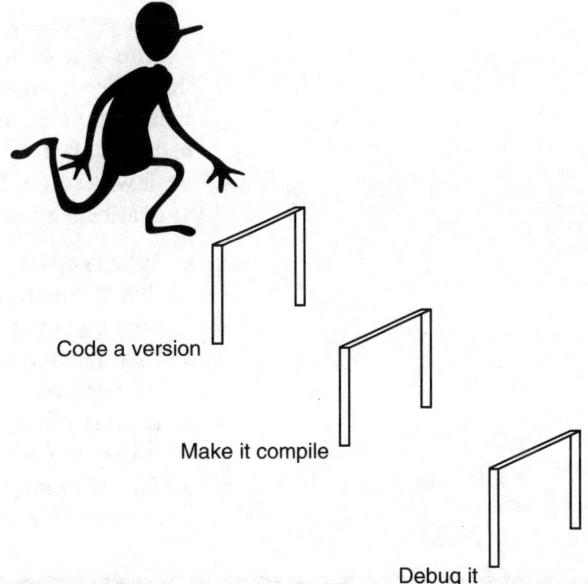

Figure 6.2 *The target*

with using the compiler's help to avoid later problems and debugging strategy to find the bugs that get written in spite of your best efforts.

6.2 Ten programming errors and how to avoid them

Our strategy in reducing debug time by increasing coding time should be to focus on the sort of errors that are most difficult to track down and concentrate on avoiding them.

The errors discussed below commonly occur in C and C++ code. Many of them apply to other languages too. They are ordered roughly by decreasing potency. In other words, the early errors are usually the worst, the most insidious and hardest to find. However, even number 10 is enough to cripple a project. Watching for them as you write is the best defence. If they get through, they may lurk undetected for long enough to avoid detection until a later phase of the development.

The invalid memory access error

An invalid access error occurs when a program tries to use memory it doesn't own. Some programming languages guard against this type of error, but one of the 'features' of C++ is that it never checks to see if a program is writing to a valid area in memory. For example, C++ is quite happy to allow:

```
char text[50];
text[300] = 'a';
```

This will compile, and probably work at the time, but that 'a' will be inserted somewhere it has no right to be in memory, overwriting the data of another object, the address of a function, an operating system variable, or something similar. Later (perhaps much later) there will be a mysterious system crash.

This kind of bug is very difficult to track down. The main causes of the error are writing beyond the bounds of an array, using an uninitialized pointer, accessing a data item after it has been **deleted**, and bad input data. These four are all special cases of other types of errors dealt with below. In general, however, the following steps can be taken to guard against invalid access errors:

- When copying from one area of memory to another, use functions that explicitly limit the length of the copy. For example, use **strncpy()** rather than **strcpy()**.
- In a function that writes to an array provided by the caller, always include a parameter which specifies the maximum index of the array.
- Beware of using more than two moving pointers in a function – it is easy to get them confused.
- *Ensure pointers point to allocated memory.* (OK, I said this one is covered below – but it bears repeating!)

How to write a virus

Most computer viruses work like this: somewhere in the program is a buffer, i.e. an array, and a function that writes to that buffer without doing bounds checking. That function reads input from somewhere – perhaps a user's file or an incoming mail message. A malicious programmer works out how the data in the program are organized, in particular what data items are beyond the end of the buffer. They then cause the program to read into the buffer data that is far too long for it. Because the program doesn't check the input data length, it copies the whole long string, obliterating other information beyond the buffer and replacing it with new data (perhaps executable code). Later, the changed data are accessed and the rogue program takes control.

Invalid memory access errors have a lot to answer for!

The off-by-1 error

The off-by-1 error appears in several forms:

- The number of times a loop is executed is off by 1.
- A test does not deal correctly with the boundary case. For example, $>$ is used where $>=$ should have been. (Not necessarily an off-by-1 error, but usually appears as such.)
- A postincrement or postdecrement in a loop test happens even when the loop terminates.
- Operations to step through linked lists (see Chapter 6) do not check correctly for the first or last item in the list.
- A string contains 1 more character than the string length (for the terminating '\0').
- Storage references are off by 1.

Here are some examples of off-by-1 errors:

```
// DO NOT COPY THIS CODE. IT CONTAINS ERRORS
// An extra character is required for the terminating '\0'
char buffer[5] = {"Hello"};
```

```
// DO NOT COPY THIS CODE. IT CONTAINS ERRORS
// <= should be <
int buffer[80];
for (i = 0; i <= 80; i++)
   buffer[i] = i;
```

```
// DO NOT COPY THIS CODE. IT CONTAINS ERRORS
// This example tries to be too clever! The programmer knows
// the maximum array index is 79, but forgets that i is
// decremented before being used to index the array.
int buffer[80];
i = 79;
while(i-- > 0)
   buffer[i] = i;
```

```
// DO NOT COPY THIS CODE. IT CONTAINS ERRORS
// Here strlen returns the number of chars in the string, so the
// copy does does not go as far as the terminating null
// character.
strncpy(array, string, strlen(string));
```

```
// DO NOT COPY THIS CODE. IT CONTAINS ERRORS
// In this case the result may be correct because of the way that
// two possible errors cancel each other out: First cin >> does
// not include the terminating carriage return in the string, so
// the number of characters in the buffer (apart from the
// terminating '\0') is one less than the number typed. But then
```

```
// the p pointer is incremented one location beyond the
// '\0' so the result of p-buffer is one more than the number of
// chars in the string. Of course, the programmer may have
// wanted to use these two facts to get the right answer, but that
// makes the code obtuse and therefore poor.
cin >> buffer;
p = buffer;
while(*p++);
cout << "The number of characters you typed was " << p-buffer
     << "\n";
```

Some of the above examples show the off-by-1 error causing an access of unallocated memory, for example an attempt to write item *n* of an *n* item array (which does not exist of course).

Incorrect initialization

Another common problem is incorrect initialization. Usually this means forgetting to initialize at all.

```
// DO NOT COPY THIS CODE. IT CONTAINS ERRORS
// i should be set to 0 before loop
int i;
while(i < 80)
   buffer[i++] = 1;
```

```
// DO NOT COPY THIS CODE. IT CONTAINS ERRORS
// p should be initialized with p = buffer; first
char buffer[80], *p;
strncpy(p, "Hello there", 80);
```

The second example also illustrates a common error among novice programmers – the belief that because a pointer is an address, pointers and arrays are the same things. Or at least that declaring a pointer mysteriously allocates array storage. A pointer is just an address and does not hold the data pointed to. Possibly, one of the reasons why this misconception arises is because the initialization:

```
char *p = "Hello";
```

is legitimate. It seems as though an array has been assigned to a pointer. What this declaration actually means is 'initialize p to point to an (unnamed) array (somewhere – I don't care where) in which is stored "Hello"'.

Copying to an uninitialized pointer is, of course, an example of an invalid access error.

There is a flip side to this error: initialization to 0 can be used as a flag to show a pointer is not pointing at anything. This is so useful a safeguard that it is worth making a rule to set pointers to 0 when they are unused, and to test against this equality before doing a sequence of operations on the pointer. For example, suppose we had a type something that contained various items including an 'optional' text field. Within the class definition there might be a line:

char *option_text;

Before knowing whether the object is to include this text, the program would explicitly set

option_text = 0;// No optional text

Then, there might be a method to put text into that field:

```
int something::add_option_text(char *text)
    {
    if (option_text)            // Previously did contain something
        delete [] option_text;  // Remove buffer that contains
                                // previous text
    option_text = new char[strlen(text)+1];
    // +1 for the null character
    strcpy(option_text, text);
    return 1;
    }
```

When operations are to be done on the **option_text** field, they are always preceded by a check against zero. The **add_option_text** member function illustrates this. Other member functions would work similarly:

```
int something::remove_option_text()
    {
    if (option_text)
        delete [] option_text;
    option_text = 0;                    // Flag to show no text stored now
    return 1;
    }
```

Using the compiler and the language to avoid bugs

When the C++ compiler and linker connect together different pieces of code they check that all the exchanged object types match. Bjarne Stroustrup compares this to plug compatibility in the physical world, where different-shaped plugs and sockets are used to prevent the wrong pieces of equipment being connected together. Without C++'s strong type checking, incompatibilities between modules that show up as compiler errors might otherwise manifest as tricky runtime errors. It should be your aim to get the compiler to do as much of this checking as possible. That means not using mechanisms that work around the type system (e.g. the use of pointers to **void** to exchange addresses of any type of object, then casting them to the assumed type) except when you must (like when using qsort() instead of the Standard Template Library alternatives). Similarly, don't use polymorphism to work out the type of a base class at runtime when it can be known at compile time. To really push the detection of errors back from runtime to compile time, you can define classes that prevent invalid inputs even being specified (see Meyers, *Effective C++*, item 46). But most of all, use **const. const** explicitly indicates what can and can't be changed by different functions. Including it ensures you think about whether your functions have to change data items or merely read them, then gets the compiler to check that the rules are kept. This has efficiency implications too. For example, suppose in our **String** class (pages 138 to 144), we wanted a member function that would return a pointer to a C-string version of the **String**. Think **const**! Don't we really just want a member function that returns a **const** C-string? If so, then the class already has a C-string that can be returned: as a **const**, there is no danger the caller will modify it.

```
something::~something()      // Destructor
  {
  if (option_text)
    delete [] option_text;
  }
```

Variable type errors

There are two kinds of errors associated with variable types. The first is using an inappropriate type of variable, leading to overflow if the range of the type is insufficient. The second is doing an inappropriate thing with a particular type of variable. For example, **float**s and **doubles** do not obey the laws of arithmetic – they inherently introduce truncation errors, which means they should not be used for counting, and tests for equality should be avoided.

```
// DO NOT COPY THIS CODE. IT CONTAINS ERRORS
// On some machines, int is short, so has range [-32k, 32k].
int tenfactorial = 1;
for (int i = 2; i <= 10; i++)
  tenfactorial *= i;
```

```
// DO NOT COPY THIS CODE. IT CONTAINS ERRORS
// Rounding in calculation of squares will cause equality to fail
// even when it should succeed
int is_triangle_right_angled(float a, float o, float h)
  {
  return(h*h == a*a + o*o);
  }
```

To fix the second of these errors, a proximity test can be done, to see whether the two quantities are 'close enough' to each other. The exact meaning of 'close enough' depends on the way that floating point numbers are represented. It is often better to be clear on what the tolerances are in the application domain, and check that differences are within those tolerances.

Loop errors

The off-by-1 loop error has already been discussed. Other examples of loop errors are:

- including within a loop any statement that must be executed in all circumstances
- relying on a loop to perform an assignment, and therefore omitting initialization
- confusion of counts and index variables in loops
- use of **break** or **continue** means that some statements that should be executed in the loop are skipped. This might lead to inconsistent exit conditions.

The first two of these errors both arise when there is a possibility that the loop might be executed zero times. If this happens, the crucial statement or assignment is not done.

```
// DO NOT COPY THIS CODE. IT CONTAINS ERRORS
// Program misleads if end == start
int lastval;
for (x = start; x < end; x++)
   lastval = array[x];
cout << "The last value in the array was " << lastval << '\n';
```

```
// DO NOT COPY THIS CODE. IT CONTAINS ERRORS
// typical example where a range from start to end is required
// but start to size is used instead.
for (x = start; x < size; x++)
   array[x] = y;
```

```
// DO NOT COPY THIS CODE. IT CONTAINS ERRORS
// End of a range confused with number of items in range
int sum = 0;
for (i = start; i < end; i++)
   sum += array[i];
average = sum/i;  // Should do division by end-start (unless
                  // end == start)
```

```
// DO NOT COPY THIS CODE. IT CONTAINS ERRORS
// At the end of the loop it is unknown whether the '\n' has
// been reached or not.
while((c = get_new_character()) != '\n')
   {
   if (p – buffer == BUFFERSIZE)
      break;
   *p++ = c;
   }
```

Incorrect code blocking

Incorrect blocking of code happens when a statement that should be part of a code block (a chunk of code enclosed in curly brackets {}), is inadvertently placed outside the block.

The code

```
if (a < b)
   cout << "a is still less than b";
   a++;
```

is indented as though the programmer intends **a** to be incremented only if the **if** statement is true, but this will not happen: **a** will be incremented anyway. The correct code for the conditional to apply to **a++**; as well as the **cout** statement, is:

```
if (a < b)
   {
```

```
cout << "a is still less than b";
a++;
}
```

Incorrect blocking is a good example of an error that creeps in during debugging. A typical scenario is:

- Pre-written code is being tested. In the above case, for example:
  ```
  if (a < b)
      cout << "a is still less than b";
  ```
- The programmer runs a test with **a** less than **b**, and discovers that **a** does not increment.
- The programmer adds a line,
  ```
  a++;
  ```
  but forgets to extend the code block by using {}.
- The test works now and the programmer moves on, missing the new bug.
- The new bug manifests later when trying to exercise a different part of the code.

This illustrates why careful testing of bug fixes is important.

So far as incorrect blocking is concerned, habitual use of indentation is a good visual cue for missing curly brackets. Many editors will help in showing matching brackets. So finding this error is not too difficult. The trick is to remember to look for it.

Returning a pointer or a reference to a local variable

A common error in C++ programming is to do something like

```
// DO NOT COPY THIS CODE. IT CONTAINS ERRORS
int *collect_values()
  {
  int input_array[3];
  cin >> input_array[0];
  cin >> input_array[1];
  cin >> input_array[2];
  return(input_array);
  }
```

The problem here is that **input_array** only exists during the function **collect_values()**. Returning a pointer to this array is returning a pointer to something that has been destroyed when the function returns. The calling function might want to access the three values pointed to by the returned pointer, but it has no guarantee that they are still there – the operating system has every right to reclaim the space used by **input_array** as soon as the function returns. In practice, sometimes, those values will still be there, because the space has not yet been reused. This can mean that a bug like this can go undetected for a long time (perhaps forever), until some change in the calling sequence suddenly activates it.

In C++, this error can occur in very subtle ways. For example, an object may use **new** to create other objects. Every **new** should have a

delete (see below), so the object's destructor will probably delete objects it has created, unless they have been explicitly handed over to other objects. Now a destructor is called implicitly whenever an object goes out of scope – for example, if an object was declared in a function, when that function returns, the object is destroyed by calling its destructor. It is easy to forget that objects created by an object during its lifetime might well be **deleted** when the 'creating' object ceases to exist.

The solution to these errors is to avoid returning pointers or references to anything created during a function, *unless* it has been created with a **new** and it is certain that the creator will not implicitly **delete** the thing when the function returns.

Other problems with **new** and **delete**

Misuse of **new** and **delete** causes some of the most catastrophic bugs in C++. Some compilers are particularly sensitive to **delete**ing something that was not allocated with **new** – typically they will crash the next time **new** is called.

The general rule is: every **new** should have a **delete**. Sometimes this is straightforward to manage: an object creates things as it goes on, keeping touch of all it is responsible for, and its destructor removes all those objects. The only thing to worry about here is the possibility that some other object will have started using one of the created objects. When an object must be created by one object and used by others, it cannot necessarily be **deleted** by its creator. In cases like this, there must be some mechanism somewhere for keeping track of what has been allocated, and some object or module is charged with the responsibility of eventually deleting the objects. 'Changes of ownership' like this should be explicitly stated in documentation (i.e. at least as comments in the program).

Making copies of objects is a particularly common place for **new** and **delete** bugs to creep in. If an object contains pointers to things it is not enough simply to copy the pointer values; the arrays being pointed to must be cloned and pointers in the new object set up to the new arrays. That way, when the first object is **deleted**, and its destructor deletes its associated arrays, the second object will retain its arrays.

Inadequate checking of input data

Program errors may not be directly caused by bugs in the logic, but by the program responding in an uncontrolled way to bad data. For fear of this it is possible to go to the extreme of having every single function check the values of its parameters. This is not usually necessary, or even helpful, because of the large amount of extra code involved. However, it is appropriate for a function to check certain kinds of expectations about its arguments. In particular if an object assumes its methods are to be called in a particular order, it should check in the later methods that the appropriate earlier steps have been done.

Debugging tips

Debugging is an acquired skill that demands patience.

Initially you have to get your program to compile. This is the easy part, because the compiler rejects syntax errors but lets through all the rest. For syntax errors, it is usually possible to work out what went wrong by looking at the line of code that the compiler flags. But remember that sometimes errors have a cascade effect: a single mistake on line 3 can cause the compiler to reject the next 100 lines. Many compilers have help facilities that will explain error messages. Borland C++ and Visual C++ both provide context-sensitive help.

Errors in the logic of your program cause it to go wrong. If you're lucky, something obvious will go wrong. You will then be able to try several things:

- Work backwards from the wrong output or clearly wrong condition to see how it could arise.
- Try different inputs to see which cases cause the error. This might point you to places in the code that differentiate those inputs.
- Check any intermediate results that are available to narrow down the source of the problem.

If these don't get you closer to the cause, then you'll have to go deeper. You can:

- seed the code with debug facilities that monitor its behaviour. These can range from print statements to a log file or standard output, through dumping an intermediate result to a file, to writing a visualization routine that will somehow animate what is going on. Debugging helps like these that can be turned on when a flag is set can be very useful. They are worth leaving in the code even when it is (supposedly) debugged because changes to specification might result in a very similar set of checks having to be run, and they point out to a maintainer what the critical points to monitor are. In this case, the debug statements have the same status as comments, and are liable to the same danger – not being kept in sync with the code. A related debugging aid is to use human-readable strings in data structures – these aid interpretation of output data files, while using a debugger to step through the program.
- use a debugger to put break points just before things start to go wrong, then trace the code, stepping line by line, examining variables as you go.

If you still can't find the problem, or if it's a bug that only shows up at obscure times and is hard to reproduce, then your next step should be:

- Read the code. 'Play computer'. Look out especially for the ten errors discussed in this chapter. Despite your extra care in writing, one may have crept in.

After spending half an hour trying to trace down a bug without results, the chances of finding it that day are slim, and get slimmer as frustration grows. Know when to give up. Many programmers have experienced entire afternoons in front of the debugger ripping their hair out without avail, followed, the next morning, by almost immediate light on the problem.

Other people can be invaluable for help in debugging. Often, the error is obvious to someone who has never seen the code before, whereas the programmer has become blind to obvious errors through familiarity. So getting someone else to look at the code is always worth a try.

When you've found the bug and worked out the fix, remind yourself that the place where bugs are most likely to be added to a program is during debugging. Faster turnaround on a single bug does not result in faster completion of the program or version. An observed problem may well be due to the interaction of several bugs, so it is worth continuing the search even after the first error has been found. Finally, make the fix, test the program using input that will exercise it and look carefully for unintended consequences of the change.

As well, some checks for bad data can be inserted almost for free. For example, in using the **switch** statement, the **default** case should always be included, even when the legal alternatives have been covered, so that illegal data can be flagged by a statement under **default**.

When a program has a number of modules written by different people, it is advisable for everyone to write their module interface functions so that they check preconditions (assumptions about the input data). Then, if module A misbehaves and calls module B's functions in the wrong order (for example), module A can issue a debug message. Usually, this practice saves a great deal of debugging time during integration, because modules watch each other's behaviour and report on problems.

There are many conventions for error reporting. Exceptions were discussed on page 160. This book uses the traditional method of returning a negative error code (or 0 for functions that return a pointer). During the debugging phase it is often helpful to make a call to a function called **DEBUG(char *message)** where **message** is a string describing the problem. For example, for bounds checking an array:

```
int access_element(int index)
   {
   // Check preconditions
   if ((x < 0) || (x >= numitems))
      {
      DEBUG("access_element() – index out of range");
      return (-1);
      }
   // Rest of function
   ....
   }
```

You can then include a header file **"debug.h"** which defines **DEBUG()** as

```
#define DEBUG(x) (debug_message(x, __FILE__, __LINE__))
```

and have a **debug_message()** function that prints out the errors, stores them in a file, pops up a debugging window or whatever. The **__FILE__** and **__LINE__** arguments to **debug_message()** are recognized by the compiler, which inserts the current file and line number values.

Different modules interpret shared items differently

A simple example of different parts of a program interpreting shared items differently is given below:

```
// DO NOT COPY THIS CODE. IT CONTAINS ERRORS
// This example illustrates using two different conventions to
// refer to an ending value or location – either the value given
// is one beyond the end or it is at the end. Here some_
// function() and some_other_function() use different
// interpretations.
some_function()
   {
   int xstart = start;
   int xend = start + size;
   some_other_function(xstart, xend);
   }
```

```
// Typically some_other_function is miles away from
// some_function
some_other_function(int xs, int xe)
   {
   int i;
   for (i = xs; i <= xe; i++)
   do_something_with(i);
   }
```

some_other_function thinks it is processing from **xs** to **xe** inclusive, but its caller only wants it to go to **xe**-1. In this case, the result of the different interpretations is an off-by-1 error.

Another common form of this error is where the caller of a function gets the order of the function arguments wrong. (The compiler will only allow this to happen if switched arguments are of the same type.)

Disagreement on what shared items mean can result in very bad errors. Here are some safeguards that can be applied:

- Keep the number of shared items to a minimum. Remember:

Rule 6

Localize information.

- Document the public interfaces of modules carefully (see Chapter 11).
- Try to avoid changing public interfaces of modules, or any other kind of shared information. Have a formal change procedure for dealing with any change that *is* required.
- In header files containing prototypes, include informative argument names as well as their types, so that programmers using the functions can easily refresh their memories about the meaning of each argument.

6.3 Style for program clarity

Kernighan and Plauger wrote the definitive book on software clarity: *The Elements of Programming Style*. The book is well worth owning; even though some of its advice is dated, most is still valid and valuable. In the back of the book, they summarize their 76 rules. Some of these have already appeared here in a slightly different context, for example 'Make sure every module hides something', and 'Let the data structure the program'.

A more recent book is Steve McConnell's *Code Complete* which has many guidelines and examples of good style in C.

The approach here is minimalist, summed up in two rules from Chapter 2 and developed as six guidelines. To begin with the more general rule derived from the literature perspective (page 17),

Rule 3

Make your code reveal its non-sequential
structure to a sequential reader.

Rule 3 reminds you of the mismatch between your code, containing objects with data and behaviour, which at different times will be in different states, interacting with each other, responding to external events, and a human reader who has to read and assimilate your program in some order. Remembering Rule 3 will remind you of this fundamental gap that your variable naming, logical structure, visual structure and comments have to bridge. Here are five key implications that turn the aspirations of Rule 3 into practical guidelines.

File structure: a commentary introduction is essential

The first few lines of a program file are the first thing a reader reads so they must provide an introduction. Comments should give the program's name, purpose, author, outline usage information, any known problems or hazards (things that *could* go wrong), and the version history (initially just the date, but with dated comments appended when the program is subsequently revised). They should also introduce any non-sequential interactions or dependencies that a reader needs to keep in mind while reading the code sequentially. This contextual documentation must not be forgotten (or put in a separate file), because it is usually exactly what is not self-evident in the code.

Explanatory structure: comment to reveal

The purpose of comments in general is to make clear what is not self-evident. Use comments only to remove ambiguity, not to repeat what the code itself says. The comments should of course be up to date – modified when the code is. The best way to ensure this is to put comments really tight against the code to which they refer, even if that doesn't look the tidiest.

Visual structure: make the program pretty

A program should show its structure visually. Vertically, a function should show its usual sequence of operations from top to bottom. Horizontally, indenting should be used to show the nesting of program blocks and the logical structure of the program. All programs should be indented, prettily, although the particular style you use for this is irrelevant. Some will write

```
function()
  {
  one();
  two();
  }
```

and others

```
function() {
  one();
  two();
}
```

What can it possibly matter? In a team project, conventions are helpful, perhaps. But they are just that – conventions.

Indentations should be clearly visible, but set tab stops close enough so that the code does not sprawl. It is good to be able to track straight down a page of code. Many levels of indentation should be avoided anyway; a function that reaches in eight tab stops probably needs some part of its functionality extracted and put into a separate function.

When a statement or a string runs on to several lines, show the continuation lines indented a couple of tab stops.

A limit of about 60 lincs on the length of any function is reasonable. If a function gets too long, it should probably be split into parts.

Verbal structure: make it possible to read the code aloud

Kernighan and Plauger sum up what is meant by this in their rule: 'Use the telephone test for readability'. The telephone test is simply asking whether someone could understand your code when read over the telephone. If yes, then the code is clear enough.

The telephone test detects overnesting, that is, too many embedded pairs of brackets. In general it is wise to avoid too much nesting of any sort – when the wrong **if**-condition parentheses start getting matched, expressions become obscure.

Variable, structure and abstract data type names should be meaningful. There have been several proposed naming conventions to guide the programmer in name choice. Windows programmers sometimes encounter the 'Hungarian convention' where a name can consist of a complicated prefix giving information about the type of the variable and a suffix that uniquely identifies it. Otherwise a good rule is to ensure that variable names are short, meaningful and mostly pronounceable. Being able to pronounce a name means it is easier to remember. If there is a way of qualifying the identity of something by a single letter abbreviation, that may be added to a pronounceable name to produce an acceptable mostly pronounceable name:

```
pnext // A pointer to the next node
ivalue, fvalue // Integer and float versions of the same quantity
getc, gets // Get a character and get a string
```

Multiple word names are fine if they help in giving meaning. Arguments about the relative value of **variable_name**, **variableName**, and **VariableName** are academic. Follow whatever local conventions apply.

Logical structure: don't be too clever

Make your code as straightforward as possible. Every trick makes the code more difficult to understand. Being too clever can include using side effects of an operation to do something, breaking typing and memory access restrictions, and using an assumed special feature of the data or the situation to cut corners on an algorithm.

Replicated structure: kill the doppelgänger

This sixth and final guideline is no more than a special case of

Rule 4

Write no more than necessary.

Good writing style means being both clear and concise. In programming, clarity can sometimes conflict with being concise, but not when redundant repetition is concerned: repeating yourself never enhances clarity; it just bloats the code and makes it difficult to maintain.

The way that needless repetition usually happens is that you realize you need to do what you're already doing plus something slightly different as well. So you copy the code you have now and make a few minor changes in the new version: much is repeated, but the two streams of code are subtly different. The maintenance problem rears its head when something that both versions rely on changes, and you have to alter both in parallel ways. Although it's a hassle, it is almost always better, right from the start to *modify not clone* your existing version so that it can be called for both cases. If you already have two similar but not quite identical code segments, take the better one and make it do the other's job too: kill the evil twin. After all, you wouldn't want to read a book that repeated the same passage with multiple minor changes just to give you the story from two characters' points of view.

Rule 4 does mean more than just removing duplication and redundant code. Look for ways to rewrite that say things more concisely without obscuring the meaning. Above all, so far as possible, make the program speak for itself (so you don't have to write additional documentation).

6.4 Multifile program structure

Rule 6

Localize information.

Integrated programming environments like Borland C++ use 'projects' to allow a program to be divided into files. They automatically keep track of which files have been updated since the program was last compiled and linked. This ensures that remaking a program only involves those parts that have changed, thus speeding up the cycle time. The same efficiency on systems without integrated environments (like Unix) can be achieved by creating a Makefile which specifies the dependencies, compile flags, etc. appropriate to the project. The program 'make' reads this file, works out what has changed, and what therefore needs to be done, and brings the program up to date.

The existence of projects, make and other kinds of automatic selective compilation is one reason for splitting programs into files. But the main reason is stylistic – to make the program clear and well structured. How then should code be organized?

The modularization of a program should drive its division into files, although, in large programs, each module will typically include several files. A file should *never* span more than one module. Where there is hierarchical decomposition into submodules, the file structure should reflect the system structure as closely as possible.

In C++, type (**class**) definitions go into header files (**something.h**) and procedures, including methods, go into program files (**something.cpp**).

Where code is common to two programs, then it should be put in a file of its own.

When a header file contains some definitions specific to one **.cpp** file, and some that are needed by other files or modules, then it should be split. Typically this happens when class definitions must be included by another module so it can call methods of objects in this module. One way of doing this is to have files such as **module1.cpp**, **module1.h** and **module1_user.h**, where the last of these contains all the definitions that are needed for other files to use the facilities in **module1.cpp**.

All non-standard code (for example, that which interfaces with a particular piece of hardware) should go in a separate file.

User-interface code should be separated from other code.

It is a good idea to separate highly volatile code (code that is likely to change – for example, that which you are currently working on) from stable code. This makes navigation easier.

The length of files is not usually a deciding issue in splitting up a program. The price for using a few long files, against many short files, is increased compilation time when a change is made. On the other hand, limiting the number of files may well make navigation through the code easier. Many programmers maintain long files, and there does not seem to be a fundamental reason why they should not.

Sometimes .h files include other .h files. In headers that do this, it is a good idea to check whether the second file has already been included. To do this, code such as the following may be used.

In **thatfile.h**:

```
#define thatfile_H
```

In **thisfile.h**:

```
#ifndef thatfile_H
#include thatfile.h
#endif
```

6.5 A program that automatically generates a multifile structure

It is often handy to start a project with an empty set of program files organized as the previous section described. Integrated development environments prepare such a framework when you open a new project. In this section we'll review a program that does the same thing, creating an empty framework, writing initial header, class implementation and test program files along with a Makefile. As a side issue, this program gives the opportunity to introduce C++ **strings**.

As discussed on page 76, the C++ standard library has a **string** data type that can often be used in preference to C-strings. It shares some of the functionality of the **String** type in Chapter 5, but has many other features too. Rather than enumerating all the features of C++ **strings**, we'll use this example program to illustrate their use.

First, here's what the program does. Given a name for a project, it sets up initial source and header files as well as a test program framework. For example, if invoked as 'mkframework example', it writes four output files:

Makefile:

```
# Makefile for example class programs, created 09/08/03
    16:42:01

all: testexample

testexample: example.o testexample.o
  g++ -o testexample testexample.o example.o

testexample.o: testexample.cpp example.h
  g++ -c testexample.cpp
example.o: example.cpp example.h
  g++ -c example.cpp
```

example.h:

```
// Header for example class, created 09/08/03 16:42:01

#ifndef example_H
#define example_H

class example {
public:
    example ();
    ~example ();
};

#endif
```

example.cpp:

```
// Implementation for example class, created 09/08/03
// 16:42:01

#include "example.h"

example::example() {
}

example::~example() {
}
```

testexample.cpp:

```
// Test program for example class, created 09/08/03 16:42:01

#include "example.h"

int main(int argc, char *argv[]) {
    example testexample();
    return 0;
}
```

Immediately this is done, it is possible to type 'make example' and the empty boring shell of a test program, testexample, will be built. You can then begin incremental development, putting the class declaration in the header file, definition of member functions in example.cpp and test code in testexample.cpp, remaking at intervals to ensure that you still have a working version that is growing in functionality.

How does mkframework work? Here is the source code:

```
1:   //
2:   // Program to make framework for a program by creating
     // makefile
3:   // newclass.cpp, newclass.h, testnewclass.cpp
4:   // John Robinson
5:   // August 2003
6:   //
7:   #include <iostream>
8:   #include <fstream>
9:   #include <string>
10:  #include <ctime>
11:  using namespace std;
12:
13:  string makecontent("# \
14:  Makefile for NAME class programs, created DATE\n\n\
15:  all: testNAME\n\n\
16:  testNAME: NAME.o testNAME.o\n\
17:   \tg++ -o testNAME testNAME.o NAME.o\n\n\
18:  testNAME.o: testNAME.cpp NAME.h\n\
19:  \tg++ -c testNAME.cpp\n\
20:  NAME.o: NAME.cpp NAME.h\n\
21:  \tg++ -c NAME.cpp\n");
22:
23:  string headercontent("\
24:  // Header for NAME class, created DATE\n\
25:  \n#ifndef NAME_H\n#define NAME_H\n\n\
26:  class NAME {\n\
27:  public:\n\
28:  \tNAME ();\n\
29:  \t~NAME ();\n\
30:  };\n\n\
31:  \n#endif\n");
32:
33:   string implementationcontent("\
34:  // Implementation for NAME class, created DATE\n\
35:  \n#include \"NAME.h\"\n\n\
36:  NAME::NAME() {\n\
37:  }\n\n\
38:  NAME::~NAME() {\n\
39:  }\n");
40:
41:  string testcontent("\
42:  // Test program for NAME class, created DATE\n\
43:  \n#include \"NAME.h\"\n\n\
```

```
44:    int main(int argc, char *argv[]) {\n\
45:    \tNAME testNAME();\n\
46:    \treturn 0;\n\
47:    }\n");
48:
49:    void replaceall(string &st, const string oldsubstr, const
              string newsubstr) {
50:       string::size_type pos;
51:       string::size_type len = oldsubstr.length();
52:       while ((pos = st.find(oldsubstr)) != string::npos)
53:          st.replace(pos,len,newsubstr);
54:       }
55:
56:    using namespace std;
57:    int main(int argc, char *argv[]) {
58:       if (argc != 2) {
59:          cerr << "Usage: mkframework classname\n";
60:          return -1;
61:          }
62:       time_t now = time(0);
63:       char c_str_now[80];
64:       strftime(c_str_now,80, "%c",localtime(&now));
65:       string strnow(c_str_now);
66:       // %c in strftime gives a vanilla date format
67:       string classname(argv[1]);          // newclass
68:       string headername(classname);
69:       headername += ".h";      // newclass.h
70:       string implementationname(classname);
71:       implementationname += ".cpp";     // newclass.cpp
72:       string testname("test");
73:       testname += classname + ".cpp";   // testnewclass.cpp
74:       ofstream makefile("Makefile");
75:       if (!makefile) {
76:          cerr << "Can't create Makefile\n";
77:          return -1;
78:          }
79:       ofstream headerfile(headername.c_str());
80:       if (!headerfile) {
81:          cerr << "Can't create " << headername << endl;
82:          return -1;
83:          }
84:       ofstream implementationfile
              (implementationname.c_str());
85:       if (!implementationfile) {
86:          cerr << "Can't create " << implementationname <<
              endl;
87:          return -1;
88:          }
89:       ofstream testfile(testname.c_str());
90:       if (!testfile) {
91:          cerr << "Can't create " << testname << endl;
92:          return -1;
```

```
 93:        }
 94:        replaceall(makecontent,"NAME",classname);
 95:        replaceall(makecontent,"DATE",strnow);
 96:        makefile << makecontent;
 97:        makefile.close();
 98:        replaceall(headercontent,"NAME",classname);
 99:        replaceall(headercontent,"DATE",strnow);
100:        headerfile << headercontent;
101:        headerfile.close();
102:        replaceall(implementationcontent,"NAME",
               classname);
103:        replaceall(implementationcontent,"DATE",strnow);
104:        implementationfile << implementationcontent;
105:        implementationfile.close();
106:        replaceall(testcontent,"NAME",classname);
107:        replaceall(testcontent,"DATE",strnow);
108:        testfile << testcontent;
109:        testfile.close();
110:        return 0;
111:        }
```

Models for each of the files that will be produced are stored in global **strings**. From line 13 to line 21 is the **string makecontent** which stores the model for the Makefile. Note that the string used to initialize **makecontent** has '\' characters at the end of every line signifying continuation (see page 98). So the initialization string stretches from line 13 to line 31. The header model is set up in **string headercontent** from line 23 to line 31, the implementation file model is set up from line 33 to line 39, and the test file content is set up from line 41 to line 47. All of these strings have embedded substrings "**NAME**" and "**DATE**", that are replaced by the program with the actual name of the framework to be created and the current date. This replacement is actually accomplished on lines 94, 98, 99, 102, 103, 106 and 107 where **main()** calls **replaceall()** to do the substitutions. **replaceall()** (lines 49 to 54) itself uses the **string** functions **length()**, **find()** and **replace()**, which work just as you would expect. The complication in **replaceall()** is its use of the type **string::size_type** and the attribute **string::npos**. **string::size_type** must be used because this is the type passed to, and returned by, the member functions. **string::npos** is the return value from **find()** when it *doesn't* find. **main()** also uses string facilities to create the actual filenames from the classname provided by the user. From lines 67 to 73, string variables are set up containing the various filenames. '+' and '+=' are both used for concatenation, just as in the **String** class exercise on page 144.

6.6 Chapter end material

Bibliography

Kernighan and Plauger's original book was:

Kernighan, B. W. and Plauger, P. J. (1978). *The Elements of Programming Style*, 2nd edn, McGraw-Hill.

A later book with the same theme (and an author in common) is:

Kernighan, B. W. and Pike, R. (1999). *The Practice of Programming*, Addison-Wesley.

The third major book referenced was:

McConnell, S. (1993). *Code Complete*, Microsoft Press.

Other references on style may be obtained from:

Thomas, E. and Oman, P. (1990). A 'bibliography of programming style literature', *ACM SIGPLAN Notices*, **25** (2), pp. 7–16.

To find out more about variable type errors, see:

Goldberg, D. *What every computer scientist should know about floating point arithmetic*, available at http://docs.sun.com/source/806-3568/cg_goldberg.html.

Hungarian notation is discussed in:

Simonyi, C. and Heller, M. (1991). 'The Hungarian revolution', *BYTE*, **16** (8).

7 Data structures

7.1 Structuring data

The structuring of data is of first importance in program design. In particular, *collections* of similarly structured data must be manipulated. In this chapter, ways of structuring collections are discussed. After a recap on pointers, the chapter introduces linked lists, which are the main alternative to arrays, then moves on to discuss array/linked-list hybrids. Trees are introduced and illustrated. The various data structure choices are demonstrated in the design of two fundamental Abstract data types, the queue and the table.

Principle 1

Software design is representation.

7.2 Memory usage and pointers

Although the amount of memory space available in today's computers is about 1000 times that available a decade ago, the objects that programs are required to manipulate have grown too. So programmers must use memory sensibly. A good rule of thumb is: *programs should be written to waste memory in moderation.* Very often, the simplest or fastest way of doing something means wasting a little memory. This is acceptable. But when the waste is substantial, or there are many copies of a data structure that wastes space, something is wrong.

For example, consider a program that processes large data structures called **hugething**s. A **hugething** is 2000 bytes big, which is typical of the size of a record in many database programs. On its own 2000 bytes is not too bad, but a crowd of **hugething**s could be. The number of **hugething** structures in use at any time depends on user activity. Typically there might be very few, but sometimes, for particular operations, the program needs to handle up to 100 000 **hugething**s.

If the maximum number of active **hugething**s is *guaranteed* never to exceed 100 000, then it would be possible to declare an array.

hugething h_t_array[100000];

is represented in Figure 7.1

Because **hugething**s are big, and because there are typically only half a dozen active at a time, this is very wasteful. Two hundred megabytes (100 000 × 2000 bytes) are absorbed in one array.

The solution is to transform this inflexible array-based collection into one whose size adjusts according to the number of active items. With **new** and **delete**, memory space can be allocated dynamically as and when needed. With pointers, dynamically allocated structures can be linked to each other, and to permanently allocated 'anchors'.

A pointer is an address, so can be used to link between objects anywhere in memory. Whenever an object has some kind of association with another object, it can represent that association with a pointer.

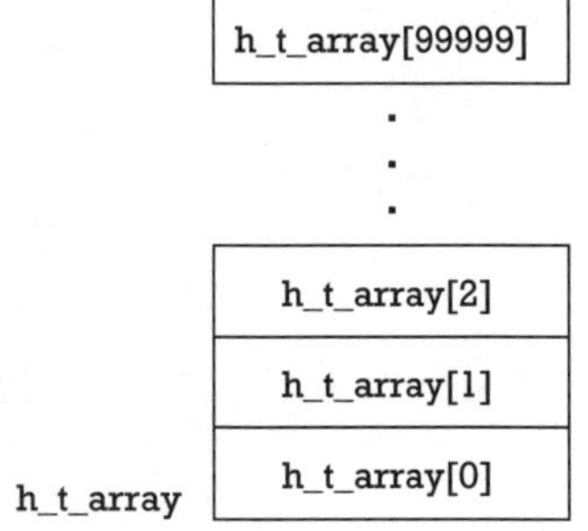

Figure 7.1 *h_t_array: an array of hugethings*

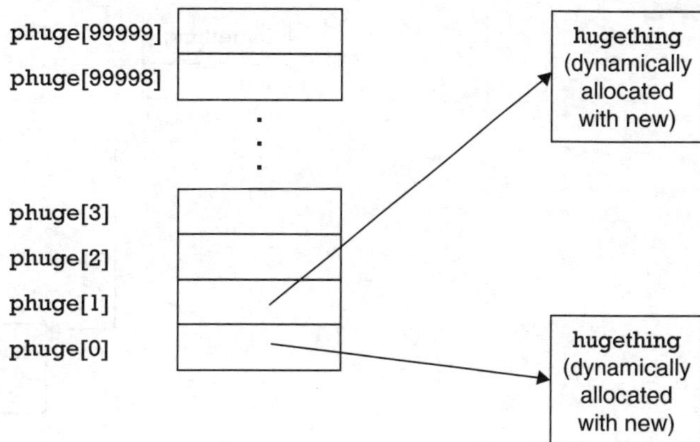

Figure 7.2 **phuge:** *an array of pointers to dynamically allocated hugethings*

A program for a theatre booking system might have customer, seat, and performance objects. A particular booking of a particular seat at a particular performance will need to associate these objects. So a booking object in the program may include pointers to a customer, a seat and a performance object.

The fact that a pointer is really an address is an accident of computer architecture – everything in computer memory has its own unique address, so the address is a general-purpose way of referring to anything. All addresses, and therefore pointers, are the same size (4 or 8 bytes usually), yet they can point to data structures of any size. For example, a **char** pointer might be pointing to a single character, or to a huge text array storing an entire book.

The simplest, and very often the best, way of using pointers to save memory is to convert a very large array of large data structures to an array of pointers. For the **hugething** example, a declaration of such an array would look like:

hugething *phuge[100000];

This still takes up some memory – 400 000 bytes if a pointer is 4 bytes – but that is considerably less than 100 000 **hugething**s, or 200 Mbytes. Now, as **hugething**s are used, they are dynamically created and destroyed with **new** and **delete**. As a new **hugething** is allocated, the next available pointer in the **phuge** array is set up to point to it. See Figure 7.2.

In the worst case, this uses more memory than the array of **hugething**s, because not only are 100 000 **hugething**s active and allocated (=200 Mbytes), but an extra 100 000-long array of pointers is used too. But most of the time, the memory usage here will be considerably less.

Now a data structure can include a pointer to an object of its own type. So another way of addressing the **hugething** array example is to add an item to each structure which points to the next in a list or chain. Then each **hugething** can be allocated when it is needed and appended to the beginning or end of the existing list. Any of the allocated structures can be accessed by stepping from one to another across the pointer links. A value of 0 in the next pointer is an appropriate way of signalling

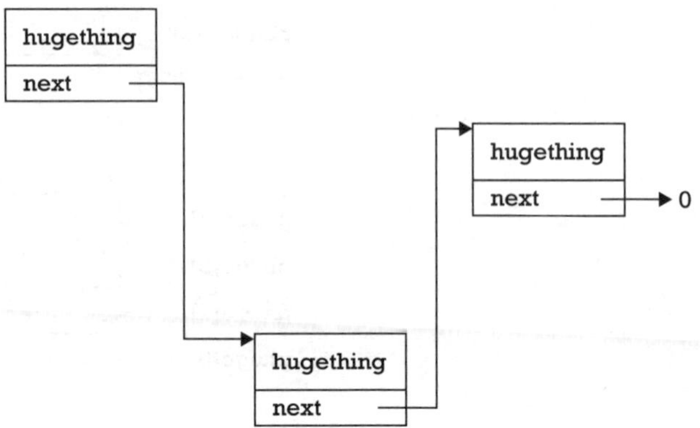

Figure 7.3 *A linked list of **hugething**+pointer structures*

the end of the list. An example is shown in Figure 7.3. The beginning of the list has to be 'anchored' with some permanently allocated item, but the remaining memory usage varies on demand. The overhead for this is an extra 4 bytes (or the size of a pointer) in each 2000-byte **hugething** structure.

This use of pointers to associate similar items into a sequence creates a *linked list*.

7.3 Linked lists

A linked list is a sequence of data structures, not necessarily contiguous (adjacent) in memory, where each one contains a pointer, pointing to its successor on the list.

The first item in the list is a permanently allocated object whose purpose is to anchor the head of the list so the program can access it. The remaining items may be anywhere in memory, but any item can be reached, by starting at the head, then stepping across the links from one structure to the next. The only way to reach the ith item on the list is to start at the beginning and step through i links.

An item may be added at any position i on a list by finding the data structures currently at positions $i - 1$ and i; making the new item the successor of the $i - 1$ structure (by putting its address in the $i - 1$ structure's link pointer), and making the old item i the successor of the new structure (by setting the new structure's link pointer to point to it).

An item may be removed from the list by setting the link pointer of its predecessor to the address stored in the removed item's link pointer (i.e. the address of the successor).

The three important properties of linked lists are:

1. Linked lists can overcome the limitations of fixed size associated with arrays. Rather than being wasteful or limiting, linked lists can provide the mechanism by which only as much memory as is required is allocated at any time.
2. Most operations on linked lists are done in three steps:
    (a) Traverse the list from the beginning to the appropriate position.
    (b) If necessary allocate/deallocate the space for the new/deleted item. Copy the data for the item appropriately.
    (c) Perform pointer changes to alter the structure of the list.

> The head of a list might be anchored by a pointer in some data structure rather than by a permanent list item.

3. Access time for an array is constant; the statement **x[i]** = **a**; takes the same amount of time no matter what the value of **i**. For a linked list, access time depends on the location of the required node in the list.

The items in a list do not *have* to be dynamically allocated. It is possible to declare an array of structures which include link pointers, then use the pointers to define an ordering different from the natural array order. This might be useful for setting up a pool of pre-structured memory elements that can be linked and unlinked at will without incurring the memory allocation overheads of **new** and **delete**. Alternatively it could be used to keep a collection of items ordered in two different ways, for example in a database of employees stored in employee number order in the array, with pointers between individual employee records linking them in alphabetical order.

Doubly linked lists

Each entry in a list can maintain *two* pointers, one pointing forwards and one backwards. It is then possible to step from item to item on the list in either direction. Such a *doubly* linked list is useful in applications that require searching in the proximity of a particular list item.

7.4 Data structures for text editing

Arrays

One of the best examples of a program that manipulates a collection of unknown size is a text editor or word processor. Assuming all the text of a document is to be stored in memory during the program, what sort of structure should be used?

Here are two possibilities:

```
char   text[MAX_NUMBER_OF_CHARACTERS];
char   text[MAX_NUMBER_OF_LINES][MAX_LINE_LENGTH];
```

A text editor needs to work on documents of many different sizes. The constants **MAX_NUMBER_OF_CHARACTERS** and **MAX_NUMBER_OF_LINES** need to be as big as possible so that large documents can be handled. In both cases, then, a large amount of memory is allocated for an array, which will rarely be full. The memory could well be needed in another part of the program, but it is tied up in the text array. The two-dimensional array is particularly inefficient, because the document will probably contain many lines shorter than **MAX_LINE_LENGTH**. However, it does preserve some of the important document structure: finding the first letter of the *i*th line with the two-dimensional array is just a case of going to **text[i][0]**; ; finding it with the one-dimensional array probably means searching through it character by character, counting the number of new lines.

Inserting text on a particular line involves data shifting. For the one-dimensional array, everything beyond the insertion point must be moved when a character is inserted. Again, the two-dimensional array is more efficient, because only the characters in the current line must be shifted.

Arrays of pointers

The two-dimensional case can be improved:

```
char *text[MAX_NUMBER_OF_LINES];
char current_line_buffer[MAX_LINE_LENGTH];
```

Now, an array of pointers is stored permanently. This has a much smaller memory requirement than either of the earlier examples. Each

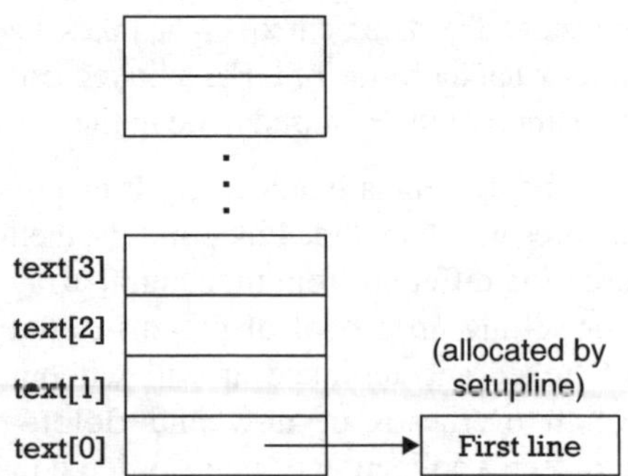

Figure 7.4 *Loading the first line into a simple text editor data structure*

line of the document will have a character array allocated for it when it is read in. The character array (created with **new**) can be exactly the right length for each line. Assume there is a function **char *setupline()** that sets up the array and copies in the text. Then all that has to be done is to set the pointers in the **text** array one by one as lines of the document are read in (i.e. by setting them to the addresses returned by **setupline**). If line 1 of a document stored in a disk file were:

First line

when the program reads the file, the first **setupline()** will allocate an array of the appropriate length (11 characters), and copy the line into the array. Then **text[0]** must be set to point to the start of the array, giving the situation shown in Figure 7.4.

When the user edits a particular line, that line is copied to the **current_line_buffer**, its allocated space is **deleted**, characters are inserted and removed while the line is in the current line buffer, then when the user moves to another line, new space is allocated, the line copied out of the buffer, and the appropriate pointer set up in the **text** array. The sequence of program instructions would be:

```
strncpy(current_line_buffer, text[current_line_number],
    MAX_LINE_LENGTH-1);
delete [] text[current_line_number];
//
// Here follows manipulation of the text in current_line_buffer
...
//
text[current_line_number] = new
    char[strlen(current_line_buffer)+1];
strcpy(text[current_line_number], current_line_buffer);
```

This solution requires limited data shifting, and provides easy access to each structurally important chunk, which in the case of a text editor is a line. For a word processor, a paragraph might be a more appropriate chunk.

Linked lists

But there is still a space-efficiency problem. *Software Design for Engineers and Scientists* contains about 20 000 lines. Should the editor always reserve space for 20 000 pointers? Perhaps that is not too bad, but the elegant solution is to adapt to the number of lines with a linked list.

Here is the building block of the solution:

```
struct line {
  line *pnext;
  line *text;
  };
```

Now the only permanent storage to be declared is:

```
linezeroline = {0, 0};
```

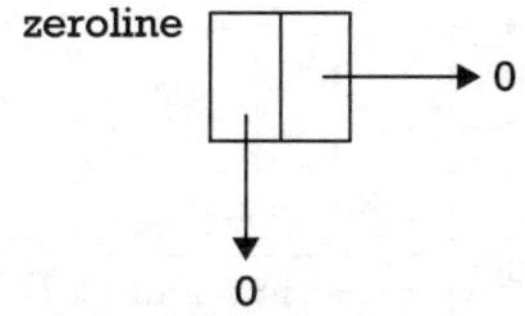

Figure 7.5 *The permanent storage in an alternative text editor data structure*

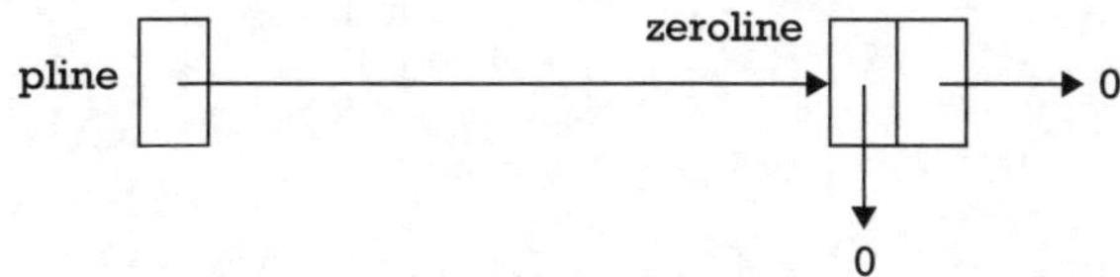

Figure 7.6 *pline initialized with the address of zeroline*

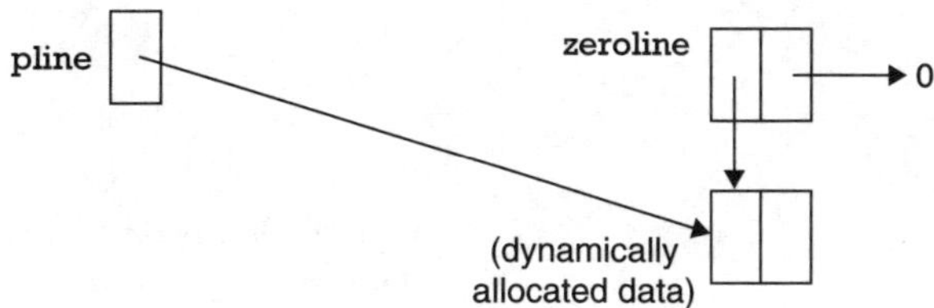

Figure 7.7 *Dynamic allocation of the first line structure. pline moves to point at it*

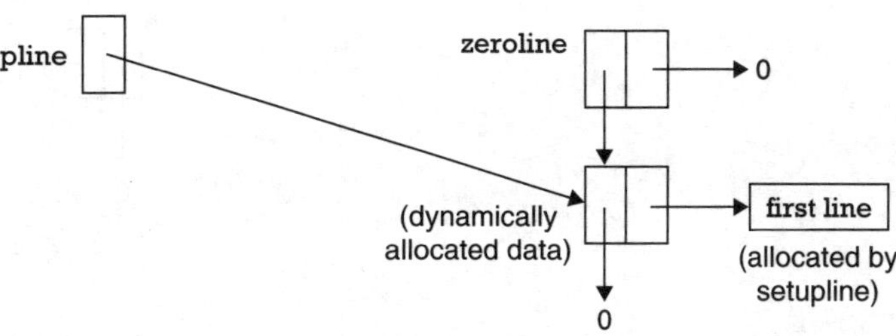

Figure 7.8 *First line loaded*

This will be line zero. It will never get filled with text. Its function is simply to hold the address of the first line of text, once it has been loaded in. Figure 7.5 shows this initial state. The address of the first line will be stored in **zeroline.pnext**. This value is initialized to 0. A pointer to 0 is a special message that means either (a) this item is the end of a list, or (b) there is an error (used to signal an error return from a function returning a pointer).

Again using **char *setupline()**, the first line could be set up with

```
line *pline = &zeroline;  // pline starts pointing at zeroline
                          // (Figure 7.6)

pline = pline->pnext = new line;
    // Allocate a new line structure, point pline->pnext at it
    // (which in this case is zeroline.pnext), then move pline to
    // point at the newly allocated line structure (Figure 7.7)

pline->text = setuptext();
    // Read in the text and set up its array; point the text pointer
    // at the array
pline->pnext = 0;
    // Set this line structure's pnext to 0 to signify the end of the
    // list. (Figure 7.8)
```

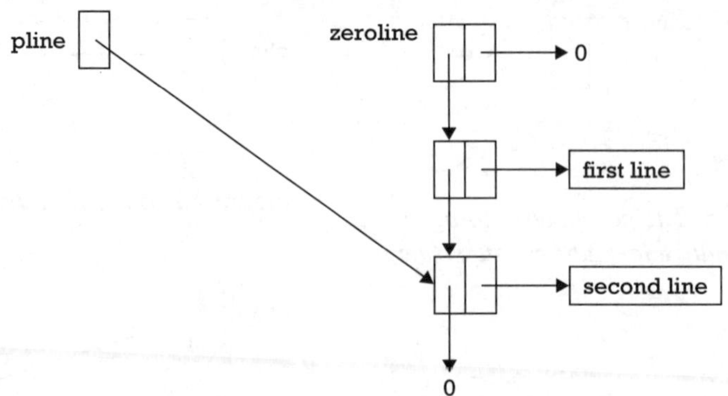

Figure 7.9 *Second line loaded.* **pline** *moved to it via the* **pnext** *pointer*

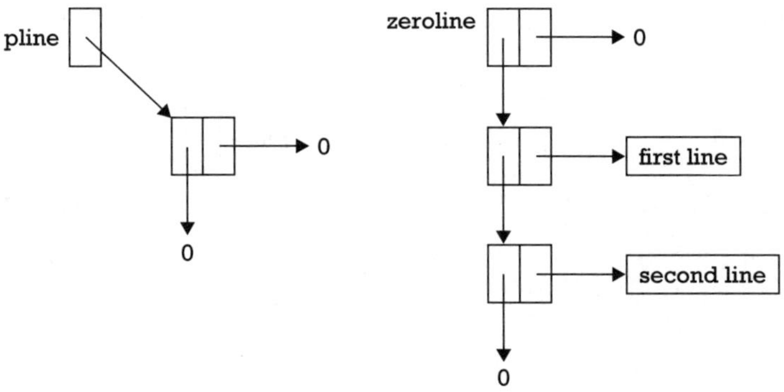

Figure 7.10 *Inserting a new line.* **pline** *points at new* **line** *structure*

The second line would be added with:

```
pline = pline->pnext = new line;
// These three lines are identical to the three immediately
pline->text = setuptext();   // above (comments changed).
pline->pnext = 0;   // (Figure 7.9)
```

So the **pnext** of any line points to the next line (and the **pnext** of the last line is 0 as a flag that this is the last line). The text '**first line**' and '**second line**' represents the first two lines obtained by **setuptext()**.

Here is some code to insert a new line *before* the existing first line.

```
pline = new line;               // Figure 7.10

pline->text = setupline();      // Figure 7.11

pline->pnext = zeroline.pnext;  // Figure 7.12

zeroline.pnext = pline;         // Figure 7.13
```

Observe how this code hooks a new first line between **zeroline** and the existing first line.

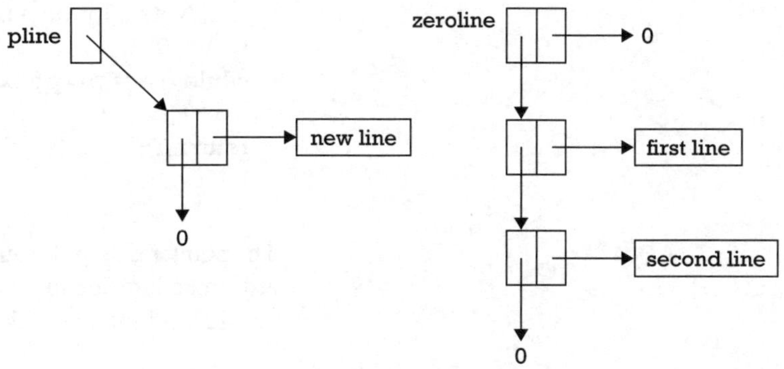

Figure 7.11 *Inserting a new line. The new **line** structure is filled*

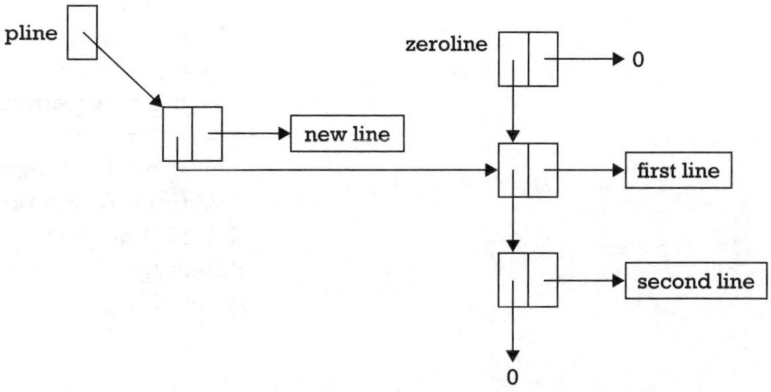

Figure 7.12 *Inserting a new line. Pointer manipulation step 1*

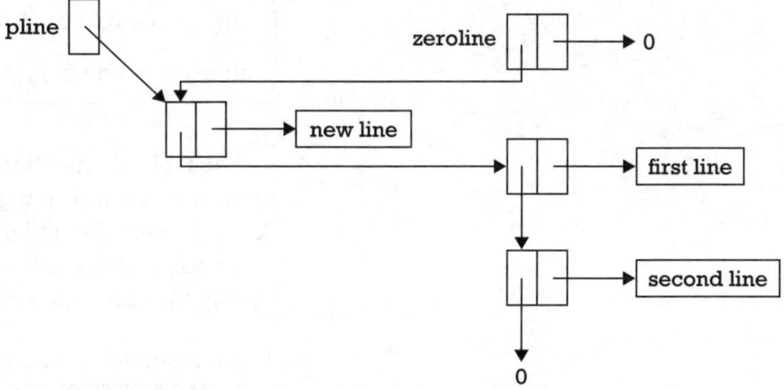

Figure 7.13 *Inserting a new line. Pointer manipulation step 2. The new line is linked in at the start of the list*

Now, suppose there was no explicit record of the number of lines. Here is a function that returns it:

```
int  numlines() {
    line*p = &zeroline; // Initializes p (a pointer to line) to
                        // point at zeroline
```

```
// Note syntax: p is being initialized, not *p
int i = 0;
while ((p = p->pnext) != 0)
   i++;
return(i);
}
```

The pointer **p** skips through from one line to the next. This is the standard procedure for moving through linked lists. As another example, consider a function to delete a given line.

```
int   deleteline(int linenum)
   {
   line *p, *q;
   p = &zeroline;
   while (linenum-->0)
      p = p->pnext;
      // p is now pointing at the line before the one to delete
   q = p->pnext;
      p->pnext = q->pnext;
      // Unhook line pointed to by q from the list
   delete [] q->text;
   delete q;
   return(0);
   }
```

> **deleteline()** has a problem. There is no check for reaching the end of the list before **linenum** lines. Tidy it up, and write a function
>
> ```
> int      insertline(int linenum, char *newtext)
> ```
>
> to insert a line before **linenum**.

Each of the functions manipulating the linked list works in the same way, by traversing the links to find the appropriate item, then doing pointer manipulations to hook items into place.

To recap, this section has considered four possible mechanisms for storing the text in an editor:

1. As a single one-dimensional array, of size **MAX_NUMBER_OF_ CHARACTERS**, which is presumably as big as possible.
2. As a single two-dimensional array, providing space for up to **MAX_NUMBER_OF_LINES**, each of up to **MAX_LINE_LENGTH** characters, and always occupying the maximum amount of space.
3. As an array of character pointers, of size **MAX_NUMBER_OF_ LINES**, where each array element points to a character array containing the text of the appropriate line.
4. As a linked list of character pointers.

The last two solutions also include a 'scratch' buffer of **MAX_LINE_ LENGTH** for storing the current line while it is being edited.

As the amount of memory available in computers expands, memory efficiency for any particular task becomes less of a problem. But at the same time, computers are used for more demanding tasks. Whereas a relatively inefficient solution may today be acceptable for storing a document in a word processing program, it certainly would not be for storing a video clip in a non-linear editing program.

Which of these mechanisms is the best? If the documents to be edited are always small in size, then the second solution may be the best. It is wasteful, but if there is a guaranteed low limit on **MAX_NUMBER_OF_LINES**, the waste can be tolerated, and the simplicity of the two-dimensional array is in its favour. But for more general use, one of solutions 3 and 4 should be chosen. Solution 4 is the best of all because it does not impose an artificial **MAX_NUMBER_OF_LINES** limit. Instead it is completely flexible.

Another linked list implementation is given in the section on queues later in this chapter.

7.5 Array/Linked list hybrids

It is possible to build a linked list of arrays, or an array of linked lists.

A linked list of arrays might be useful when a collection has a nominal size, but might grow in exceptional cases. For example, in the case of the text editor discussed above, a single line structure might be defined:

```
struct line {
   line *pnext;
   line *pcontinuation;
   char  text[NOMINAL_MAX_LENGTH];
   };
```

This would be used as a component in a linked list of lines as before, but now the text of a line is stored in an array of characters within the **line** structure. If a particular line is too long for the **NOMINAL_MAX_LENGTH** number of characters allocated to it, a new line structure could be linked in. This would lead to situations such as that shown in Figure 7.14.

Although linked lists of arrays are sometimes useful, they are less important than the other alternative – an array of linked lists. The usual way of creating an array of linked lists is via a *hash table*, and as it is such an important structure, it deserves a subsection of its own.

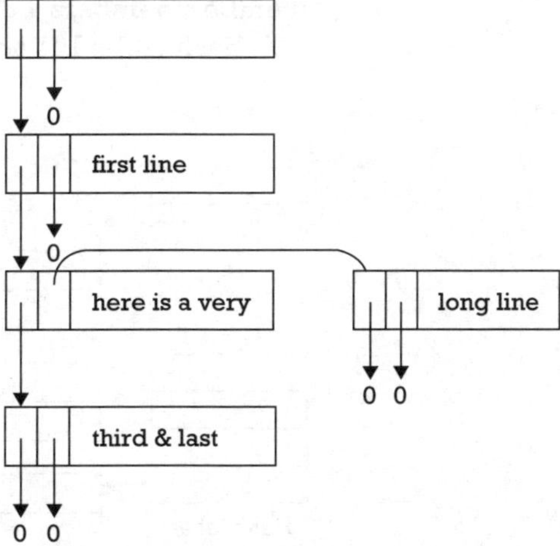

Figure 7.14 *Extending the fixed size* **text** *buffer in* **line** *using* ***pcontinuation***

Hash tables

The main reason for using an array of linked lists is to trade off space and time efficiency. A linked list adapts its memory usage as required, whereas an array has a fixed (and therefore large) memory usage. To find the *i*th item on a linked list takes on average *n*/2 operations, where *n* is the number of items; to find the *i*th item in an array takes only 1 operation. So linked lists are better so far as space is concerned, and arrays are better so far as speed is concerned.

Using *m* linked lists, rather than just 1, means that each list will have, on average, *n*/*m* items on it. Thus the search time is reduced by a factor of *m*. An array of linked lists, as in Figure 7.15, is therefore a way of using some permanent storage (as much as it takes to hold *m* start anchors for the linked lists) to gain speed; by adjusting *m*, the programmer can find a space/time trade-off appropriate to the problem at hand.

Usually, some data field of the structures to be stored is chosen as a key; a *hash function* is applied to the key which maps it into a number between 0 and *m* − 1, i.e. into an index for the array.

If the hash function works well it will distribute the entries uniformly through the table (assuming the original values were uniformly distributed). A generally effective hash function is to take the value of the key modulo N, where N is a prime. This is so that patterns in the data do not result in most of the entries hashing to the same table location. There is no guarantee against this of course, but the modulo-prime approach is generally successful.

Here is an example for a hash table and hash function to store a data structure called **member**.

```
struct member
    {
    member *pnext;    // To link next member in
    int key;
    int other_stuff;    // Some arbitrary other stuff
    };

member members_collection[101];
    // Hash table: 101 anchors for 101 linked lists
```

Ideally different hash keys should map into different addresses, but hash tables are always smaller than the range of values the key can take, so some keys hash to the same address. This is called *collision*, for which there are several *collision-resolution* policies. Chaining of hash table entries in linked lists, as here, is called *hashing* with *separate chaining*. An alternative is *open addressing*, where a collision causes *re*hashing, until a vacant space is found.

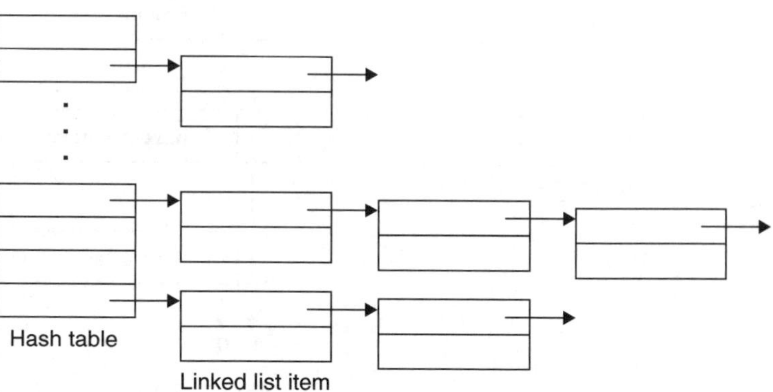

Figure 7.15 *Structure of a hash table with separate chaining*

```
int hash(int key)
  {
  return(key%101);
  }
```

Here 101 linked lists are maintained, the start node of each stored in the permanent **members_collection** array. An item with a key value 480, for example, will be appended to the list which starts at **members_collection[76]**.

Sometimes the key is long, as when strings are being stored. If, for example, a hash table is being used to store a dictionary, it would be good to use all the characters in a word to compute a table value. This would ensure that 'the', 'this', 'there', 'where', etc., all stood some chance of hashing to different values. Typically this is accomplished in an iterative way such as:

```
unsigned int hash(char *p)
  {
  unsigned int hash_value;
  hash_value = 0;
  while(*p != '\0')
    hash_value = (hash_value << 6) + *p++;
  return(hash_value%N);// N is prime - size of hash table
  }
```

The hash table provides a good solution to some abstract data type implementation problems. However, it does not allow some things that an ordered array gives for free: to find the highest or lowest key value, or the closest stored key to a given one, the hash table must be searched as if it were a collection of linked lists – the hash function is useless.

7.6 Trees

The **line** linked list created in section 7.4 is more than a simple list. Each **line** structure has two pointers – one used to join the structure in the list of lines, and one pointing to the text of the line. Clearly, using pointer links allows more flexible data structures than simple linear lists. To represent a graph, such as Figure 7.16 (taken from Chapter 13), for example, a multiple link structure could be used as follows:

```
struct arc
  {
  char label[20];
  struct node *destination;
  };

struct node
  {
  char name[20];
  struct arc arcs[3];
  };

node nodes[5] = {
  "PRECOUNT",{"T2,T3", &nodes[0], "T1", &nodes[1], "", 0},
  "BATCHING",{"T1", &nodes[2], "T3", &nodes[3], "T2",
      &nodes[4]},
```

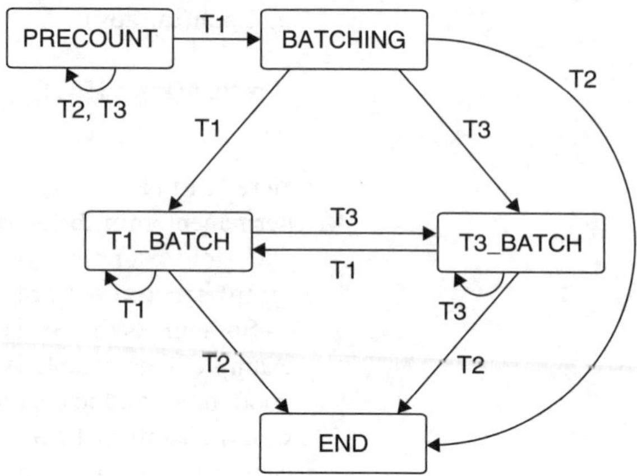

Figure 7.16 *An example of a graph*

```
"T1_BATCH", {"T1", &nodes[2], "T3", &nodes[3], "T2",
    &nodes[4]},
"T3_BATCH", {"T1", &nodes[2], "T3", &nodes[3], "T2",
    &nodes[4]},
"END", {"", 0, "", 0, "", 0},
};
```

Of course, this could be made even more complicated by allowing arbitrary numbers of links from each node – i.e. keeping a linked list of arcs with each node.

Graphs have wide application in computer science. But they have an important subset, which is as widely used as all other types of graph put together: trees. A tree is a graph with no cycles in it; that is, there is no way of traversing a series of arcs and arriving back at the starting point (without going backwards the way you came). Figure 7.17 shows a tree (*not* an inheritance tree, but a parse tree).

Clearly a tree can be represented with the same sort of data structure as a graph. The lack of cycles does not change that. The node at the top is called the root. If every node in a tree has the same number of children (nodes connected to it from below), then it is a multiway tree. Nodes at the bottom without children are called terminal nodes or leaves. A multiway tree with two children to each non-terminal node is called a binary tree – an especially important data structure.

Figure 7.18 shows a binary tree that is *completely full*. That is, every node except the leaves has exactly two children.

A way of representing this in code might be:

```
struct b_tree_node {
    char label[40];
    b_tree_node *left;
    b_tree_node *right;
};

b_tree_node example[7] = {
    "Suits", &example[1], &example[2],
    "Red", &example[3], &example[4],
```

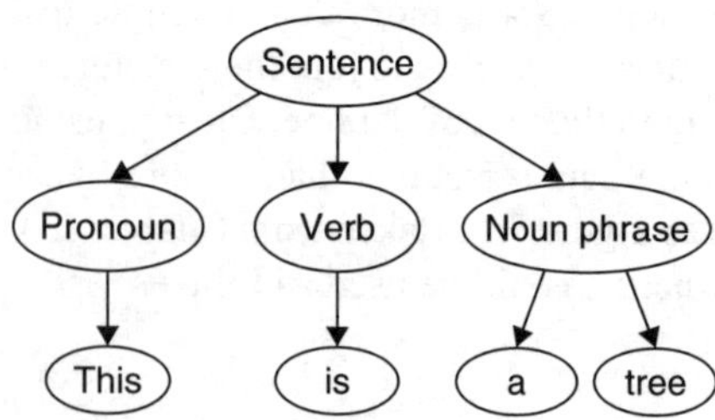

Figure 7.17 *An example of a tree*

Figure 7.18 *An example of a completely full tree*

```
"Black", &example[5], &example[6],
"Hearts", 0, 0,
"Diamonds", 0, 0,
"Clubs", 0, 0,
"Spades", 0, 0
};
```

It is always possible to have a pointer in a node data structure that points back to the 'parent' of the node. What is more, a general tree can always be represented in a binary tree structure. Thus it is possible to define a very general-purpose node structure that works for all trees:

```
struct tree_node {
    char label[40]; // or whatever stuff is to be associated with
                    // this node
    tree_node *parent; // pointer to my parent
    tree_node *sibling; // pointer to my adjacent sibling
    tree_node *children; // pointer to linked list of my children
};
```

This data structure implements a binary tree with the **parent** pointer pointing to the parent, and the **sibling** and **children** pointers acting like the **left** and **right** pointers in the example above. Figure 7.19 shows the relationship between an original tree and its binary tree representation, where the thick arrows represent **children** pointers, the thin arrows represent **sibling** pointers. **parent** pointers are not shown, but point up the lines on the top diagram.

Note that this binary tree is *not* completely full: some of the non-terminal nodes have just one child. But it's still a binary tree.

It is important to be able to navigate around trees. The most fundamental operation is *traversal* – visiting every node in the tree. For a binary tree, there are four basic methods of tree traversal.

- Preorder traversal recursively visits the root, then the left subtree, then the right subtree of every node.
- Inorder traversal recursively visits the left subtree, then the root, then the right subtree of every node.

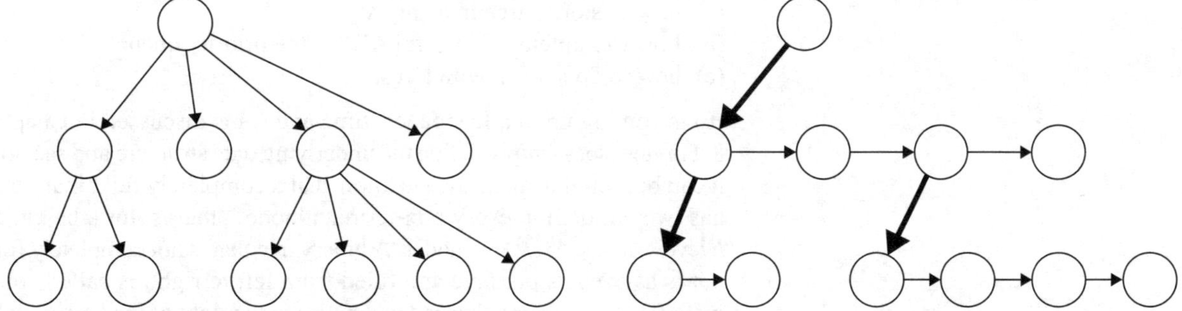

Figure 7.19 *An example tree (left) and its representation as a binary tree with children and sibling pointers (right)*

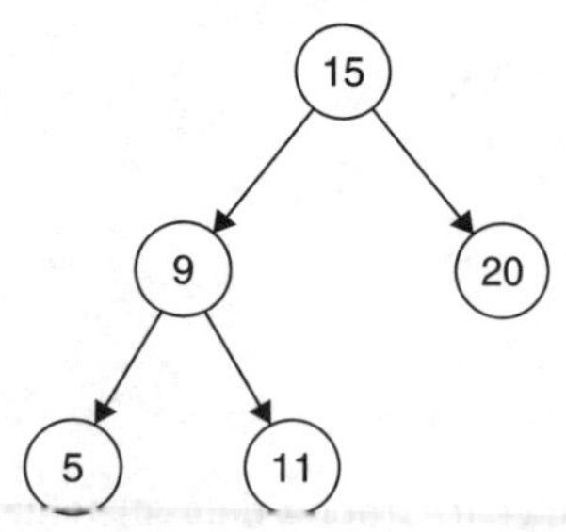

Preorder traversal	15	9	5	11	20
Inorder traversal	5	9	11	15	20
Postorder traversal	5	11	9	20	15
Level order traversal	15	9	20	5	11

Figure 7.20 *Traversal order for an example tree*

- Postorder traversal recursively visits the left subtree, then the right subtree, then the root of every node.
- Level-order traversal visits the nodes from top to bottom and left to right, going downwards a level at a time.

Figure 7.20 illustrates the four types of traversal on an example tree. Here is some code for inorder traversal.

```
struct b_tree_node {
   char label[40];
   b_tree_node *left;
   b_tree_node *right;
   };
void in_traverse(b_tree_node *t)
   {
   if (t == NULL)
      return;
   in_traverse(t->left);
   visit(t); // visit is the function that does whatever
            // needs to be done at each node.
   in_traverse(t->right);
   }
```

> Write code for pre- and postorder traversal of a tree.

> Harder: write code for level-order traversal. (A solution is given on page 213.)

Trees are widely used data structures: binary trees provide one of the fundamental search mechanisms of computer science; data compression, computational geometry, cryptography, and artificial intelligence all make heavy use of trees (and, to a lesser extent, graphs). A good data structures and algorithms book will introduce binary search trees, red–black trees, splay search trees, B+ trees, and many more. Many ingenious algorithms use trees for jobs that you wouldn't naturally suspect. There is insufficient space here to catalogue these. Instead, we are going to look at a particular example that illustrates:

(a) how to store a tree in an array
(b) how to implement a useful ADT – the priority queue
(c) how to do sorting with trees.

(b) is coming up in a few pages' time and (c) is discussed in Chapter 8. For now, let's simply define the underlying data structure and see how it can be stored in an array. We know that a completely full binary tree has two children to every non-terminal node – that is, for a height of h levels, it has $2^{(h+1)} - 1$ nodes. A binary tree that is not completely full, but is as even as possible and filled from left to right, is called complete. That is, you only have to add nodes to the right of the lowest-right node to make the tree completely full. So a level-order traversal of a *k*-node complete binary tree is the same as the first *k* steps of a larger, completely full tree.

Figure 7.21 shows an example of a complete binary tree. Its lowest level is incomplete, but is being filled from left to right.

This particular complete binary tree has another property: the value in every node is greater than (or equal to) the value in that node's children. This does not mean that all the values at a particular level are greater than all the values lower down: only that the node at the top of the tree (and at the top of any subtree) has the greatest value in the tree (subtree). A complete binary tree with this ordering property is called a *heap*. We will see how useful heaps are shortly.

Complete binary trees (including heaps) can be stored in a linear array. To do this we simply number nodes from 1 at the root, in the order of level-order traversal, and use those numbers as indices into an array. The zero index is not used. Figure 7.22 shows the mapping of a simple complete binary tree to the equivalent array.

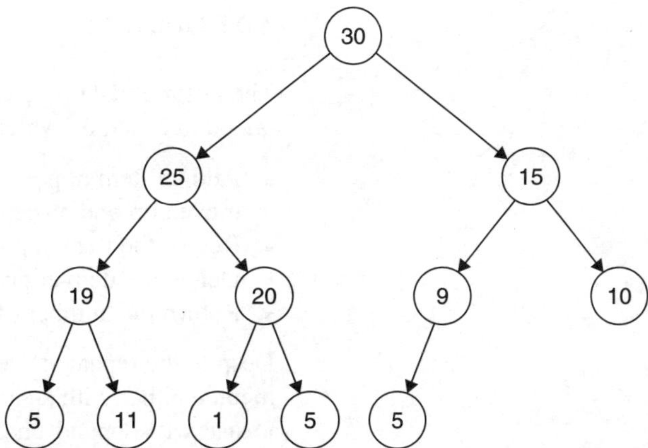

Figure 7.21 *A complete binary tree*

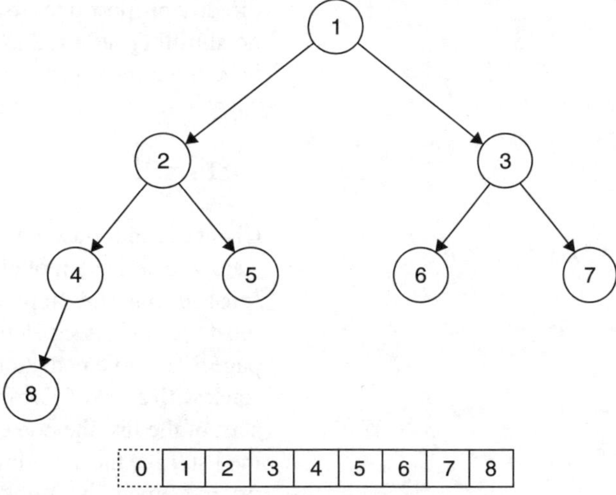

Figure 7.22 *Storage of a complete binary tree in an array*

Now it is easy to move around the tree via the array. The children of node (index) k are at locations $2k$ and $2k + 1$, and the parent of node k is at location $k/2$ (rounded down, i.e. in integer arithmetic).

7.7 Elementary abstract data types

Arrays, linked lists, their hybrids, and trees are all data structures for collections. Getting at a particular item of a collection requires an algorithm matched to the data structure. Before considering algorithms, we must decide what kinds of access to items a collection might provide. This means attaching behaviour to the data structure – moving into the realm of abstract data types. Abstract data types hide information, and provide a well-defined interface between the 'outside world' (the rest of the program) and the data structures. Because collections are so important, there are many kinds of fundamental ADT associated with them. In this section, the ordered list, the stack and the queue are briefly discussed, then in the next, the ADT table is developed.

ADT Ordered List

The abstract data type Ordered List is a collection of elements maintained in order, on which the following operations are defined:

- Add an item at position i on the list. All subsequent items will be moved up one place.
- Return the item at position i on the list.
- Delete the item at position i on the list.
- Return the number of items on the list.

Despite the repeated use of the word 'list', this ADT might be implemented either with an array or a linked list. The decision about which to use will depend on the size of the items, the amount of storage available, and the relative frequency of operations. If the ADT is implemented as an array, adding and deleting items may require memory moves, but access to each element will be very fast. If the list is represented as a linked list, the memory requirement will be directly proportional to the number of items on the list, there will be no shuffling required as items are added and deleted, but the list will have to be traversed every time any element is accessed.

ADT stack

Chapter 5 introduced a C++ implementation of a stack (page 123). A stack is a collection of elements ordered by insertion time: elements are 'pushed' onto the 'top' of the stack and 'popped' from the top. So the most recently added element is the one which will be retrieved first. The page 123 implementation used an array. We could instead implement a stack with a linked list: the push operation attaches a new element to the start of the list, the pop operation removes the element from the start of the list. Like the array implementation, the linked list does not do memory movement; but unlike the array, it grows and shrinks with the stack, *and* the list does not have to be traversed to get at the item to push or pop.

ADT queue

A third important ADT is the queue. This is another linear collection of elements ordered by insertion time. Again, only two operations are defined: items are 'enqueued' at the back of the queue and 'dequeued' from its front. Again, either an array or a linked list can be easily used. Here is code for both array and linked list implementations of a queue of pointers. The pointers are declared as **itemtype** * but the definition could be templated so the pointers could be to any other type.

The array implementation is first:

```
class queue
   {
                                 // First come the private data items
   int queue_length;          // To allow the caller to set max length
   itemtype **queue_base;     // Points to where queue is in
                                 // memory
   int head;
   int tail;
public:
   queue(int size)
      {
      queue_base = new itemtype *[queue_length = size];
      head = tail = 0;
      };
   ~queue()
      {
      delete [] queue_base;
      }
   int enqueue(itemtype *);
   int dequeue(itemtype **);
   };

int queue::enqueue(itemtype *value)
   {
   if ((tail == head-1)||((tail == queue_length-1)&&(head == 0)))
      return(-1);             // Queue full
   queue_base[tail++] = value;
   if (tail == queue_length)        // Wrap around to start of array
      tail = 0;
   return(0);
   }
int queue::dequeue(itemtype **value)
   {
   if (head == tail)
      return(-1);  // Queue empty
   *value = queue_base[head++];
   if (head == queue_length)        // Wrap around to start of array
      head = 0;
   return(0);
   }
```

Here is a linked list implementation that does the same thing, and has the same external interface.

```cpp
class queue
  {
  struct item
    {
    itemtype *value;
    item *pnext;
    } *head, *tail;
  int queue_length;    // To allow the caller to set max length
public:
  queue(int size)
    {
    queue_length = size;    // Unused in this implementation
    head = 0;
    tail = 0;
    };
  ~queue()
    {
    item *p;
    while(head)    // Step through, deleting items one by one
    {
    p = head;
    head = head->pnext;
    delete p;
    }
  }
int enqueue(itemtype *);
int dequeue(itemtype **);
};
int queue::enqueue(itemtype *value)
  {
  item *p = new item;          // Make new item
  if (tail)                    // Already something on queue
     tail = tail->pnext = p;   // so tack new item on end
  else
     head = tail = p;          // Otherwise this is single item
                               // on queue so far
  tail->pnext = 0;             // New end of queue
  tail->value = value;
  return(0);
  }
int queue::dequeue(itemtype **value)
  {
  if (head == 0)
     return(-1);               // Queue already empty
  *value = head->value;        // Get value in head item
  item *p;
  p = head;
  head = head->pnext;          // Move head pointer along one
  delete p;                    // and delete former head item
```

```
    if (head == 0)              // Last item just removed
        tail = 0;
    return(0);
}
```

> Compare and contrast the two implementations. 'Play computer'
> and see what happens in the programs as items are enqueued and
> dequeued. Make sure you understand how the array version
> 'wraps around'. Observe how the size parameter is unused in the
> linked list version. Modify the code so the two implementations
> are exactly compatible.

Now, with a queue, it is possible to go back to tree traversal (page 208). Recall that level-order traversal visits the nodes from top to bottom and left to right, going downwards a level at a time. Using the **queue** ADT, level-order traversal can be accomplished:

```
void level_traverse(b_tree_node *t)
    {
    queue traverse_queue(1000);    // Size depends on size of tree
    traverse_queue.enqueue((void *) t);
    while (traverse_queue.dequeue((void **) &t) == 0)
        {
        visit(t);
        if (t->left)
            traverse_queue.enqueue(t->left);
        if (t->right)
            traverse_queue.enqueue(t->right);
        }
    }
```

Level-order traversal of trees is made possible using queue ADTs, just as pre-order traversal implicitly uses a stack ADT because it is recursive.

> Explain why recursion involves implicit use of a stack.

ADT priority queue

Let us return to the *heap* data structure. Recall that a heap is a complete binary tree with the property that every node's value is greater than or equal to the value of its children (page 209). This immediately sets up the heap as being an appropriate data structure for getting at items in order of their value. In particular it is suitable for a priority queue.

A *priority queue* is an abstract data type to which priority-tagged items may be added in arbitrary order, but which are retrieved highest priority first. The priority queue determines internally how the prioritizing is done – for example, each call to **pq_put()** could simply add the item to a list, and each call to **pq_get()** could search through that list for the current highest priority item. A linear search is, however, time consuming, as we shall see in greater detail in the next chapter. A heap looks like a better approach because **pq_get()** has simply to fetch the item currently at the root of the tree. The only problems are:

(a) We have to preserve the heap property when removing an item from the root.
(b) We have to preserve the heap property when adding an item.

How to do these two things efficiently is an algorithmic question, but we are going to anticipate the next chapter here by specifying a pair of algorithms. This will allow us to continue discussing the data structuring of heaps, even though their efficiency won't be tackled until page 232. Here then are the algorithms:

(a) Call the item currently in the bottom right position the 'dropper'. After removing the item from the root, move the dropper into the root, vacating the bottom right node. While the dropper is smaller than at least one of its children (i.e. until the structure is a heap again): exchange the dropper with the larger of its two children.
(b) Call the new item the riser and put it in the next available leaf position (filling up the bottom layer from left to right or starting a new layer if the tree is completely full). While the riser is greater than its parent (i.e. until the structure is a heap again): exchange the riser with its parent.

Here's an example:
To **pq_get()** the root from the heap shown on page 209, the 30 at the top is removed, and the 5 in the bottom right position replaces it. Figure 7.23 illustrates.

Next the 5 is swapped with its larger child – the 25 (Figure 7.24).

The swaps ripple down the tree (Figure 7.25) until it is a heap again.

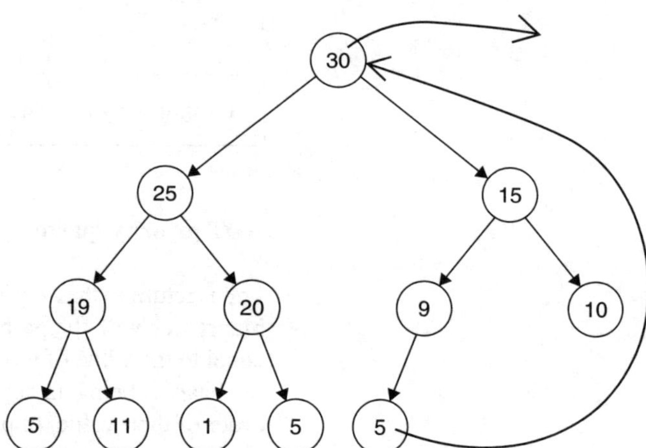

Figure 7.23 *Maintaining a heap: Removal of the root; substitution with bottom right node*

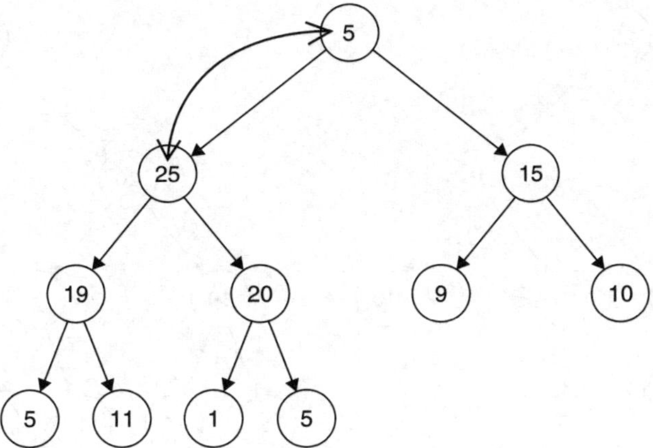

Figure 7.24 *Maintaining a heap: first exchange of the 'dropper' with a child*

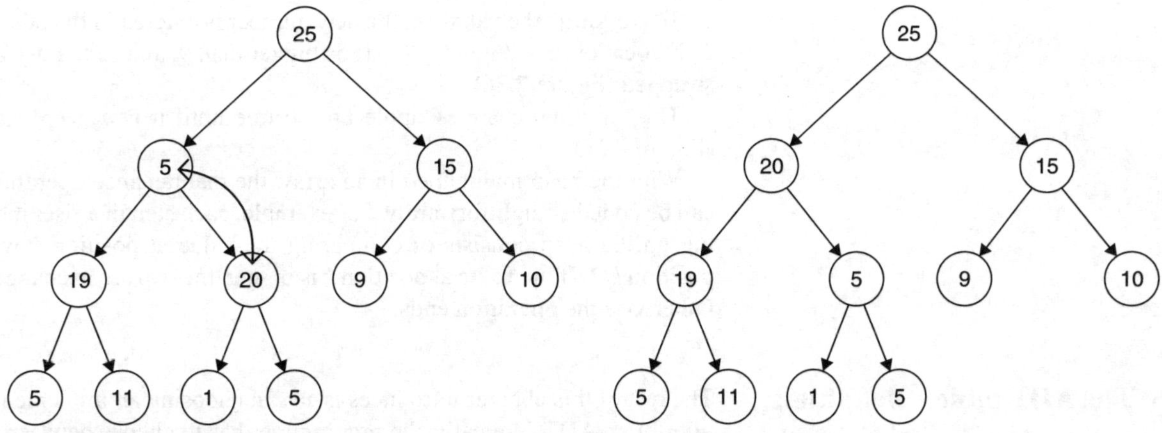

Figure 7.25 *Maintaining a heap: the dropper moves down until the heap property is restored*

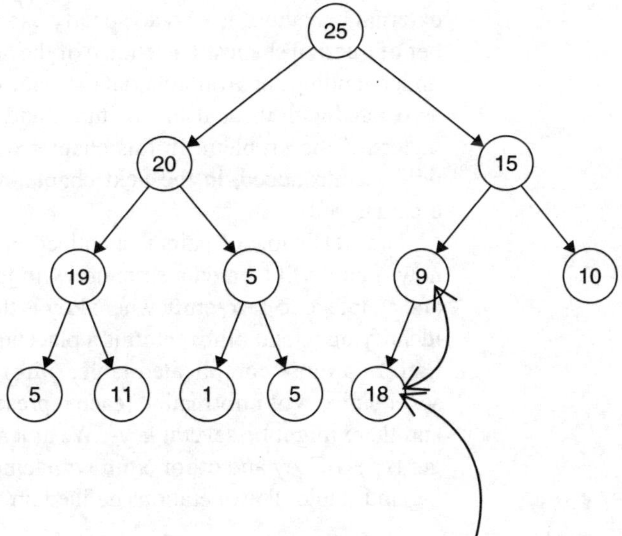

Figure 7.26 *Maintaining a heap: insertion of new element (the 'riser') at bottom right, first exchange with parent*

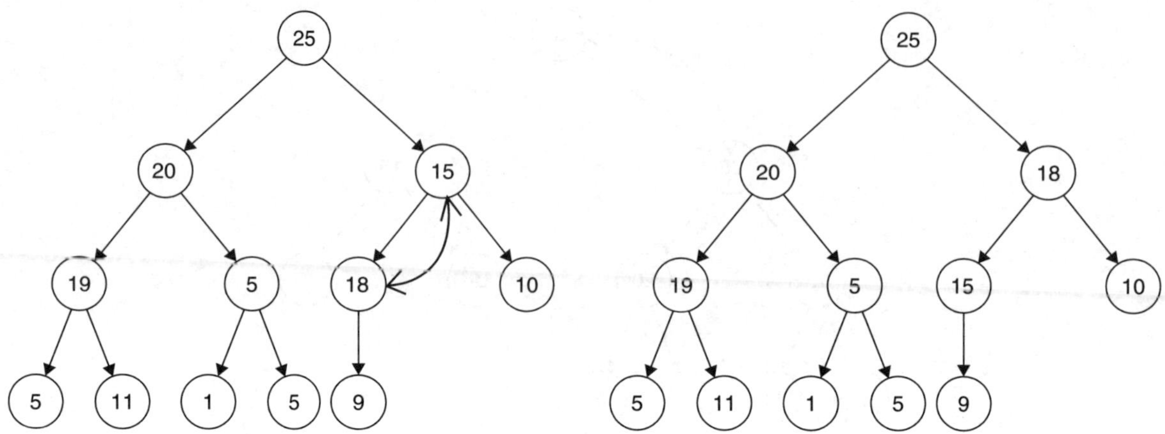

Figure 7.27 *Maintaining a heap: the riser moves up until the heap property is restored*

To **pq_put()** the value 18, the new number is entered in the bottom right location. See Figure 7.26. 18 is bigger than 9, and so the two are swapped (Figure 7.26).

This time the changes ripple up the tree until it is a heap again (Figure 7.27).

With the heap maintained in an array, the maintenance operations can be coded straightforwardly. For example, each step in a riser moving up the heap consists of comparing the value at position k with position $k/2$. If the value at position k is bigger, the two are exchanged. Otherwise the operation ends.

7.8 The ADT table – definition

The rest of this chapter introduces issues in choosing an implementation for an ADT. Typically, the programmer has to choose between an array, a linked list, a hybrid of the two, or some tree-like data structure. To illustrate and motivate the discussion of how this choice is made, we are going to define an abstract data type table (of pairs). Although its external behaviour is easy to specify exactly, it can have a large number of internal shapes. The choice of the appropriate shape depends on understanding the available data structures, analysing the algorithms associated with those data structures, and knowing the particular parameters of the problem. In this chapter the alternative data structures will be introduced, in the next chapter we will see how algorithms are analysed.

The ADT table of pairs is a collection of pairs. What is a pair? For now, a pair will be a data structure with just two items, an integer **key** and an integer **other_stuff**, where **key** is the label of the record used for identifying it, and **other_stuff** is a placeholder for some other information. In a more complicated table, **other_stuff** might be replaced by several pieces of information (each represented as a structure member), and there might be several keys. We at least expect to replace the integer types of **key** and **other_stuff** with templates once the table is working and stable. The operations defined on the Table ADT are:

- Add an item to the table.
- Remove an item from the table.

- Find the 'other stuff' associated with a particular item's key.
- Return the number of items in the table.

For any **key**, there may be only one entry in the table. This means that the add item operation may involve replacing the **other_stuff** for an existing **key**, rather than adding a new item.

The ADT makes these operations available by access functions (C++ member functions), which are as follows:

add_to_table(int key, int other stuff)

remove_from_table(int key)

get_other_stuff(int key, int& other_stuff)

number_in()

At first sight, this looks like a simpler ADT than the ordered list, stack or queue – the lack of explicit ordering means it does not matter how the data are stored. But this simplicity is illusory. We seek an efficient implementation that makes the access functions fast. In consequence, there are many possible design solutions depending on the particular constraints and performance expectations of the situation.

7.9 Implementing the ADT table with an unordered array

Given below is a program that implements this ADT and allows the user to manipulate an instance of table interactively.

```
//
//
// C++ Program to implement ADT table of pairs
// using unordered array
//
//
#include <iostream.h>
#include <stdlib.h>
#include <stdio.h>

struct pair {
  int key;
  int other_stuff;
  };

class table_of_pairs {
  int table_size;
  pair *ptable;
  int num_in_table;
public:
  table_of_pairs(const int size) {
    num_in_table = 0;
    ptable = new pair[(table_size = size)+1];
      // Set up an array of the requested size
    };
  ~table_of_pairs()      {delete [] ptable;};
  int add_to_table(const int, const int);
  int remove_from_table(const int);
  int get_other_stuff(const int, int&) const;
```

```cpp
    int number_in() const  {return(num_in_table);};
    void print_table() const;
    };
//
// Methods for the class table_of_pairs follow
//

int table_of_pairs::add_to_table(const int key, const int other)
    {
    remove_from_table(key);
        // Remove any existing entry with this key value
    if (num_in_table == table_size)
      return(-1);       // table is full
    ptable[num_in_table].key = key;
        // Add new item at the end of the list
    ptable[num_in_table++].other_stuff = other;
    return(0);
    }
int table_of_pairs::remove_from_table(const int member)
    {
    int offset = 0;
    ptable[num_in_table].key = member;  // sentinel
    while (ptable[offset].key != member)
       offset++;     // Keep going until find something with
                     // this key value
    if (offset == num_in_table)    // The one I found was
                                   // the sentinel -
       return(-1);   // - so member is not in table
    // If member is in table, copy down final element in its
    // place (to overwrite it)
    ptable[offset] = ptable[num_in_table - 1];
    num_in_table--;
    return(0);
    }
int table_of_pairs::get_other_stuff(const int member, int&
       other) const
    {
    int offset = 0;
    ptable[num_in_table].key = member;  // sentinel
    ptable[num_in_table].other_stuff = 0;
       // set up, so that if key is not in table, this zero
       // value of other_stuff will get put into other
    while(ptable[offset].key != member)
       offset++;
    other = ptable[offset].other_stuff;
    return(offset != num_in_table);
       // returns 0 if the one found was the sentinel
    }
void table_of_pairs::print_table() const
    {
    int offset;
    for (offset = 0; offset < num_in_table; offset++)
```

```cpp
            cout << ptable[offset].key << ","
                 << ptable[offset].other_stuff << " ";
      cout<< "\n";
      }

//
// The main program here is just a simple test loop
// The commands q, a, r, c, p are available, to quit, add element,
// remove element, check for element, and print table repectively.
//
void main()
   {
   char c;
   int key, other;
   table_of_pairs mytable(8);
   // Here saying that 8 is max num of elements I will use in
   // any run of this test program (A bit restrictive).
   while(1)
      {
      cout <<":" ; // Prompt
      cin >> c;
      switch(c)
         {
         case 'q':
            return;
         case 'a':
            cin >> key >> other;
            mytable.add_to_table(key, other);
            break;
         case 'r':
            cin >> key;
            mytable.remove_from_table(key);
            break;
         case 'c':
            cin >> key;
            if (mytable.get_other_stuff(key, other))
               cout << "Item labelled " << key << " has value "
                    << other << " in mytable\n";
            else
               cout << "There is no item labelled " << key <<
                    " in mytable\n";
            break;
         case 'p':
            mytable.print_table();
            break;
         default:
            cout << "Use q, a, r, c or p only\n";
            break;
         }
      cout << mytable.number_in() << " items in table\n";
      }
   }
```

In this program, the table is implemented as an array of pairs. New entries are made by appending to the array, and searches are done by scanning through the array. Note that three of the four accesses functions require searching. The reason for making the array one element longer than the maximum number of elements in the table is so that the last location in the array is always available for use as a *sentinel*. When the array is being searched for a particular value, that value is placed as a sentinel in the position after the last element of the array. This ensures that the search loop will always terminate – all that needs to be checked after coming out of the loop is whether the termination is within the table array or at the sentinel. This function is illustrated in the **remove_from_table** and **get_other_stuff** methods.

> If **add_to_table** merely added a new item, it would not involve searching. However, the requirement that there be only one item per key value means a search for any existing item with this key must be made first. In the code this is done by **add_to_table** calling **remove_from_table**.

7.10 Alternative implementations

Implicit in the code is the assumption that an array is an appropriate (good) way of storing the pairs. Suppose we had instead chosen to use a pure linked list:

```
struct member
   {
   member *pnext;
   int key;
   int other_stuff;
   };
class  table_of_integers
   {
   member  zeromember;
   int   num_in_table;
public:
   ...
   };
```

The cost of implementing the ADT in this way is a pointer for each pair element.

We might compromise with a hybrid scheme, allocating chunks of storage:

```
struct chunk {
   chunk *pnext;
   int keys[100];
   int other_stuffs[100];
   int num_in_this_chunk;
   };
```

Initially a single chunk is allocated, then every time the number of elements exceeds the amount of storage available (100 per chunk in this example), an extra chunk is chained in. This is very like the continuation line example on page 203.

A hash table provides another possible implementation solution. The table is accessed according to a pair's key value, so hashing seems a natural choice. Furthermore, only one item with any particular key value is allowed. If it were possible to have many items with the same key value, then, when this happened, all those items would be attached to the same linked list. As it is, collisions occur only when

Binary search is like looking up a name to find a number in a phone book. Open in the middle and check the names at the top of the page. If the names are earlier in alphabetical order than the one you're looking for, split the back half in two and try again; if they are later, split the front half. Eventually you'll home in on the name. It sounds simple, but coding it right is tricky. The first publication of a binary search algorithm was in 1946. The first publication of a *correct* binary search algorithm was nearly 17 years later!

two different key values hash to the same index. Unless the data have unusual structure, the hash function should be able to distribute the items fairly well.

So far, then, we have considered four alternatives for storing the table: an array, a linked list, a linked list of arrays, and a hash table. But it is not that simple!

Suppose the table of integers were stored in an array that was always maintained in sorted order. Adding a new element would mean finding the appropriate position, moving everything from that position to the end upwards one place and inserting the new element. This obviously would take longer than just appending the element to the existing (unsorted) array. But keeping the array sorted means it can be searched more efficiently. Binary search works by successive partitioning of the array; recursively checking the middle element to see if it is smaller than or greater than the element sought, then applying the same check to the half of the array in which the element is now known to lie.

As Chapter 8 will show, binary search requires $\log_2 n$ comparisons to find an element, where n is the number of items in the array. On average, sequential search through an unordered array requires $n/2$ comparisons. If n is 2048, say, that means 11 comparisons for binary search versus 1024 comparisons for linear search. And the difference increases with n. So keeping the array in sorted order might be well worthwhile.

Given that binary search is so much more efficient than sequential search, we might consider whether there is a way of applying it to linked lists. There isn't. The nature of a linked list is that it is accessed through its links, which means visiting each item in sequence. But is there another way of linking things together that *does* allow searching in a binary fashion? There is! A binary tree – a data structure whose very shape means it is searched in a binary fashion. Finding an element would involve walking down the tree from the root comparing the values in non-terminal nodes with the key. Adding and deleting elements are problematic though. Its shape has to adapt to the particular values that are added and removed – there is a need for 'tree balancing'.

Now we have six possible data structures: an ordered array, an unordered array, a linked list, a linked list of arrays, a hash table and a binary tree. Surely it is time to stop! With information hiding, there is at least the reassurance that the external interface will be the same, whatever the choice of structure. But it is the implementer's responsibility to make that choice the best possible.

Before a reasonable decision about data structure can be made, the associated algorithms must be analysed. We turn to this issue in Chapter 8. The final consideration of the ADT table of pairs' implementation is deferred until Chapter 13, when we can consider the particular parameters of example problems as well as the capabilities of the possible data structure/algorithm options.

7.11 Chapter end material

Bibliography

Data structures and algorithms come as a package. See Table 8.1 at the end of Chapter 8 (Algorithms) for a structured list of references.

8 Algorithms

A program should be correct, reliable, maintainable and efficient. In this chapter we focus on the last. The other three are important too. Never underestimate the virtues of a simple solution – it is likely to be easy to program correctly, reliable and maintainable. But it may not be efficient. Does that matter?

Yes, efficiency matters. Efficiency *always* matters because you always have to ask the question, 'Is efficiency an issue?' Often the answer may be 'No' and that's the end of it. But you can't just assume 'No'. People who claim that efficiency is less important than correctness, reliability and maintainability change their minds when the correct, reliable, maintainable program takes days to run.

In short, it is important to have a good appreciation of algorithm efficiency, so you can make a judgement about how much efficiency considerations matter for the problem at hand. If they matter at all, you need to go further in analysis, to quantify computational complexity, and so make informed decisions between simpler, less efficient solutions, and complicated, more efficient solutions.

Any well-defined procedure is an algorithm. But some algorithms are so important they form the focal point for an introductory study of algorithm analysis. Books on algorithms usually divide them into a few major types: Sorting, Searching, String Processing, Geometric, Graph Theoretic, Arithmetic, etc. Here we will concentrate on Searching and Sorting as representative and important algorithms.

8.2 Searching algorithms

Here are three ways to search for an integer **x** in a collection of integers:

Unordered linked list – sequential search

```
struct item {
   item *pnext;
   int value;
   };
int search(int x)
   {
   extern item zeroitem;
   int i = 0;
   item *p = &zeroitem;
   while ((p = p->pnext) != 0)
      {
   if (p->value == x)
      return(i);
```

```
      i++;
    }
return(-1);
}
```

Unordered array – sequential search

```
int search(int x, int array[], int num_in_array)
  {
  int i;
  for (i = 0; i < num_in_array; i++)
    {
  if (array[i] == x)
      return(i);
    }
return(-1);
}
```

Ordered array – binary search

```
int search(int x, int array[], int num_in_array)
{
int middle;
int start = 0;
int end = num_in_array - 1;
while(end >= start)
  {
  middle = (start + end)/2;
  if (array[middle] == x)
    return(middle);
  if (array[middle] < x)
    start = middle + 1;
  else
    end = middle - 1;
  }
return(-1);
}
```

> Hand execute these three alteratives on a simple search problem:
> find the letter 'j' in the alphabet. Treat 'a' to 'z' as symbols for
> consecutive integers starting at 0. Sketch out the data structures
> and 'play computer'. How many operations does each alternative
> take?

Clearly the choice of data structure affects the choice of algorithm,
and vice versa. The best choice depends on:

- the maximum number of items in the collection
- the usual, or typical number of items in the collection

- the relative frequency of different operations
- required optimization trade-offs (memory vs speed)
- how much code complexity is allowed or desired.

To make these judgements it is important to understand the memory requirements and performance of each solution. Memory requirements are usually straightforward to calculate; for assessing performance, we need to consider how to measure algorithm efficiency.

8.3 Expressing the efficiency of an algorithm

An algorithm's efficiency is expressed as the order of magnitude of the algorithm's running time as a function of the problem size. Problem size usually means the number of items involved, denoted N or n. For example, sorting algorithm efficiency is expressed in terms of the number of items to be sorted. If we say that the running time, or computational complexity, is order $f(N)$, we mean that $f()$ describes how the number of operations depends on N. For example, we might say an algorithm is order $\log N$, or order N^2, which means that the number of operations required will be some multiple of $\log N$ or N^2 respectively. It is the functional relationship that matters; algorithms whose runtime differs by just a multiplicative constant are equivalent.

Different functions of N have different growth rates. Figure 8.1 shows some typical examples encountered in algorithm analysis. In the curves shown the scale factor is 1 (and the logarithm base is 2 – the base is just a scale factor too). If an algorithm consists of parts with different growth rates, for large N it is the fastest growth rate that predominates. So, for example, an algorithm that has an order N part and an order N^2 part is simply an order N^2 algorithm.

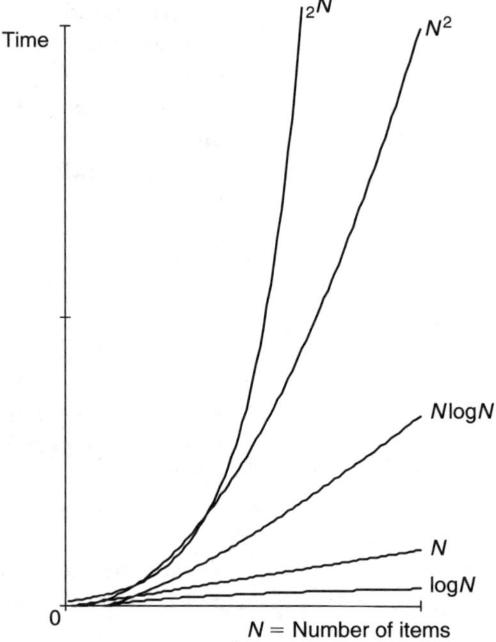

Figure 8.1 *Comparative algorithm growth rates*

The actual coding of the algorithm is usually not considered in algorithm analysis; there may be subtleties in programming that can change the running time by a scale factor, but these are of secondary importance compared with understanding how the time depends on the number of items. For example, an algorithm that takes 2^N operations is order 2^N, whereas one that takes $1\,000\,000\,000\,\log_2 N$ is order $\log N$. The graph in Figure 8.1 shows an order 2^N algorithm growing rapidly, whereas an order $\log N$ one grows very slowly. But surely that $1\,000\,000\,000$ counts for something! Well, yes, for N less than 30 it makes the fast $\log N$ algorithm take longer than the 2^N algorithm. But for N larger, even slightly larger, the 2^N runtime is soon outstripping the age of the universe on any conceivable computer, whereas $\log N$ is just growing very slowly.

However, when you *do* know N for a particular problem, and it is small, it is worth calculating the actual running time (in number of operations) for each alternative design. Even $1000\,\log_2 N$ is greater than 2^N for all N from 0 to 10.

Many algorithms are data dependent. In the worst case their order of magnitude may be one function of N, say N^2, but in the best case, they follow a different function, say $N\log N$. Worst case performance is written with 'big-oh' notation, e.g. $O(N^2)$ and best case with 'omega' notation, e.g. $\Omega(N\log N)$. These are upper and lower bounds on runtime, but, when they differ, we are usually interested in the average case too. This is often harder to work out. In this book, we go through most worst case calculations and a couple of average cases to illustrate the differences.

8.4 Search algorithm analysis

The analysis of basic search algorithms (page 222) is straightforward.

Unordered linked list – sequential search

In the worst case, the item to be found is at the end of the list. N items have been traversed to reach this item. So the worst case performance of sequential (linear) search on a linked list is $O(N)$ (order N).

If the items to find are uniformly distributed over the list, they will take $N/2$ operations to find on average. The average running time is therefore also order N. Scale factors are ignored in expressing orders of magnitude.

Unordered array – sequential search

The analysis for worst case and average case accesses is the same as for linked lists. Instead of stepping from one item to the next via a pointer, we successively index array locations. But the visiting of all items in the worst case and $N/2$ in the average case is identical to lists. Sequential search on arrays is an order N operation.

Access of the ith element in an array is an order 1 operation; an array affords immediate access to any given index. This is different from access to the ith item on a linked list.

> What is the order of magnitude for accessing the ith item on a linked list? Why?

Ordered array – binary search

Binary search has its worst case when the item to be found is at the bottom of the recursive subdivision tree. To get the bottom of a binary tree takes $\log_2 N$ operations, and the worst case performance of binary search is therefore O($\log N$). The base is irrelevant, because to convert between logarithms of different base, you simply multiply by a constant.

What about the average case? Sometimes the algorithm finds the appropriate element before traversing down to the bottom of the tree. But in the 'best' case (that is, when there are most likely to be 'hits' before reaching the bottom of the tree), there are still half of the items that take $\log_2 N$ operations to find. The average case performance is therefore order $\log N$.

Show that in the 'best' case for binary search, half of the items take $\log_2 N$ operations to find.

What is the efficiency of searching in a hash table with separate chaining?

8.5 Sorting algorithms and their efficiency

We now turn to analysing sorting algorithms. To make the analysis consistent, we will assume that the sort is always to be done on an array of integers. In practice it is often arrays (or linked lists, etc.) of more complicated data structures that are to be sorted. Typically, a particular item of a data structure will be used as a sort key.

The five sorting algorithms we will consider are selection sort, insertion sort, mergesort, quicksort, and heapsort. Why five? (And why these five?) The first two are simple and worthwhile solutions when the value of N is small (below 50). They are easily described and analysed. The others are good general purpose sorts for large N. They illustrate three ways that inventive people have reorganized the sorting problem, and thereby optimized particular qualities in their solutions. Other contenders for inclusion would have been shell sort and radix sort. The purpose here is not to be exhaustive but to illustrate calculation of efficiency.

In each case, the operation of the algorithm will be illustrated on the array

int a[6] = {37, 48, 22, 49, 37, 26};

and the progress of the sort will be shown with a diagram where each successive row represents a step in the process.

Each sorting algorithm swaps elements of an array using the **swap** function:

```
void swap(int *a, int *b)
  {
  int temp;
  temp = *a; *a = *b; *b = temp;
  }
```

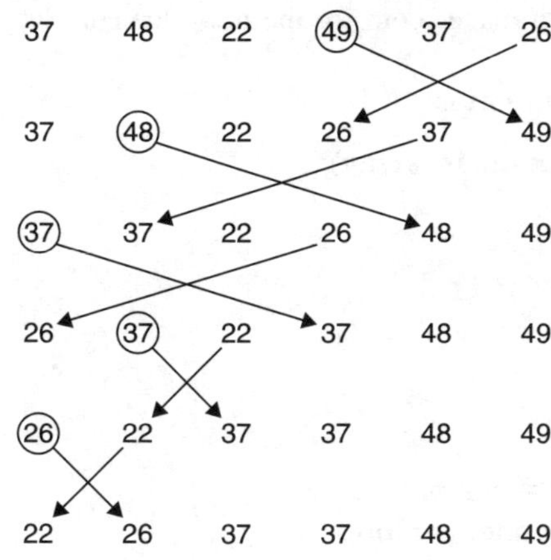

Figure 8.2 *Selection sort on an
example array*

Selection sort

Selection sort works by finding the largest element, exchanging it with the element in the last position, finding the next largest, exchanging it with the element in the second-last position, and so on. The progress of the algorithm on the example array is illustrated in Figure 8.2.

The C++ code for selection sort is:

```
int sort(int array[], const int num_in_array)
  {
  int max;
  int position;
  int i, j;
  for (i = num_in_array-1; i >= 1; i--)
    {
    max = array[i];
    position = i;
    for (j = 0; j < i; j++)
      {
      if (array[j] > max)
        {
        position = j;
        max = array[j];
        }
      }
    swap(&array[i],&array[position]);
    }
return(0);
}
```

There are two loops in the selection sort. The first time through the **i** loop, the **j** loop runs **n-1** times; the second time it runs **n-2** times, and so on. The last time the **j** loop runs twice. The total number of times the inner loop runs is therefore $(n - 1) + (n - 2) + \cdots + 2$ which is $n(n - 1)/2 - 1$. The swap operation at the end of the program only runs n times. Therefore the total running time is n^2. This amount of time is required in every case so both average and worst case performance is order n^2.

As an $O(N^2)$ algorithm, selection sort should only be used for fairly small N (less than 50). The same is true of insertion sort (coming up next). The advantage of these algorithms over more efficient alternatives is their simplicity.

Insertion sort

Insertion sort works by considering each element in turn and inserting it in the correct place relative to the elements that have already been sorted. At the beginning only the first element of the array is considered sorted, the second is extracted and put in front of or after the first element, as appropriate, the third is extracted and put in its proper place relative to the first two, and so on. So each insertion means shifting some of the

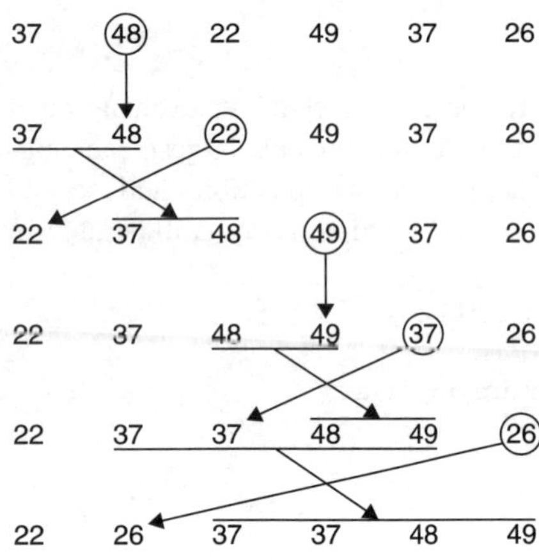

Figure 8.3 *Insertion sort on an example array*

already sorted elements right to make room for the new element. See Figure 8.3.

The C++ code for insertion sort is:

```
int sort(int array[], const int num_in_array)
{
  int i, j, temp, current;
  for (i = 1; i < num_in_array; i++)
    {
    current = array[i];
    for (j = i-1; j >= 0; j--)
      {
      if ((temp = array[j]) <= current)
        break; // out of the inner for loop
      array[j+1] = temp;
      }
    array[j+1] = current;
    }
  return(0);
}
```

The algorithm always executes $n - 1$ outer loops and between 1 and $n - 1$ inner loops. The number of times the inner loop is executed depends on how the data are organized. The worst case running time happens when the inner loop has to go all the way from j=i-1 down to j=0. This is similar to selection sort, so worst case performance is $O(N^2)$. In the best case, when the array is already sorted, selection sort is linear in time (order N) because the inner loop is executed only once for each outer loop. What then is the average case performance?

The critical operation in insertion sort is comparison. That is, the number of times the inner loop is executed depends on the number of times the (((temp = a[j])) < current) comparison test fails. For a particular element a[j], the test fails as many times as there are elements before a[j] in the array that are greater than it. This number of elements is the same when a[0] to a[j-1] is sorted as before it was sorted. So, defining an 'inversion' to be any pair of elements a[i], and a[j] with i < j and a[i] > a[j], then, over the whole array, the number of comparisons insertion sort must do is equal to the number of inversions in the original unsorted array. In the example array {37, 48, 22, 49, 37, 26}, there are eight inversions: (37,22), (37,26), (48,22), (48,37), (48,26), (49,37), (49,26), (37,26).

Now, we can calculate the average case performance of insertion sort if we know the average number of inversions in an array. Ignoring duplicate entries (as in the example array), we can label each element according to its size with the first n integers. So element 1 is the largest, element n the smallest. The question now is, what is the average number of inversions in the permutations of the first n integers? If we pair permutations that are mirror images of each other, for example 2 4 5 1 6 3 would be paired with 3 6 1 5 4 2 (the same thing backwards), then for any ordered pair of elements, say (1,2), these will occur in one permutation in order, and in the other, as an inversion. Therefore, between the two paired permutations, there must be a total of $n(n - 1)/2$ inversions. The average number of inversions over the whole set of permutations is therefore $n(n - 1)/4$.

We have shown that in the average case, the critical part of insertion sort happens $n(n - 1)/4$ times. Therefore, in the average case, as in the worst case, insertion sort is an order n^2 algorithm.

Mergesort

We now come to the order $N\log N$ algorithms. Here we will discuss three of the most important: mergesort, quicksort and heapsort.

Mergesort is order $N\log N$ in both the average case and the worst case. It is not as fast as quicksort on average, but it does have nice features that justify its inclusion here. In particular it is useful when a large sorted file must be merged with a set of unsorted new entries.

Mergesort has an initial 'downward' phase, where the array is recursively divided into halves. So it is first split into two, then each half is split into two, and so on, until the array has been divided into subarrays each containing one element. There is then an 'upward' phase where pairs of arrays are merged. Starting from the one-element subarrays, each merging creates an array where the elements are correctly ordered. The sorting problem becomes how to merge two already-ordered arrays into one ordered array. Figure 8.4 illustrates the process on the example array.

In the code below, the merging of two sorted arrays is done by the function **merge**; execution is controlled by the recursive function **sort**.

```
void merge(int *astart, const int *aend, int *bstart, const int
*bend, int *out)
  {
  while ((astart < aend) && (bstart < bend))
    {
    if (*astart < *bstart)
      *out++ = *astart++;
    else
      *out++ = *bstart++;
    }
```

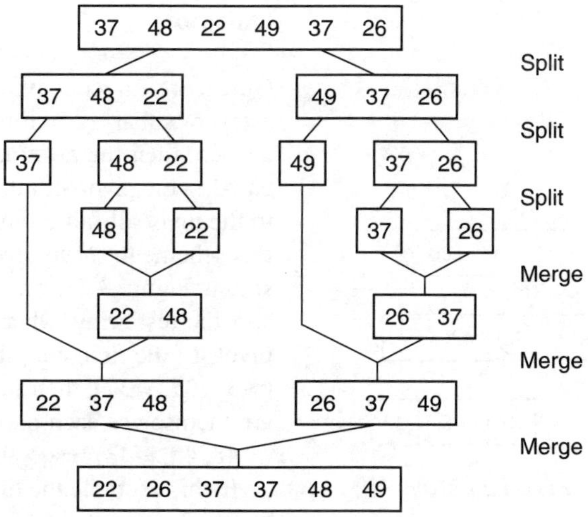

Figure 8.4 *Mergesort on an example array*

```
    while (astart < aend)
        *out++ = *astart++;
    while (bstart < bend)
        *out++ = *bstart++;
    }
int sort(int *array, int n)
    {
    int mid;
    extern int *buffer;
    // Temporary global storage, same size as original array
    mid = n/2;
    if (mid == 0)
        return(0);
    sort(array, mid);
    sort(array+mid, n-mid);
    merge(array, array+mid, array+mid, array+n, buffer);
    memcpy(array, buffer, n*sizeof(int));
    return(0);
    }
```

Note the heavy use of pointers in this program. It is possible to code **merge** using array subscripting rather than moving pointers. The pointer version is usually up to 40% faster though.

At each level of mergesort there are N operations in total, although these are divided up among the subarrays of that level. There are $\log_2 N$ levels. Therefore the running time of mergesort is always order $N\log N$; that is, its worst case and average case times are identical.

Mergesort requires an auxiliary array of the same size as the array to be sorted. There is rarely a space problem, but all the copying back and forth slows the algorithm down. This can be avoided by swapping the roles of the temporary **buffer** array and the original array **a** at alternate recursion levels, but that makes the code messier.

Mergesort is not usually as fast as quicksort, but it is an elegant algorithm, and one that is used in adapted form for *external* sorting – that is, when the whole array will not fit into main memory.

Quicksort

Quicksort is a recursive sorting scheme that at each step partitions the array, or subarray, in three parts, about a 'pivot' element. The pivot is chosen, then the array is reorganized so that elements lower than or equal to the pivot all fall on its left and elements greater than or equal to the pivot all fall on its right. The total sort is achieved by applying this scheme to successively smaller subarrays until a single element is sorted. Figure 8.5 shows how quicksort uses just two recursive steps to sort the test array, when the first element in the array is used as the pivot. In the first step, the whole array is partitioned into three parts ($\leqslant 37$, 37, $\geqslant 37$), then in the second step the $\leqslant 37$ and $\geqslant 37$ subarrays are themselves each partitioned into three parts: ($\leqslant 22$, 22, $\geqslant 22$) and ($\leqslant 48$, 48, $\geqslant 48$) respectively.

In this example the first element of the array has been chosen as the pivot. This is not always the best strategy, but here it works quite well.

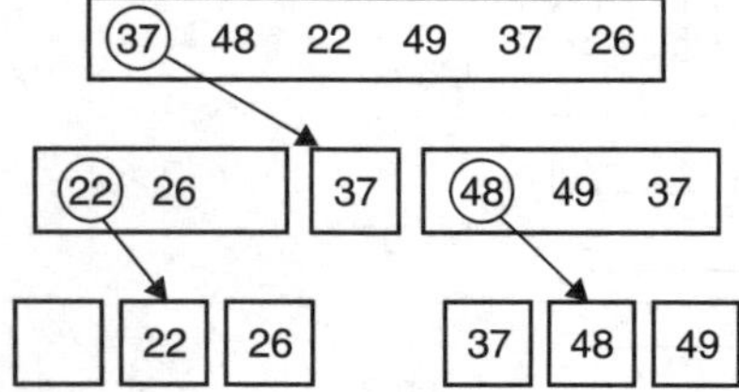

Figure 8.5 *Quicksort on an example array*

It turns out that quicksort performs best when the pivot is the median of the array to be sorted. Of course, without sorting the array, there is no way to know what the median is, but an effective strategy is to take the first, middle and last elements of the array, find their median, then use that. The C++ code below adopts this strategy.

```cpp
void quicksort(int *start, int *end); // Just the prototype
int sort(int *array, const int num_in_array)
// Wrapper for quicksort
  {
  quicksort(array, array + num_in_array - 1);
  return(0);
  }
void quicksort(int *start, int *end)    // The real thing
  {
  int *p, *q;
  int *centre;
  int pivot;
  if (end <= start)
    return;
  if (end == start + 1)
    {
    if (*end < *start)
      swap(end,start);
    return;
    }
// At least three elements; do median of three to get pivot
centre = start + (end-start)/2;
if (*start > *centre)
  swap(start,centre);
if (*start > *end)
  swap(start, end);
if (*centre > *end)
  swap(centre, end);
// Now with *centre as pivot, *start and *end are certainly
// in the correct partitions.
if (end == start + 2)
  return;
// Get pivot value, then put pivot element out of the way
pivot = *centre;
swap(centre,end-1);
p = start;
q = end-1;
// Now ready to partition the rest of the array
while(1)
  {
  while(*++p < pivot);
  while(*--q > pivot);
  if (p >= q)
    break;
  swap(p,q);
  }
```

```
// Finally put pivot element in its place
swap(p,end-1);
// And call recursively to do subarrays
quicksort(start,p-1);
quicksort(p+1,end);
}
```

In the worst case, quicksort is an $O(N^2)$ method. To see this, imagine what happens if the choice of pivot is unlucky: one partition of the array contains zero elements, then there is the pivot, then there are all the other elements in the array. Quicksort then calls itself recursively on the $N-1$ elements that are left. If this happens repeatedly, there could be full N recursive call to quicksort. Each one of these has $O(N)$ operations, so the total running time is $O(N^2)$. At the same time, those N recursive calls to quicksort are using up lots of space on the stack. Quicksort always requires this auxiliary storage to keep track of which subarray is being processed, even if it is converted from this recursive structure into an iterative form.

It is worth noting that if the pivot element was chosen to be the first element in the array, then an already-sorted array would cause the worst case to happen (repeated partitioning into one empty half below the pivot and one full half above it). This is why median-of-three is used to choose the pivot. (It turns out, however, that for the example array we have been using, choosing the first element is better than choosing the median-of-three! To see this, work through the sorting of the test array with quicksort using the median-of-three.)

Although quicksort is $O(N^2)$ in the worst case, its average performance is order $N\log N$. Showing this is quite involved, so will not be done here. Suffice it to say that quicksort is the fastest known sorting algorithm for large N.

Heapsort

Recall the *heap* data structure discussed in Chapter 7. We saw there how a heap could be used to implement a priority queue. Simply adding all the items to be sorted to a priority queue, then removing items until the queue is empty accomplishes sorting. So how efficient would that be?

First, heapsort is space efficient. The complete binary tree in which the data are stored can be represented in an array. The operations of adding to the heap and removing from it work 'in place' – they just involve exchanges of data positions in the array. So heapsort requires no auxiliary memory. Second, to determine complexity, we must consider the steps involved in the elementary operations of adding to and removing from a heap. The algorithms given on page 214 work up or down the binary tree to maintain the heap property. A binary tree with n nodes has $\log_2 n$ levels, so each of these algorithms takes $O(\log n)$ steps. To add n nodes is $O(n\log n)$, and then to remove those n nodes, in order, is also $O(n\log n)$. Heapsort is therefore an order $n\log n$ algorithm – the same as mergesort. On average, it is not as fast as quicksort, but some people believe that its space efficiency and its order $n\log n$ *worst* case performance (as opposed to quicksort's n^2 worst case) make it the best choice for general-purpose sorting.

8.6 Exploiting existing solutions

There is a growing body of standard algorithms in key areas of computer science. Often there are several sources for these. Several library packages are available – you can link these with your code without ever knowing how the algorithms work. Numerical libraries such as those published by the Numerical Algorithms Group (NAG) are typical. But there are times when you need to know how an algorithm works, and might need to include it in your own source code. Remember:

Principle 4

Every program has surprises.

Table 8.1 shows recommended sources for some important classes of algorithms.

Be careful to check the source's ground rules before copying an algorithm. Numerical recipes in C, for example, use unit-offset vectors.

Table 8.1 *Useful sources for algorithms*

Algorithms	Reference	Comments
General: Theory, sorting, searching, string manipulation, graph algorithms	Knuth, D. E. (1968, 1973, 1999). *The Art of Computer Programming*, Vols 1, 2, 3, Addison-Wesley.	The standard, classic reference.
	Sedgewick, R. (1998). *Algorithms in C++: Parts 1–4*, Addison-Wesley.	Sedgewick is a favourite for DS&A courses.
	Weiss, M. (2003). *Data Structures and Algorithm Analysis in C*, Addison-Wesley.	Weiss is also popular.
	Skiena, S. S. (1998). *The Algorithm Design Manual*, Springer-Verlag. Online version at www2.toki.or.id/book/AlgDesignManual/.	Skiena has very good and wide coverage, especially of graph algorithms.
Linear algebra, numerical methods, statistics, random numbers and encryption	Press, W. H. *et al.* (2002). *Numerical Recipes in C++*, 2nd edn, Cambridge University Press.	Very comprehensive.
Graphics	Foley, J. *et al.* (2003). *Computer Graphics. Principles and Practice in C*, Addison-Wesley.	Standard text, recently converted to C.
Data compression	Bell, T., Cleary, J. G. and Witten, I. H. (1990). *Text Compression*, Prentice-Hall.	Algorithm descriptions and pseudocode.
Compilers	Pyster, A. (1998). *Compiler Design and Construction*, Van Nostrand Reihold.	Uses C and pascal.
	Fraser, C. and Hanson, D. (1995). *A Retargetable C Compiler: Design and Implementation*, Benjamin Cummings.	Fraser and Hanson's compiler is a Literate Program (see page 15).
Operating systems	Tanenbaum, A. S. (1992). *Operating System: Design and Implementation*, Prentice-Hall.	Discusses design of a Unix-like system.
Miscellaneous	*Collected Algorithms of the ACM*, Association for Computing Machinery. www.acm.org/calgo	Various languages and pseudocode. Mainly numerical.

That is, it assumes that the first component of a vector, stored in a $1 - D$ array is in **a[1]**, up to the **n**th component in **a[n]**. This is contrary to the zero-offset thinking of C++ arrays which run from 0 to **n-1**.

Remember, too, that you may be required to pay for a licence to use code in products.

Finally, a word of warning. *Always* think about data structures and algorithms together. How do you sort a linked list? Of the five sorting algorithms here, mergesort is perhaps the obvious contender. Mergesort always steps through lists in order, it never needs random access to an arbitrary element, so it can deal with linked lists almost as efficiently as it deals with arrays. This might be the answer, but if there is temporary storage available, the best answer may very possibly be: convert the data structure into a temporary array, sort the array with quicksort and convert back. Always consider a data structure change before a once-off algorithm run.

9 Design methodology

9.1 Introduction

The next six chapters introduce a software design methodology suitable for scientific and engineering programming – one that is concise but that can be applied to problems of different scales.

This chapter is about the overall shape of design methodology. After introducing a comprehensive linear process, we'll sidestep briefly to consider how reliable design techniques are. We will then look at a minimalist methodology specifically for software design; this will form the framework for the following chapters.

9.2 Generic design methodologies

Since the 1920s, design has been seen as a unified art – something that can be understood and taught in the abstract. The subject has many subdisciplines, including graphic design, architectural design, machine design, and software design, but there is a single essence, and it is possible to talk about generic attributes, methods and procedures of design. Antoine de Saint-Exupéry's famous words 'A designer knows he has arrived at perfection not when there is no longer anything to add, but when there is no longer anything to take away' have as much to do with software design as with aircraft design. (If you're already wondering if that statement is strictly true – good! Questioning aphorisms is a major theme of this chapter.)

Successful design has been observed to happen in a variety of ways. Some great designs have been the result of a flash of inspiration. Many more arrived slowly, step by step, converging by theory and experiment on a solution. Is there a single underlying model that captures the process in all cases?

The answer, from advocates of design as a discipline, is yes. Pragmatically, the answer is more or less. Real-life examples can be given as exceptions to any simple model you can think of. On the other hand, a single model that is sufficiently complicated *can* capture the process – if there is no path through the model for a particular example, just add it. Such a model is so general as to be useless. But the useful models

---

The emergence of design as a discipline

The German design centre/school, the Bauhaus (Building House), came to prominence in the 1920s. The vision of its founder Walter Gropius was the 'reconciliation of art and life by abstract artists creating universally relevant forms for industry'. A philosophy of abstract design was taught and practised in an atmosphere of great creativity and intellectual energy. Understanding design in the widest possible sense led to Bauhaus excursions into literature, music and politics. The last, together with the fact that many of the Bauhaus faculty were Jews, led to its closure by the Nazis in the 1930s. Although the design output of the Bauhaus was not ultimately as significant as that of contract industrial designers in the US and elsewhere, its philosophical and educational contribution was profound. The Bauhaus has inspired several successors – the New Bauhaus in Chicago, the Hochschule für Gestaltung at Ulm in Bavaria, and the Systems Design Engineering program at the University of Waterloo, Canada.

Design and art

Design and art have the shared goal and responsibility of making artifacts that are *good*. Neither discipline is satisfied with mere 'existence proofs'; they both seek optimal 'solutions' given the situation at hand. For the artist the criteria of quality may be unquantifiable, perhaps even inexpressible. But they exist, making the art legitimate, rather than meaningless. For the designer the criteria might be measures of accuracy, economy, safety, and robustness, but there may be aesthetic criteria too. In both cases, finding a good solution is a search that calls for experience and insight as well as knowledge and skill. Both disciplines demand clear thinking and planning, the ability to concentrate, painstaking work, determination, and practice.

do not claim to capture the essence of the design process for every case. Rather they provide a structure that, if followed, systematizes the process in a helpful way. They do not guarantee success, but they do increase its chances.

The earliest and simplest model for structuring the design process is linear, typically going from beginning to end of a list such as this:

Problem identification
Problem statement with design criteria
Specification of requirements
Ideation: devising possible solutions
Feasibility analysis
Optimization
Selection of solution
Detailed design
Implementation
Test
Release
Maintain

There are many variants, but this list is fairly comprehensive.

Many authors have suggested the linear model is inadequate, instead preferring a cyclic model, where operations are repeated many times on a more and more detailed level. Cycles may be introduced anywhere in the model, denoting repetition of any subsequence of activities, or of the whole process. A cyclic model can emphasize the hierarchical nature of complex designs, where decomposition into subproblems is one level, and solution of each subproblem follows the same sequence of steps but at a lower level.

Whatever design methodology model we adopt, it only tells us what sort of things happen at each stage, not how to do them. Explaining how to do things is what teachers and writers on design offer. But you need to know that much of the advice is provisional, debatable and sometimes even wrong. To see this, we take a brief look at how the scientific method has tested a not-so-scientific design method.

9.3 Reliable steps to a solution? – a sceptical interlude

The literature on design is full of suggestions about how to think up good solutions ('ideation'). Often you'll be urged to adopt design or problem-solving methods on the basis of inspiring anecdotes and enthusiastic endorsements. But few methods have been carefully evaluated in controlled experiments. When they are, the evidence may not support the claims. To illustrate this, consider one of the best-known methods for thinking up ideas – brainstorming.

Brainstorming

Brainstorming was developed in the 1930s by the advertising executive Alex Osborn, and published in his book *Applied Imagination* in 1953. It is often taught (and used) in the context of a design methodology, as a way of generating many alternatives from which the best can later be picked. Here is what a popular design textbook, *The Universal Traveler*, has to say about brainstorming:

> Over the years, Osborn's technique [brainstorming] has become the definitive basic method for finding ideas. Today it continues, with several variations, to be just as valuable … Observing an experienced Brainstorming team at work is a sure way to sell yourself on the power of this proven method. Even first-time, inexperienced groups generally boggle the minds of viewers by producing an average of *ten ideas per minute*. It certainly beats the normal production rate of struggling for an hour or more to think up a mere two to three alternatives.

The 'rules' of brainstorming may be summarized:

- The quantity of ideas is more important than their quality. Aim to generate as many candidate solutions as possible. It doesn't matter if ideas seem silly or off-the-wall; indeed, crazy ideas are encouraged.
- Suspend judgement. While generating ideas, do not analyse, critique or judge them. If working in a group, this goes for both your own ideas and other people's.
- Try to develop, combine, rethink previous ideas to get new slants. If working in a group, use other people's ideas as stimuli to your own imagination.

Does brainstorming really work? Stop for a moment and question whether this procedure makes sense. Is more better when it comes to generating ideas? Does withholding judgement really enable you to avoid blind spots? Is it better to do brainstorming alone or in a group? What is the experimental evidence?

Let us begin with the question of groups versus individuals first. The reason is that here the experimental evidence is clear cut: more and better ideas are generated by having people work by themselves then pooling the ideas, than by brainstorming as an interacting group. This is a very robust result, confirmed in over 20 studies (Diehl and Stroebe, 1987). There is no scientific basis for the use of groups when brainstorming. There may be a social advantage to having all parties present when ideas are suggested (Bouchard, 1971) but there is no evidence for this. Strangely, brainstorming in groups seems to have a hold on educational consciousness – people are still taught it as if it works better than solitary thinking.

The issue of whether (solitary) brainstorming as a methodology is of value is still undecided. Osborn's book provided mainly anecdotal evidence. The work of Parnes and others in the late 1950s and early 1960s substantiated the belief that under some circumstances better ideas are generated by people trained in brainstorming than the untrained. In a 1964 study, Gerlach *et al.* replicated Parnes' results, then showed that the superior performance of the brainstormers could be matched by untrained subjects simply by rewriting the test instructions so that the criteria used to judge the test were made explicit. Later contradictory studies followed. One that seems of particular value is that of Sappington and Farrar (1984). They showed that subjects who used brainstorming produced better ideas than subjects told to judge each idea as good or bad as soon as they had thought of it. This result seems to support the principle of withholding judgement. At least, being forced constantly to evaluate one's own thinking slows up idea production. Nevertheless, there is surprisingly little evidence that non-disciplined mental exploration of the space of all possible solutions as in brainstorming, actually helps in finding good ideas. And it definitely hinders when there is just one correct solution (Johnson *et al.*, 1968).

In fairness to brainstorming and other ideation methods, it is worth highlighting the importance (and reality) of their common thread. Many, perhaps all, methods aim to offset human rigidity. Pattern recognition, which humans are very good at, can blind a person to what is really happening, or to some possible alternatives. The phenomenon of 'set' is well documented – humans tend to adopt a rigid response to a repetitive situation. In this respect, a method like brainstorming aims to switch off the effect. On the plus side, human pattern recognition allows very powerful analogical thinking – seeing relationships between things that are superficially quite different. A technique like lateral thinking aims to capitalize on this positive mode of human pattern recognition. The value of these techniques is that they ask not 'How should we think?', but 'How do we think?', and then try to channel that behaviour into problem solving.

In summary, it is wise to be wary of overenthusiastic claims for ideation techniques like brainstorming. Usually the experimental evidence is inconclusive, and the anecdotal evidence lacks comparison with alternatives. Where there are firm results, as in the groups versus solitary question, they often contradict the popular wisdom.

Ideation and judgement

The idea that ideation (thinking up possible solutions) and judgement should be kept separate isn't confined to brainstorming. Even in the full design methodology given on page 236, ideation is separated from solution selection by two intermediate steps, encouraging the belief that analysis and criticism should be deferred until after you've released your imagination. Unfortunately, this causes exactly the opposite problem to the technology-driven developer who is determined to make every problem look like a nail so they can use their hammer. Instead, with the ideation/judgement split, the question of hammer size is not allowed to intrude until we've listed the possibilities of knocking in the nail with a wrench, screwdriver or another nail (a particularly stimulating and elegant candidate solution).

In software design it is usually wrong to separate ideation and evaluation. Candidate solutions are developed through analysis and knowledge of what the technology has to offer. Consequently you will not see the ideation/judgement split in this book.

Sceptical design for designers

The dubious and arguable issues in design methodology do not stop there. What should the writer on design do?

(a) Adopt a motivational enthusiasm that captures readers' imagination, or

(b) Adopt a scientific realism that promotes scepticism and perhaps pessimism about the process?

For the most part, this and the coming chapters present techniques that are supported by evidence. Some have been confirmed by experiment. Many rest on the collective experience and wisdom of good programmers. Some are little more than common sense (which is also worth questioning from time to time). You are encouraged to take most of the advice at face value. Habitual scepticism is very time consuming (and can lose you friends), though it is important now and then to challenge and question. The discussion of brainstorming above is not because it is unusually dubious, but to make you aware that things are not always as clear cut as they are sometimes presented.

9.4 Design methodology for software

Discussion on the shape of design methodology has been echoed in the software engineering literature. Some books begin with the 'waterfall' model of the software design process (and many then go on to say how wrong it is). The waterfall model is linear, and consists of most of the steps in the list on page 236, slightly modified so as to apply directly to software. It can include cycles, which signify what happens when problems have to be fixed. Backtracking is represented as something to be avoided, and the larger the cycle the greater the cost.

By contrast to the waterfall model, some authors prefer a spiral model of design, where steps in the methodology are repeated, with refinement at each iteration. Here, backtracking is not just a response to problems, but a legitimate part of the total design process. The spiral model was popularized by Boehm (1985), and has been seen in several variants.

In this book, a linear/all-at-once methodology is presented. It is linear because its five steps fall in a natural sequence, used for the next five chapters of the book. Practical design is non-linear – it has iterations. But trying to codify the place and time of those iterations clutters the presentation of methodology. It is far better to understand the steps of the process, and be mindful of them all from beginning to end, than to try to draw all the links between them. That's why the methodology is all-at-once as well as linear. It's meant to be minimal enough that you can keep all five steps in mind throughout, not so much *re*visiting design phases as continually maintaining them.

Here, then, are the steps of a software design methodology. The coming chapters will introduce these, and show how they apply to real problems.

Understand the problem
Research possible solutions
Modularize
Program
Test

At this stage, it may seem that the model is not all that minimal – it has five steps. But each of these is fundamental. Chapters 9–13 will show how, with very little further decomposition, the methodology guides you through the crucial stages in design.

Each of these steps in the design process is, to some extent, the generation of a solution. Understanding a problem may go a significant way towards solving it. Researching possible solutions provides the insight to make the right choices in the final design. Modularization and programming have good claims to being the 'problem solving' step because they determine the shape of the program. And testing is so fundamentally linked to the problem statement and the implementation that it has a causal role in reaching the final solution.

In fact, the whole process constitutes the solution of the problem. This is not just a truism, but is a necessary consequence of ...

Principle 1

Software design is representation.

Three of the five steps – problem statement, modularization, and programming – are about representing the situation in different ways. The other two – researching solutions and testing – are about getting the representations truly to conform.

9.5 Design team organization

So far as possible, the minimalist methodology given here is as applicable to large programs as to small ones. But, of course, there is a huge difference between a program that's the work of one individual, and a large software system produced by a team. This text takes no stand on the appropriate way to organize a programming team. One model that has proved successful is that of the 'chief programmer team', where members have well-defined roles such as:

- Chief programmer (the 'surgeon' who designs the program, codes it, tests it and documents it).
- Tool builder (creates and maintains the specialized tools needed for development and test of the program).
- Tester (designs test cases, sets up frameworks for tests and runs them).
- Language lawyer (an expert in the language who is adept at answering specialized questions about the low-level implementation).
- Librarian (maintains all the technical records, tracks status of software and documentation, and controls the integrity of the product as it is integrated).

Many teams fall into this pattern almost by default, as the star programmers take control of the project. The chief programmer structure emerges more because of big differences in programming ability, than because of its original rationale, which was to reduce the number of minds working on a project and thereby cut down the chances for miscommunication and poor interfacing.

Where there is near equality in the abilities of the programmers, it might be just as effective to organize the design team by having everyone working on different parts of the code, or to use pair programming (see page 37). The overriding consideration for the project manager is not the particular team structure, but the principle:

Principle 12

Productivity depends on ability, morale and environment.

In a successful team, members can leverage each others' ability to produce more than the sum of the parts. This is the key to team morale.

9.6 Documentation

As a programmer, you will conscientiously keep your own logbook. Your programs will be good stylistically and thus self-documenting. But sometimes, more documentation is required. For a large project, each step of the design methodology has a corresponding document:

Understand the problem	Requirements specification
Research possible solutions	Intermediate design documents: Class and state diagrams
Modularize	Module interface specification
Program	The program
Test	Test plan (and results)

What are the characteristics of real-world software documentation? Typically we find that it is non-existent or late or out of date. Of these, the best is non-existence. As you know, this book advocates,

Rule 4

Write no more than necessary.

For small projects, making the code self-documenting, and properly finishing programs, should make further documentation unnecessary. But for larger projects, you do need more.

Perhaps the best strategy is to insist that documents are helpful to their writers. That is, design ideas, developments and changes can be worked out effectively through changes to documents. The upshot is the need for notations that are easy to work with and support the expression of good design ideas. The Unified Modelling Language (UML) has been expressly developed to support design through documentation with a range of notations. Chapters 11 and 12 include simplified versions of two UML notations to aid in design. The coming chapters will include other examples of 'minimalist' documentation. At each stage the aim is to keep the documentation job small enough and relevant enough that it will be done. Larger projects will probably have more formal processes.

If you do have to produce extensive documentation, it is important that, at the close of the design process, the documents are up to date. Parnas and Clements (1986) present a version of design through documentation (more complicated than that introduced here) where they argue that even if documents are not written until the end of the project, they should be structured as if they had been used for design.

> On large, well-organized projects, the software documentation is often maintained by dedicated writers (dedicated in the sense that that is all they do). Provided there is good communication between developers and writers, this can work very well. Medium-sized projects are more problematic: they need documentation beyond the program and the programmers must write it.

9.7 Chapter end material

Bibliography

Here are some general references about design. They are all interesting; most of them take the '(a) adopt a motivational enthusiasm that captures readers' imagination' approach to the subject (page 239). Read with appropriate care!

Koberg, D. and Bagnall, J. (2003). *The Universal Traveler*, Crisp Publications. (Previous editions were published by William Kaufmann.)

Wycoff, J. (1991). *Mindmapping: Your Personal Guide to Exploring Creativity and Problem-Solving*, Warner Books.

Higgins, J. M. (1994). *101 Creative Problem Solving Techniques*, New Management Publishing Company.

Papanek, V. (1991). *Design for the Real World*, 3rd edn, Thames and Hudson.

Papanek, V. (1983). *Design for Human Scale*, Van Nostrand Reinhold Co.

Evans, B., Powell, J. and Talbot, R. (eds) (1982). *Changing Design*, John Wiley.

Polya, G. (1990). *How to Solve It: A New Aspect of Mathematical Method*, Penguin Science. Reissue of 2nd edn, Princeton University Press, 1971.

The references for the 'Sceptical Interlude' are:

Osborn, A. F. (1963). *Applied Imagination*, Scribner.

Diehl, M. and Stroebe, W. (1987). 'Productivity loss in brainstorming groups: toward the solution of a riddle', *Journal of Personality and Social Psychology*, **53**, pp. 497–509.

Bouchard, T. J. (1971). *Journal of Creative Behaviour*, **5**, pp. 182–189.

Gerlach, V. S., Schutz, Baker, and Mazer. (1964). 'Effects of variations in test directions on originality test response', *Journal of Educational Psychology*, **55**, pp. 79–84. This paper gives many references to earlier work, including Parnes' main results.

Sappington, A. A. and Farrar, W. E. *Journal of Creative Behaviour*, **16**, pp. 68–73.

Johnson, D. M. *et al.* (1968). *Journal of Educational Psychology*, **59**, Supplement.

References to books on UML and other notations are provided in the bibliography for Chapter 11.

Design methodology for software is treated at length in most software engineering texts. See the references in Chapter 2 for examples.

The Chief Programmer Team model is due to Harlan Mills, and is developed in Chapter 3 of *The Mythical Man Month*. The original reference is:

Mills, H. (1971). 'Chief programmer teams, principles, and procedures', IBM Federal Systems Division Report FSC 71–5108, Gaithersburg, MD.

A paper by Parnas and Clements was referred to in the section on Documentation. Here it is, plus another from the same angle:

Parnas, D. L. and Clements, P. C. (1986). 'A rational design process: how and why to fake it', *IEEE Transactions on Software Engineering*, **12**, pp. 251–257.

Hester, S. D., Parnas, D. L. and Utter, D. F. (1981). 'Using documentation as a software design medium', *Bell System Technical Journal*, **60**, pp. 1941–1977.

10 Understanding the problem

10.1 Problems

Principle 6

**Good design relies on understanding the problem
and its context.**

Designers make artifacts – machines, buildings, circuits, or programs – in response to needs. The current state of things has a deficiency, a lack, or an encumbrance, and a better state can be envisaged. The designer is presented with the start state and the goal to be achieved, and given the job of getting from one to another. The gap between the start and the goal states is called the 'problem', and the process of achieving the goal is called 'solving the problem'.

Problems come in many different sizes. Big problems are usually 'solved' by a high-level design that turns them into sets of small problems. But, large or small, a problem needs to be understood. Solving the wrong problem spells defeat.

The first difficulty with problems is that they aren't always fully definable. (The second difficulty with problems is they keep changing – more on that later.) So the first step in understanding a problem is working out how slippery it is. For this you should try to write a problem statement.

Ideally, a problem statement is a specific and searching definition of requirements. It reserves judgement on solutions, so as to map the farthest boundaries of the space of possibilities. It focuses you, the designer, on the success criteria, while allowing the greatest freedom in achieving them. Unfortunately, it isn't always possible to be specific and searching at the initial phases of a project. Trying to write a problem statement anyway codifies what you *do* know, and is there for filling in as soon as the problem becomes more focused.

This chapter deals first with writing a problem statement, then with problem domain research. Research starts while the problem statement is being developed, but it continues throughout the project. The first part of the chapter is therefore about a step in the process – defining the problem, while the second part is about continual learning that only ends when the project does. A final part of the chapter deals with understanding human users – a common requirement of just about all software systems. Users should be thought of as part of the problem domain; neglecting their role in the existing state or the goal state is like forgetting the system's most critical component.

10.2 The problem statement

A problem statement is a written way of capturing the key issues concisely and systematically, making assumptions, limitations and goals explicit.

Here is a four step recipe for writing a problem statement:

1. Describe the present state of things.
2. Describe the new and better state of things.
3. List the constraints upon a solution that will move you from the present state to the new state.
4. List the criteria that will determine when the new state has been successfully reached.

The new state should be described only in terms of what it enables, not how it is done. The object is to avoid even implying a solution, until the problem is well defined.

The constraints are the economic, physical, time and accuracy limitations on possible solutions, and on the design process itself. Constraints are things which must be satisfied in a design if the solution is to work at all. They are non-negotiable.

The criteria are ways of measuring the value and worth of the engineered product. They are the explicit answer to the question 'How will I know I have succeeded?' Success has to be defined in measurable terms at the outset. This is not to say that criteria will not change during the project. But even if they do, it is important to know at each stage what the rules for success are.

Although the problem is stated independently of possible solutions, this does not mean that problem definition is done in a vacuum, devoid of scientific knowledge or insight. It's your responsibility as designer to judge whether the problem is of a scale that can be practically solved. This may mean ensuring that the criteria are reasonable given the current state of the art.

By carefully following these four steps, you can set out the framework for the design, and take the first (big) step towards solutions.

But what if the problem is slippery, that is, hard to define? Sometimes it is near impossible to express a software project in the form of a single problem, or to do so would be trite. For example, what problem is a word processor aimed at? The gap between people's ability to compose text, and their handwriting skill? The gap between their creative goals in writing and the way the process is encumbered with details like spelling and grammar? The gap between unformatted typewriter output and typeset print? None of these is a good, concise problem statement.

If you have a problem that is hard to define, you should still follow the four steps of the problem statement recipe and *try* to be as specific as possible. Summarize what the context is under item 1, provide some *use cases* for item 2 and list any constraints and criteria that you know about under items 3 and 4.

Use cases are examples of what a user should be able to do with your program. A user in this context may be either a human or some other piece of software that uses your code. By merely giving examples of the sort of thing the program should do, you provide information that can be used to steer design.

Whether the problem is well defined or slippery, the problem statement will be a living document, changing as more is understood or as the customer needs change. At each stage, it should be as explicit as possible about the success criteria because they are the main beacons to guide development.

Rule 10

Steer design by clear criteria.

Examples

The following examples demonstrate the value of explicitly stating a problem, whatever the scale of design.

A. In his book *Programming Pearls*, Jon Bentley gives an example of elegant problem solving. Lockheed engineers wanted to transmit engineering drawings between two plants, about 25 miles apart. About 12 drawings were to be sent each day. How was this to be done?

The full problem statement needs more data, but first the information given so far needs fixing. Already, in two sentences, there is a potential limitation on the solution space – one that is unnecessary. The word 'transmit' usually carries connotations of electrical communications (radio or wire), so even though it does have a wider meaning, it is best to avoid it. Let's replace it with 'transfer' in the full problem statement.

1. Drawings are generated at A; they are required at B. About 12 must be transferred each day. Currently there *is* a solution: a car courier service which takes over an hour and costs $100 per day. The drawings are originally generated on a CAD system. There is a microwave data link between the two sites. Both sites have photographic processing facilities.
2. A cheaper and faster reliable transfer method is desired.
3. There is no appropriate plotter or printer available at B. (The date of this example is 1981, so the cost of such a printer was high.) At least 99% of drawings should be transferred correctly.
4. A service providing transfer time less than one hour, at a cost of less than $100 per day, will be judged a success, since it betters current performance. The faster and cheaper the service, the better. The sooner the service starts the better, so as to curtail the $100 daily charges.

Note that by thinking explicitly about constraints, we have considered reliability, and made a judgement about how many errors are tolerable. If the current system is error free (as presumably it is), we might otherwise not have considered how important errors are.

> See if you can go from this problem statement to a solution.

B. Recently, I was asked for advice on the solution of the following problem.

A multinational manufacturer of household products wishes to improve the productivity of product marketing managers who are involved in preparing TV commercials. In particular, these managers are to be given much faster access to videos of earlier commercials of their products, as advertised around the world.

Let's formulate the problem statement:

1. There are libraries of videotapes scattered about the world, in the various national and regional offices of the corporation. To gain access to a particular commercial, the manager submits a requisition, which is processed by the local library, then sent on to remote libraries. The appropriate clips are copied at the remote libraries then sent back to the originator. The turnaround time is typically two weeks. Project managers have personal computers on their desks, with multimedia facilities. Broadband communication is being introduced throughout the corporation: this will allow MPEG-coded video to be distributed to people's desktops. Very often a manager does not know exactly what is required and can only formulate the requisition vaguely. A lot of knowledge resides in the librarians' heads – they are a key resource in finding and suggesting clips.
2. A faster delivery of videos is required. (The assumption is that creation of new material is indeed helped by reviewing previous commercials, and those from other countries and markets.)
3. The amount of time the manager spends in requesting the information should not be increased. The specialized advice of librarians should not be sacrificed.
4. A system which delivers videos within the day will be a success. A system which enhances the value of librarians' expertise will be a success.

Note that when originally posed to me, the key role of librarians was not clear. This information came out by asking questions to get at the constraints and criteria. The statement of success in the criterion 'within the day' was crucial. Had success been on the criterion of immediate access, then an interactive online database system would be essential. (People who work in multimedia systems often believe the interactive online databases *are* essential.) But providing this would possibly be outside the limits of what can be afforded. It would also be an open question whether a database access system could be made sufficiently powerful that the manager's time would not be increased.

See if you can go from this problem statement to a solution. It might be different from mine, but I hope that by having the statement explicated formally, you avoid some of the traps.

C. Teachers often answer questions by questioning the assumptions of the questioner. In a way this is equivalent to formulating a problem statement. Typically what the teacher is doing is asking:

1. What do you know already, and what do you think of it?
2. What piece of wisdom do you believe there to be and why do you think I have it?
3. What is the context?
4. What type of answer will satisfy you?

I have learned to apply this technique in concrete situations. For example, students often ask me a question like, 'How do you do XYZ with a linked list?' It took a long time for me to learn not to try to answer the face-value question. Instead I now handle this with questions like:

1. Why are you using a linked list for that? Why not an array, for example?
2. What will be the benefit of doing XYZ? Why not ZYX?
3. Is this a language issue, an algorithm issue, or what?
4. No seriously, why *are* you using a linked list?

(Incidentally, almost all problems like this are posed as procedural/ algorithmic problems, and almost all turn out to be data structure problems. There are different ways of expressing a problem, and teaching yourself to question your perspective is good policy.)

Some solutions

A. The solution adopted by the Lockheed engineers was to photograph the drawings at site A, send the film by carrier pigeon to B, where it was enlarged and printed. Only two pigeons were lost in hundreds of flights, the average time was 45 minutes and the cost was a few dollars per day.

B. I suggested replicating the entire video database at all sites and enhancing the librarian's communication facilities. Assuming that there are 50 operating regions, that two unique commercials per month per region are produced, that a 30 second commercial has 3 or 4 minutes of outtakes preserved with it, and that TV commercials have been produced at this rate for 40 years, the total database is less than 4000 hours in length, easily stored compactly on Super 8 cassettes. Librarians would still be expected to be most familiar with material produced in their region, and would be called to advise about what to extract from the local library. Librarians would be given opportunities to get to know each other, and would collectively develop protocols for exchanging information quickly by phone. Communication between librarians would always be fully available (no blocking).

10.3 Researching the problem domain

During and after problem definition a designer spends time researching the problem domain. The aim is to know the problem context well – to become a (temporary) expert in the application area. A slippery problem or incomplete information doesn't relieve you from that task. Your knowledge of the problem context will enable you to tailor your program to its domain.

Every problem domain has a literature that you can use for background research. Part will be in books, periodicals and other kinds of hard copy, part will be online. Aim to be proficient in your use of both: you'll need hard copy and online resources for researching possible solutions as well.

Library research

The main resource that an academic or technical library has for the problem solver is periodicals. Much that you need may be in books, but if you can find the *right* journal or magazine reference, it will probably be more helpful. First, the journal article will deal specifically with a problem like yours; second, it will be relatively concise; third, it will probably yield a good set of references to related and relevant work.

How do you find that correct article?

It usually takes at least three steps to get to the right article. First, find some fairly recent article somewhere that talks about the problem of interest. The only criterion for this publication is that it should have a good length citation list (or list of references). Next, use that list to start looking for a 'pivot' reference. This is usually a paper from the early days of a young field, or, if you are working in an old field, about ten years old. Such a publication is like a fulcrum on which rest two other sets of references. Those that the publication cites obviously predated it and give not just the development of the subject, but probably more detail on some things the pivot reference left out. More recent works which cite the pivot form the other set of references. Through the invaluable Science Citation Index (and other citation indices), you can follow the influence of a particular paper forward from its publication date. If you do not know how to work the Science Citation Index (nowadays on CD-ROM or through an online service), you should learn the skill immediately. Almost certainly, if an author developed a significant technique, or perhaps the first solution to a significant problem, the forward references that the Science Citation Index gives will lead you to important improvements or additional relevant evidence. The process is illustrated in the Figure 10.1.

You will rarely be lucky enough to get from the start reference to the pivot to the one you want in just two jumps. But as you practise using the library you will gain skill in judging paper titles and journal titles so that the number of false trails you follow is minimized.

What if you do not have any literature reference to start from? My advice is to find a periodical in the general area of the problem, and browse through the last five years' publications. If there is anything to be found, you have a good chance of hitting an article of direct relevance or a reference to work in the area which will send you towards a more

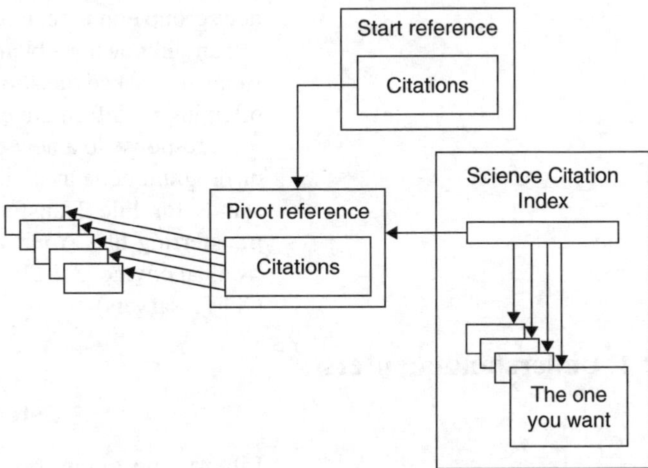

Figure 10.1 *Backward and forward references from a Citation Index*

appropriate source. In any event, this browsing will help you become familiar with the problem domain.

Getting information from the network

Almost all programmers have direct access to the resources of the Internet. Many problems can be researched without leaving your computer, simply by careful and informed use of network facilities. However, some problems can't, which is why libraries came first in this chapter. In problem solving the main three resources are websites, anonymous FTP sites (which are usually accessed directly through a web browser) and network news.

It isn't necessary here to discuss how to surf the web or do a Google search. Web directories and search engines *are* the number one tools for online information, but I will assume you are very familiar with them. The only thing to emphasize is that, in addition to general-purpose directories like Yahoo and search tools like Google, the scientific and engineering programmer has a number of other valuable web resources. For code, there are sites like the Open Source developers network, osdn.com. For algorithms, the SIAM and ACM sites are excellent. For insights into technical problems and solving them, there is Citeseer – which effectively automates the library search process described above, but with limited scope. Assuming that you are familiar with these tools, we can turn to a resource you may not know as well.

The huge number of newsgroups carried on the Internet includes many dealing directly with both the technology and the typical problem domains of software designers. Examples like comp.graphics, sci.math.num-analysis, sci.engr.mech illustrate this. (There are of course many recreational newsgroups too.) Because readers and posters to most newsgroups include some who are reasonably expert, there is a good chance of getting an answer to specific questions. Asking vague questions ('Can anyone tell me about fractals?') is not worthwhile; although asking for references often is ('Can anyone tell me where to find out about fractals?'). There are several rules of newsgroup etiquette, but the most important one in terms of researching an unknown problem domain is read the FAQ. Newsgroups have frequently asked questions lists just as many websites do and they are often highly informative. See the collections at www.faqs.org.

A response to a newsgroup question often refers to documentation or program code available at a website or anonymous FTP site. FTP stands for File Transfer Protocol, and is an old-fashioned way of transferring files. Your browser will be able to browse and fetch files from anonymous FTP servers as easily as it does from websites (HTTP servers).

10.4 Understanding users

Rule 17

Determine what users need.

Library and online resources are valuable, but working with users who understand the problem domain is invaluable. So far as you are able, involve the people who will use your program at every stage of the design, *especially* in tapping their expertise to understand the problem.

Rule 18

Understand how users will understand the program.

The user interface now comprises the majority of code in a majority of personal computer applications. Understanding users and usability is therefore a prerequisite for software design.

Understanding users is such a large subject that it is impossible in a book of this size to give adequate coverage. The references at the end of the chapter point towards good treatments.

What we really seek in defining a problem is measurable criteria. Unfortunately criteria related to the interface are usually only measurable by subjective tests which can be long and expensive. When they *can* be afforded, the design process iterates as alternative techniques are tried and tested. (Part of the research effort in user interface design is in developing models that can give quantitative predictions of interface performance. Such models might avoid some of the subjective testing iterations – but they are still tentative and difficult to apply.) Because quantitative analysis is so hard to do, it is worth searching for and applying qualitative guidelines.

Guidelines should be well defined. For example, the advice 'Make the interface simple' is so ambiguous as to be useless. Which is more simple: a controller with two buttons that are heavily overloaded with functions, or a controller tightly packed with single-purpose buttons? It depends on your definition of 'simple'. Unfortunately too many interfaces have been justified as being 'simple' according to a private definition of the designer. A similar complaint can be made about 'Make the interface logical'. The logical structure of the interface may not be visible to the user, who wants structure but not abstraction. There are few excuses for the designer who says 'It's simple and it's logical'. After all, people are neither.

A much better guideline is 'Make the interface consistent'. Although still vague, it has the virtue that it follows directly from observations of human performance. The source for this and most other valuable guidelines is cognitive psychology, and Table 10.1 summarizes results from that discipline that can be directly applied to interface design.

Shneiderman (see the references at the end of the chapter) distils results such as those in Table 10.1 into 'Eight Golden Rules of Dialog Design' that can be applied to user interfaces in general. These are:

1. *Strive for consistency*. Use consistent terminology in menus, prompts, error messages, etc., use consistent screen layouts; similar actions should do similar things in similar situations.
2. *Enable frequent users to use short cuts*. For example, provide command key equivalents to menu selections.
3. *Offer informative feedback*. Give feedback for every action, appropriate to its importance – modest feedback for minor actions; substantial feedback for major actions.
4. *Design dialogs to yield closure*. Every sequence of actions should have a beginning, middle and end; it should be clear when a particular group of actions is complete.
5. *Offer simple error handling*. Try to prevent serious errors; make error messages clear and concise and error correction easy.
6. *Permit easy reversal of actions*. If actions are easily reversed, the consequences of error are not bad and the user has confidence.

Table 10.1 *Psychological phenomena and their application to interface design*

Phenomenon	Application to interface design	Examples
Rigidity: tendency to adopt single approach to similar problems even if it is inefficient (or incorrect) for some.	When there are several ways to do something, help the user towards the most appropriate. Flag exceptional situations.	(i) Global replace in a text editor is done by many users by repeated find-change steps, because the more powerful command is hidden. (ii) Responding to a particular command with 'Are you sure?' will usually evoke a conditioned response ('y') and not help in preventing errors.
Reasoning is often done according to the similarity of the situation to remembered instances. Humans are poor at abstract reasoning.	Use similar sequences of commands where similar operations are to be done. Do not rely on the user doing chains of logical deduction. They are more likely to try something that worked in a similar situation before, without thinking through the differences.	Consistent use of menus and keyboard equivalents between different applications.
Bias to confirmation: humans tend to seek and interpret evidence to confirm hypotheses they already have. Humans are very good at transferring skill between analogical situations.	Make the effects of an operation obvious – if they are misattributed, a chain of flawed cause-and-effect reasoning will follow. Use metaphor to get leverage on user's existing skill.	Desktop metaphor used, e.g. on the Macintosh. Note that deviations from the metaphor, like dragging a disk icon to the trash to eject it, are not easily accepted.
Humans identify analogical situations by their surface similarities. Humans expect any apparent structure to be real. (Any observed pattern is interpreted as structure.)	A metaphor should not only have parallels in structure, but should 'look like' the source context. Do not suggest pattern in the interface where there is none. When there is a change in the structure of the interface, i.e. a change to a different mode, signal it.	Use of pictorial icons within a desktop metaphor. Give feedback about which mode the system is in.
Humans are very sensitive to temporal exceptions. Humans have multiple input channels.	Make response time fast and provide a special flag to show when something is taking longer than normal. Use visual, aural, and kinesthetic channels to communicate to user.	The various busy icons (hourglass, watch, busy bee, etc.) reassure the user that things are still working. Visual: default channel – use spatial organization. Aural: error beeps. Kinesthetic: Shift key held down (muscular tension feedback) for as long as upper case characters are to be typed.
Human working memory small – about seven chunks of information. Memory is episodic – we remember the gist, not the specifics. (Meaning is necessary for memory.)	Provide as much visual context as possible (user cannot remember it). Precise command names, filenames, etc. are hard to remember. The user needs cues so that gist can be converted back to a specific name.	Windows allow user-configurable visual context. Menus and form-filling interfaces give good cues.

7. *Support internal locus of control.* (This is Shneiderman's term, you may prefer to think of user-centred control.) Make users the initiators of actions rather than the responders.
8. *Reduce short-term memory load.* Provide cues to the user about what their options are. Menus and form-filling interfaces do this. Windows allow the user to maintain their own visual context.

10.5 Documenting a specification

If you are part of a medium-sized or large project, a simple problem statement won't be enough. If you're delivering a mission-critical component to an outside customer, you will be faced instead with a formal specification. This section outlines what can be done with a specification to make it useful as a design document.

The specification tells the designers what it is they are doing, what problem is to be solved, what constraints apply, what criteria will be used for evaluating the product. It is their repository of answers about users' needs, behaviour, etc., about the technology, about standards to follow. Although this sounds like the problem statement, it is more complete and more exact – it represents the top-level solution to the problem.

The specification tells the sponsor/customer what it is the designers think they are doing. It is their check on the development process, and after review, their commitment to accept what the designers provide, if that product meets the specification.

The specification is the project manager's source text for estimating resource requirements at the beginning of the development, and for tracking progress throughout. Also from the start, the specification is the tester's map for developing test cases. Indeed the ability to test for certain things may determine whether they are included in the specification at all.

The specification contains everything the designers need to know to write software acceptable to the sponsor, and no more. If the product satisfies every statement, it should be acceptable. This is why the tester is involved even at the stage of specification writing – to ensure that the specification does not require anything that cannot be tested.

Because the requirements specification has this formal structure, it is best seen as a reference document, not as an introduction. In small projects, it's usually best to avoid the temptation to write an introductory document during development – the reference documents, plus oral tutorial help, are usually the best means for getting new project participants up and running.

The requirements specification usually contains all of the following:

1. Description of formalisms/mathematical models used in the specification.
2. Description of the hardware/software platform.
3. Description of the input/output interfaces of the whole system.
4. Timing, space and accuracy constraints and criteria.
5. Description of how undesired events will be handled.
6. Identification of high-level objects in the system.
7. Specification of event/state/action behaviour for each object.

In order to use the specification as a medium for design, two special sections should be included: Open and Closed Extensions. The Open Extensions section is initially drafted at the beginning of the project.

To understand its role, think first of it as a 'Freedoms' section – a statement of all the things that are possible in the design, but the customer does not care about. That is, while the Requirements say what must be done, the 'Freedoms' say where the designers can make choices without violating the specification. Of course, if there really were a 'Freedoms' section, it could be cluttered up with all sorts of marginally relevant material. That is why we have an Open Extensions section instead. In it go those things that, with today's understanding of the project, are not explicitly required or cannot be explicitly stated, but have some risk of being required in the future.

Another way of looking at the Open Extensions section is to say that it describes where the Requirements are incomplete or likely to change.

The other special section of the specification is the Closed Extensions. This is written in the same format as the Requirements, but what it contains are designed-in behaviours. That is, as design decisions are made, and freedoms exploited, the product takes a particular form that the Requirements have not insisted upon. This form is recorded in the Closed Extensions section, by writing in the specification that would have been fulfilled by the particular implementation.

10.6 Chapter end material

Bibliography

For problem statement formulation, see the general books on design cited in Chapter 9.

The Lockheed example is adapted from Bentley, J. L. (1999). *Programming Pearls*, Addison-Wesley.

Guidance on harnessing the resources of a library is available in many study methods books. But the best strategy is to go to an information session hosted by the library, and use it.

Users and user interfaces

The following books give good coverage of some of the important topics in user interface design. The first three are from a traditional human–machine interaction background, whereas the remainder are focused on user interfaces for computer applications.

Bailey, R. W. (1989). *Human Performance Engineering*, Prentice-Hall. This is also good on design methodology in general.

Wickens, C. D. and Hollands, J. S. (2000). *Engineering Psychology and Human Performance*, Longman.

Boff, K. R. and Lincoln, J. E. (eds) (1988). *Engineering Data Compendium: Human Perception and Performance*, Imprint of Wright-Patterson A.F.B., Ohio: Harry G. Armstrong Medical Research Laboratory, 1988.

Norman, D. A. (2002). *The Design of Everyday Things*, Basic Books. (Originally appeared as *The Psychology of Everyday Things*, Basic Books, 1988.) Readable introduction to human-centred design.

Shneiderman, B. and Plaisant, C. (2003). *Designing the User Interface*, 4th edn, Addison-Wesley.

Laurel, B. (ed.) (1990). *The Art of Human–Computer Interface Design*, Addison-Wesley.

11 Researching possible solutions

11.1 Introduction

In this chapter we discuss the route from understanding a problem to having a clear idea of how to solve it. This is an exciting phase in software design, where you explore alternative approaches and work out the overall shape of the solution. As usual, it overlaps and interleaves with other phases.

Part of researching possible solutions is researching existing solutions, and the methods discussed in the previous chapter for finding out about the problem domain apply equally to finding prior art. Here we concentrate on the three tools of analysis, experiment and simulation, then go on to discuss notations. A notation is a way of writing down a model of the design, which might be used in documentation. But notations are also important for developing ideas, whether by discussion around a whiteboard, or filling in a growing sketch in your logbook. This chapter concentrates on one kind of notation that is valuable in a wide range of circumstances: state diagrams. We look at simple state machines first then develop in two directions: first, towards adding more expressiveness through statecharts, and second, towards state sketches for procedural programming which help to clarify algorithms.

11.2 Basic analysis

Analysis starts with the quantitative things you know about the problem domain – the sizes of the entities involved, the response times required, the money available. If you are going to process data, you will need to know immediately how much, how quickly, and in what sized chunks. Sometimes you will have to make guesses, and then multiply your estimate by some scale factor to reflect your uncertainty. The scaling provides some leeway for the specification to change incrementally, but more importantly, it reflects how much you do not know yet about the problem and its possible solution. In most branches of engineering scale factors of about 10 are common, but this is usually overkill for software, where estimates can usually be made fairly accurately.

If you have written a problem statement and researched the background of the problem domain, possible solutions will already have suggested themselves. They may not be very good solutions just yet, but they will be old solutions or modifications of solutions used in other contexts. If the problem has not been solved before, you can generate candidate solutions for yourself simply by saying 'Suppose everybody thinks the problem is insoluble; what counter-example would prove them wrong?'

Next apply your knowledge of the technology – possible platforms, algorithm analysis, past implementations. For whatever ideas you have at the outset, start assessing their requirements and performance. In doing this, you should aim to modify the original ideas, and look for ways of using analogical approaches, perhaps with different technologies. It is also worth taking a step back from the problem and applying Polya's Inventor's Paradox: 'The more general problem may be easier to solve'. This is very often true for software. A good example at the programming level is where a module is implemented to read in constant values and text strings from an initialization file, rather than those values being hardcoded. This will mean structuring the storage for those constants in a consistent way and probably making the rest of the program easier to write and understand. The result is of course also much easier to debug and maintain.

Basic analysis quickly achieves two things: you are able to reject some possible solutions because they do not meet the constraints, and you can identify parts of the problem domain that are not well enough understood to decide between some of the candidate solutions. Then you must turn to experiment and prototyping.

11.3 Experiment

Rule 7

Test to discover information.

Experimentation is devising tests for the problem domain, or for the users, based on your candidate solutions. It may involve writing code, but need not. The purpose is to get the information that will enable you to choose between candidate solutions and to optimize.

The scientific method proceeds by developing a hypothesis, devising experiments that have the potential to disprove it, conducting the experiment to gather data, analysing the data and drawing appropriate conclusions. The designer can follow just the same strategy, with one additional level of 'indirection'. Instead of hypotheses there are candidate solutions. The first question is to decide what direct implications each of those solutions might have if used to solve the problem. Then return to the problem domain with a simulation of the implications (the 'hypothesis') and do tests to see what the payoffs are.

Example

Consider the problem of telecommunication for deaf people – how to provide for the deaf an analogue of the telephone. There are several potential solutions to the problem: text-based communication over computers, fax, video. But the problem constraints point to difficulties with each of these: text (in English) is not the native language of many deaf people (Sign is), fax does not give real-time interaction, video is much too high bandwidth to send over telephone channels. One solution that might work is transmission of highly compressed video – where only the most relevant information is sent.

Such was the nature of the deaf communication problem at the end of the 1970s. Although no one knew at that time what algorithms were appropriate to extract and code the relevant information, or how such a service could ever be provided affordably, it was possible to go back to the problem domain and test candidate solutions. George Sperling and associates conducted experiments with deaf people wearing black with white spots on the ends of fingers, the nose, and other linguistically important features. The subjects signed in appropriate illumination conditions in front of a video camera and were taped. The result was a set of videotapes of moving white spots on a black background. Another set of deaf subjects viewed the tapes and was able to understand what was being said. Sperling showed that sign language communication was possible with very little visual fidelity provided the important moving points were transmitted.

Sperling's experiment is a beautiful example of probing a problem domain with a candidate solution in mind, before the detailed work on that solution was begun.

11.4 Prototyping and simulation

A prototype that provides a working solution to part of the problem can be valuable in choosing a solution. When doing this the developer has to be prepared for two eventualities. The first is that the prototype will effectively address some of the crucial issues in the design, and so become part of the final system. It must therefore be robust and maintainable. The second is that the prototype may have to be thrown away completely. It is important to have the freedom to throw away prototype ideas as soon as they have fulfilled their role.

Solution selection often has much to do with the user interface. Prototyping a proposed user interface, even without functionality, can help clarify the problem and point towards a solution. Direct manipulation creation of Graphical User Interfaces (GUIs) is now supported in many integrated development environments, so allows you to put together an interface very quickly.

More generally, simulation can be used to mock up any aspect of design and see what happens. It pre-releases ideas for test, provided the differences between the virtual context (the simulation) and reality are monitored.

Simulation can be used to converge on an optimal solution. This is design-by-successive-approximation, and there is nothing wrong with it for detecting problems and making incremental improvements. There is some temptation though to let simulation supplant design – to rely on the loop between you and the simulation to grow the design from nothing. This is sure to lead to frustration and bad designs. Simulation accelerates solution optimization. The space of all possible solutions probably has many local optima, but you will not find the global one unless you get close to it by analysis before starting simulation.

11.5 Notations and languages for developing designs

Principle 1

Software design is representation.

Once we have a promising candidate solution, we want to develop it further, to bring it closer to implementation, to understand what it entails. A simple solution can be coded immediately, but for more complicated cases we need intermediate representations. We need an expressive notation that helps us work out the states, actions, objects and types that best describe the solution.

Among the many formal and semi-formal methods for diagramming software systems, the most prominent today is UML – Unified Modelling Language – which provides nine different kinds of diagram within a formal object-oriented framework. In this book we will consider just three different types of diagram, and all are simpler than the nearest equivalents in UML. Moreover, the orientation will be towards using these as sketchable design media – ways to understand what's going on through re-representation – rather than formal documentation. In this chapter we discuss data flow diagrams briefly, then concentrate on representing state-based systems. In the next chapter we will revisit class diagrams.

The references list at the end of the chapter includes several good books on UML, its predecessors and alternatives.

11.6 Dataflow diagrams

A generation ago the main role of computers was centralized data processing. The computer was in control, and once the system started on a particular job it was left to do its work on passive information: payrolls, accounts, inventories. Data processing still accounts for a significant proportion of processor cycles and lines of code, and similar kinds of transformational processing are important in science and engineering. The transformations may be complex, but they can be adequately described in terms of relations between outputs and inputs. Techniques such as the Jackson System of Design (JSD), Structured Analysis (DeMarco, 1978) and Logical Construction of Programs (LCP) were developed for specifying and describing transformational systems.

Perhaps the simplest kind of transformational language is the dataflow diagram. Dataflow diagrams consist of ovals representing processes, rectangles for sources and sinks of data, parallel lines above and below labels for databases and arcs representing the flow of data. They omit procedural details, emphasizing the central role of data and its movement through the system. They can also represent hierarchies of detail: a process can be expanded into subprocesses, and the flow of data between these can be shown in the same way. Figure 11.1 shows a dataflow diagram for a flight reservations system operated by a travel agent.

The transformational approach to program specification is inadequate for event-driven real-time systems. More and more programs now fit into the category of event-driven rather than transformational. Certainly, the personal computer applications with which most people are familiar are event driven – an interactive user interface means that the program must respond on demand to unpredictable input. In this chapter, therefore, the emphasis is on specification/analysis techniques for event-driven systems. These subsume the transformational category, although they are perhaps not as appropriate for large transformational systems (e.g. large payroll programs) as the more traditional techniques.

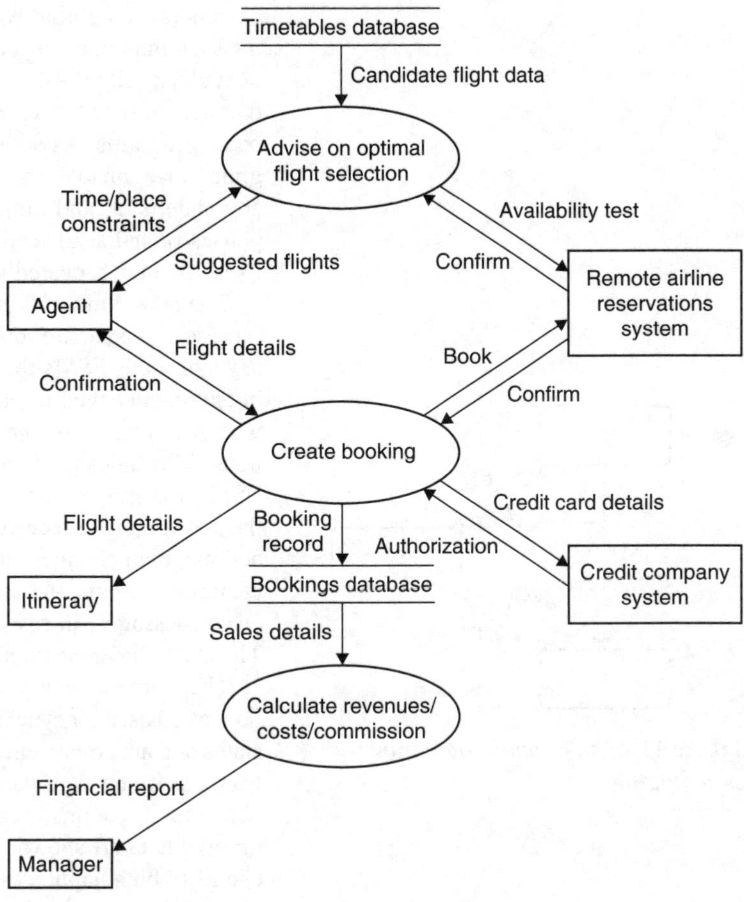

Figure 11.1 *A dataflow diagram for flight reservations*

11.7 Specifying event-driven systems

Rule 15

Design your program around these six sufficient concepts: states, events, conditions, actions, objects and types.

It is important to understand what event-driven systems are and why they are important. Programs that handle unscheduled events – for example, mouse and keyboard input to a user interface – are 'event driven'. The program doesn't dictate the flow of control, rather it responds with actions according to an unpredictable sequence of inputs or events. Such systems are becoming increasingly common, not just in the context of user interfaces. The natural, procedural, way of programming step by step has led to the unfortunate idea of an 'event loop'. Here, a repeated sequence of instructions polls for input, then executes a branch instruction depending on what the user has done. But this is not a natural description of what's required. It is better to think in terms of finite state machines.

11.8 Table-driven finite state machines

A Finite State Machine (FSM) is a system that is always in one of a finite number of states. There are two types. First, the 'synchronous' FSM undergoes a state transition at every 'clock' event. Digital circuit designers are familiar with this kind of FSM. During any clock period the system is in a particular state; at the end of the period – i.e. at the next clock 'edge' – the system changes state according to its inputs at that instant. The 'synchronous' FSM also underlies conventional computer programs. As soon as a statement has been executed, the program moves on to the next. Inputs can be interrogated (or polled) during any statement, and outputs can be changed. But the program never pauses or halts. Instead, to wait for something, the program busies itself looping, repeatedly looking at the appropriate input or variable.

The other kind of finite state machine, and the one of interest to us, represents asynchronous behaviour. Digital circuit designers use asynchronous FSMs too, and in principle theirs are the same as ours, but in practice their implementations of FSMs truly don't have a clock, whereas ours do. At the level of representation though, an asynchronous FSM hides the looping and polling of waiting: an asynchronous FSM stays in the state it is in until some external event occurs. When the state machine receives notification of the event, it executes certain actions, then changes state. So, as opposed to monitoring inputs and changing outputs, the asynchronous FSM sits dormant until it is sent *event* messages, in response to which it carries out *action* methods. The state change is internal to the FSM, hidden from the outside.

There are many ways of representing an FSM. A simple representation, often used for synchronous FSMs, is a graph where nodes represent states and arcs represent events. This can be adapted for asynchronous FSMs as in the example in Figure 11.2.

A graph like this is called a state transition diagram (or a state diagram). States are shown as rectangles with rounded corners (S1, S2, S3), events (which happen asynchronously) label the arcs (e1, e2, e3, e4). If a particular transition depends not just on the start state and the event, but also on a precondition, that condition is put in parentheses next to the event (p1). If actions are associated with a particular state transition, then they are written after the event label, separated from it by a slash '/' (a1). Finally, the start state of the system is indicated by the black spot arrow. The state transition diagram in Figure 11.2 therefore specifies a system where:

- the start state is S1
- if in state S1 and event e1 occurs, the system transfers to state S2
- if in state S1, event e4 occurs, and p1 is true at the time, the system transfers to state S3
- if in state S2 and event e2 occurs, the system transfers to state S1
- if in state S2 and event e3 occurs, the system does a1 and transfers to state S3
- if in state S3 and event e1 occurs, the system transfers to state S2.

Figure 11.3 shows a more concrete example which specifies how Macintosh transient menus work. It should be read anticlockwise from the top left. (A state transition diagram for Microsoft Windows menus would be similar, but have an extra 'clicked-open' state.) States and actions are written in upper case and events in lower case.

There are several other diagrammatic methods for specifying FSMs. Some are more powerful than the simple graph representation, for

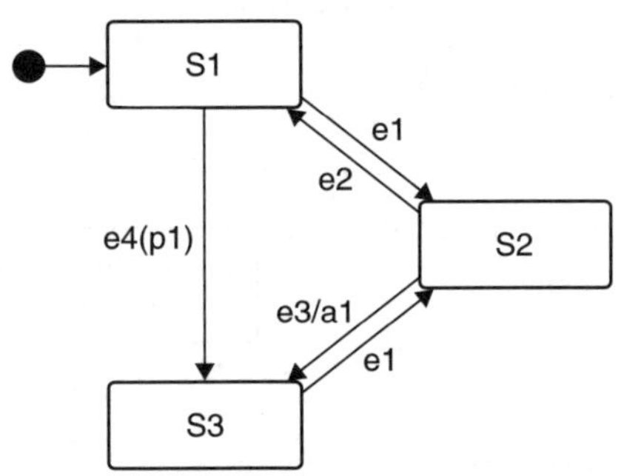

Figure 11.2 *An example state transition diagram*

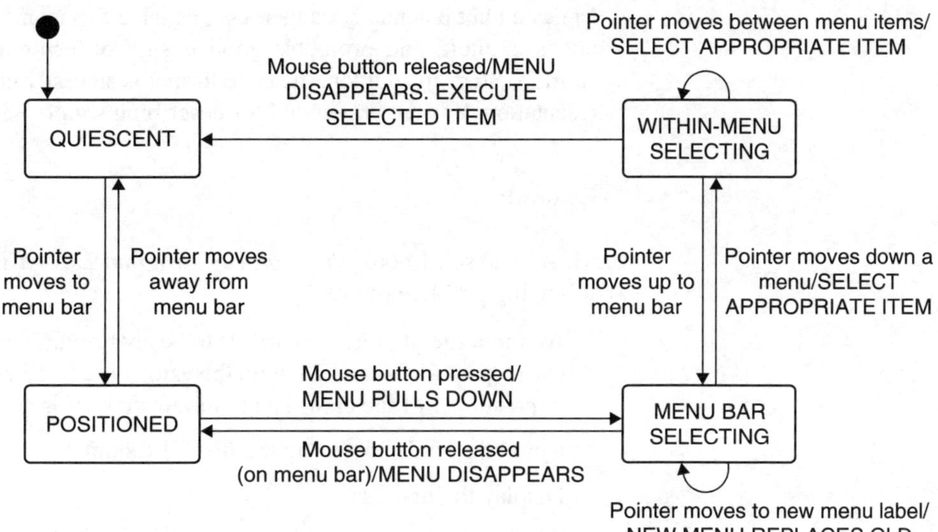

Figure 11.3 *State transition diagram for a Macintosh menu*

Table 11.1 *Structure of a finite state machine table*

Current state	Event				
	1	2	3	4	…
A					
B					
C					
…					

example the CCITT Specification Description Language (SDL). We will introduce another – statecharts – shortly. But all are concerned with representing event-driven systems. FSMs can also be described textually, for example with a table. Table 11.1 shows the structure.

Each entry in the table gives (i) the actions to be taken and (ii) the next state to go to. So when an event occurs, it is cross-referenced with the current state to get to the relevant actions and the next state.

A table is by no means the most compact way of representing an FSM. For any state, it is probable that most possible events are just ignored – i.e. they result in no actions and next state is the same as current state. But it translates very easily into a data structure and code (see page 289). By representing the state vs event combinations in a table, you think about what to do in every possible combination. Although many combinations are not expected to happen, some demand that an error flag be raised, because they are unusual – not irrelevant but potentially dangerous. The table format makes these situations explicit, and promotes good design on error handling right from the start. In addition, the table format is an easily understood representation. It is recommended for describing simple FSMs.

Example

M. A. Jackson's book *Principles of Program Design* introduces the following problem (p. 49).

> An input file of punched cards is to be analyzed. There are three card types, T1, T2 and T3, with the values 1, 2 and 3 respectively in position 1 of the card. The required analysis is as follows:
>
> Count the cards preceding the first T1 (count A)
>
> Display the first T1
>
> Display the last card, which is always the first T2 following the first T1
>
> Count the batches following the first T1, where a batch is either an uninterrupted succession of one or more T1 cards or an uninterrupted succession of one or more T3 cards (count B)
>
> Count the T1 cards after the first T1 card (count C)
>
> Count the batches following the first T1 card which consist of T3 cards (count D)
>
> All counts are to be displayed following the display of the last card. The file is known to be in correct format; that is, there is at least one T1 card, the last card is a T2, and no T2 intervenes between the first T1 and the last card.

Jackson gives this problem as the first example of serial input processing. It clearly dates his book, being concerned with card reading, but the idea of structured records being read from a file is still important – only the media have changed.

As stated, Jackson's problem is hard to understand, so we work through example. Here is an illustrative sequence of cards, showing the interpretation the program is to make of them:

T3	
T3	
T2	A = 4
T2	
T1	Start counting batches
T3	
T3	Batch B = 1, C = 0, D = 1
T3	
T1	Batch B = 2, C = 1, D = 1
T3	Batch B = 3, C = 1, D = 2
T1	
T1	Batch B = 4, C = 4, D = 2
T1	
T2	Count ends

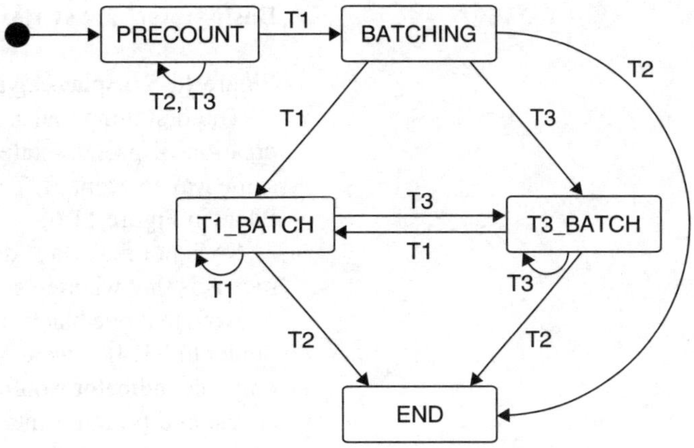

Figure 11.4 *State transition diagram for Jackson's problem*

Table 11.2 *FSM table for Jackson's problem*

	T1	**T2**	**T3**
PRECOUNT	*Display T1*	*Increment A*	*Increment A*
	BATCHING	PRECOUNT	PRECOUNT
BATCHING	*Increment B and C*	*Display T2, A, B, C, D*	*Increment B and D*
	T1_BATCH	END	T3_BATCH
T1_BATCH	*Increment C*	*Display T2, A, B, C, D*	*Increment B and D*
	T1_BATCH	END	T3_BATCH
T3_BATCH	*Increment B and C*	*Display T2, A, B, C, D*	
	T1_BATCH	END	T3_BATCH
END	*Signal error*	*Signal error*	*Signal error*
	END	END	END

Jackson solves this problem in nine pages using data structure diagrams – i.e. in a transformational way. Although the heart of his book, that data should structure programs, is also the heart of much of the advice here, it seems incongruous to treat this problem as a data-structuring problem. Surely what is needed is a finite state machine.

Figure 11.4 shows this problem expressed as a finite state machine. This converts easily into Table 11.2.

And this table gives the complete specification. In Chapter 13, we shall see how this is turned into code.

11.9 Statecharts

The main problem with state transition diagrams is that their complexity grows as the product of the number of states and the number of events. Because they are flat and non-hierarchical, any new behaviour in part of the system means extra states with an associated bundle of event transitions. Statecharts bring structure and economy to state machine representation. In short, they provide a hierarchical structure; they allow independent parts of the state machine to be appropriately decoupled and they provide a state machine-wide mechanism for communication.

Basic statechart syntax

Figure 11.5 displays again an earlier state transition diagram.

The first thing that statecharts do is to abstract commonality by the creation of pseudo-states. In this example, S1 and S3 behave in the same way to event e1. This can be abstracted by creating a 'superstate' S4 as in Figure 11.6.

S4 is not really a state, but for convenience we call it one. What it means is that when in state S4, the system is really in either S1 or S3.

Note that one black spot arrow points to S4, then within S4 another points to S1. If it were simply a case of indicating the system's state, only one indicator would be necessary, presumably starting outside all states, and pointing into S1. But the black spot arrow symbol means 'default state'. So the default state of the system is S4, within which the default state is S1. Arrows can point at superstates, with the meaning 'if this transition is taken, go into the default substate'. In Figure 11.7, if e4 happens in S5, then the system goes to S1 (because the explicit transition is to S4, within which S1 is the default).

The H-in-a-circle symbol in Figure 11.7 means 'go into the most recently visited substate of this state' or 'enter by history'. In this

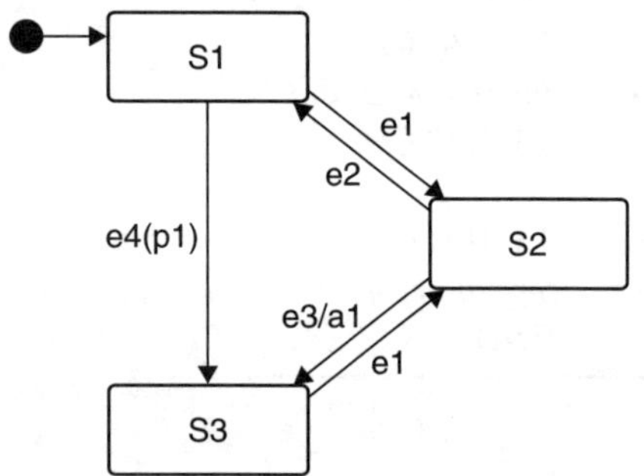

Figure 11.5 *Example state transition diagram*

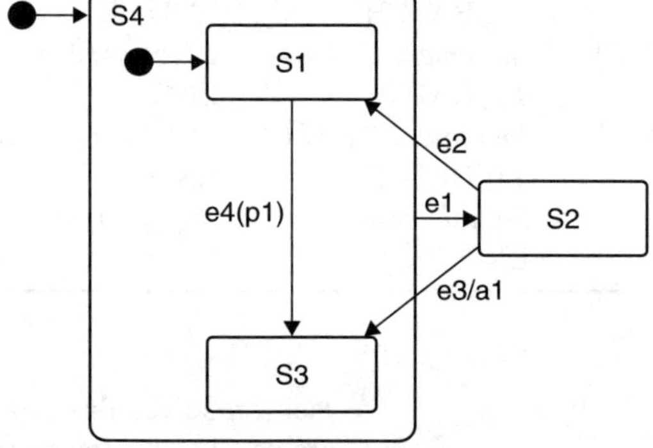

Figure 11.6 *Statechart with the same meaning as the state transition diagram in Figure 11.5*

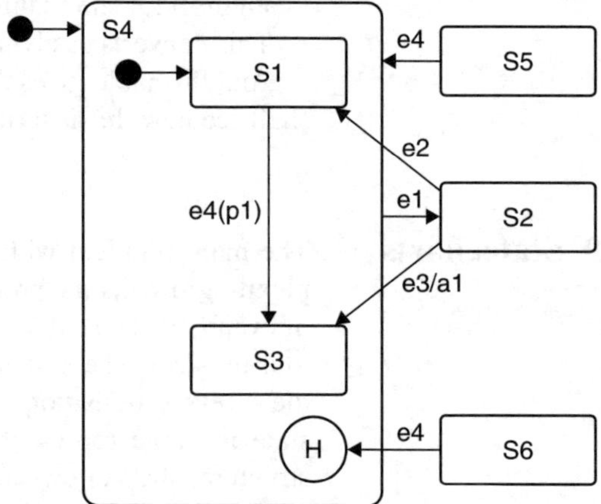

Figure 11.7 *An expanded statechart, including enter-by-history*

example, if e4 happens in S6, the system will go back into whichever one of S1 and S3 it was in most recently.

Superstates can be nested. When a lot of states are involved and the nesting goes deep, it may be appropriate to 'zoom out' so as not to show all the details in a superstate. For example, we could zoom out from Figure 11.6 to Figure 11.8.

The stubs at the end of the two arrows mean that those transitions terminate in a substate of S4.

Because there are potentially many levels of state nesting, it is handy to have a way of indicating whether enter-by-history applies just to the level in which it appears or all the way down to the lowest level (the 'real' states). The H-in-a-circle on its own means the history applies only to the current level; appending an asterix to it means the history applies to all levels. So in Figure 11.6, if S1 and S3 were really superstates, then the particular state to go to, when moving from S6, would be the default substate of whichever one of S1 and S3 was most recently visited. If the H symbol had had an asterisk, then the state to go to would be the most recently visited lowest-level state of all the states in S4.

Superstates encapsulate the idea of being in one of several substates. This corresponds to XOR (exclusive-OR) decomposition of superstates. They can be extended to capture the idea of AND decomposition. That is, the system must be in all of its AND components. This is represented with a dashed line dividing the superstate box. Figure 11.9 gives an example.

Entering the superstate simultaneously enters both of SA1 and SA2. If the transition arrow just touches the outer box, then the system simultaneously enters SB and SF, since these are the two default substates. We write this combination (SB, SF). Then, if e1 happens, the system moves to (SC, SF). Note that the substate of SA2 is going to stay as SF until event e4 occurs with the precondition SD; that is, the system must be in SD when e4 happens for the transition from SF to SE to take place.

We now have all the basic syntax of statecharts. To consolidate this information (and add just a little extra syntax) we turn to an example.

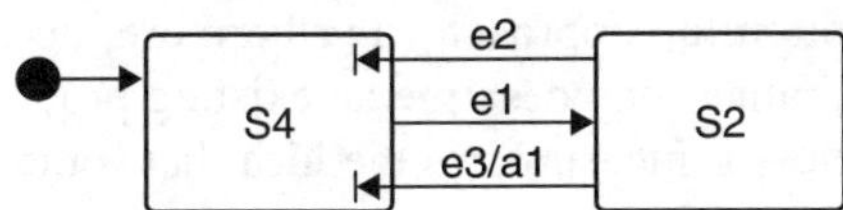

Figure 11.8 *A 'zoomed-out' statechart*

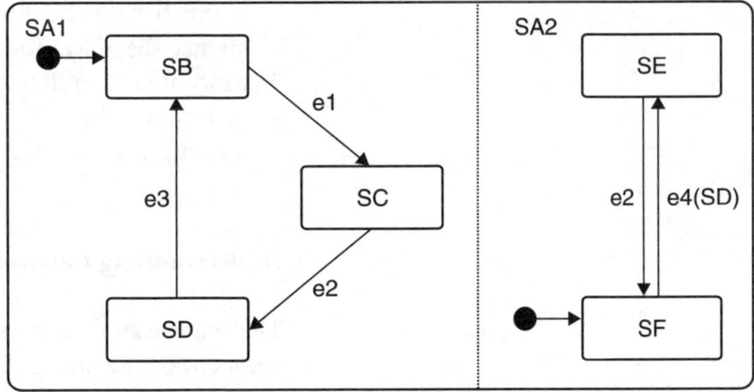

Figure 11.9 *A statechart showing AND decomposition of a superstate*

11.10 Using statecharts – a clock radio example

David Harel's excellent 1987 paper (see references) introduced statecharts by means of a digital watch example. It is hard to find another application that is as good for illustrative purposes (in terms of being at the right level of complexity). A clock radio is a slightly simpler system, but still captures most of the key features of statecharts. In contrast to Harel's paper, the treatment here begins at the top level and moves down (he followed a bottom-up approach). Furthermore, our concern is to develop a system rather than describe an existing product. This is a little more realistic – it incorporates the idea that some issues are unspecified.

Problem statement

I already have a clock alarm, but it is small, tacky and exasperating. I know what I dislike about it, so I'm in the ideal position to write a problem statement. Recall the recipe from Chapter 10 (page 245):

1. Describe the present state of things.
2. Describe the new and better state of things.
3. List the constraints upon a solution that will move you from the present state to the new state.
4. List the criteria that will determine when the new state has been successfully reached.

For my clock alarm, the problem statement is:

1. The time is set with a slide switch: left = increment hours, centre = run, right = increment minutes. I often find that I get the right minute set, try to move the switch into the run position and accidently overshoot and increment the hours. Another three position slide switch controls the alarm: left = off, centre = on, right = set. When this switch is in the right position, the other switch increments the hours and minutes for the alarm time. I often confuse the two switches. When the alarm goes off, I have to find the second switch and slide it left: half the time I find the other switch and start incrementing hours. After a power outage, the clock flashes 00:00 until it is set.
2. My new clock will be easy to set. It will have a radio as well as an alarm buzzer. It will have a 'snooze' button that I can press to silence the alarm for a few minutes.
3. Since this is an illustration, there are no constraints.
4. A design where I press the wrong button (or move the wrong switch) less than one time in ten will be judged a success.

Understanding the problem domain

The main way to understand the problem domain is to see what the competition are doing. A quick survey of the Argos catalogue revealed that almost all clock radios have a battery backup for the time and alarm setting. Of course! I did not write that into the problem statement, because it did not occur to me how useful it would be. Having done this minimal research, the problem statement step 2 can be modified to include:

> The system keeps time and retains the alarm setting during power outages.

Developing and specifying a solution

Clearly the important part of this design will be getting the user interface right. So let's abstract the user interface away from the rest of the system, so that if it changes, the effect on the rest of the specification will be minimal. To do this, we can define messages like Increment_minutes and Increment_hours that are assumed to be generated by the user interface. These messages then become events for the non-user-interface parts of the statechart.

At this stage, we might map out a preliminary design for the user interface. Based on what is known of the competition, it might be good to start with a button that increments hours for as long as it is held down. This is a bit simplistic though. The way the button should work is like a typematic key on a typewriter: increment the hour once as soon as it is pushed, and then every second thereafter. Figure 11.10 shows an appropriate state diagram.

In this example Idle just refers to the function of this part of the user interface. Tick is a clock signal generated once a second. Check that you understand what this miniature state machine is doing.

Having drawn the diagram, we might realize that this will only work if the particular button does not have any other roles, that is, if there are no modes to the interface. Modes in general are things to be avoided, but it may be that we will have to overload the button with more functionality later in the design. So that should be borne in mind.

Let's go up to the top level now, and consider the system as a whole. Clearly the behaviour of the system depends on power, which suggests that the top level should look something like Figure 11.11.

The one-level of nesting in Figure 11.11 is not really necessary because at present there is no commonality between Backup and Working. But of course there should be. In both states the clock should keep 'ticking' (that is keep time) and store the alarm setting. In the working state, the displays, buzzer and radio should be enabled too. What we really need is a way of capturing exactly what is different between the two states. The next diagram (Figure 11.12) does this.

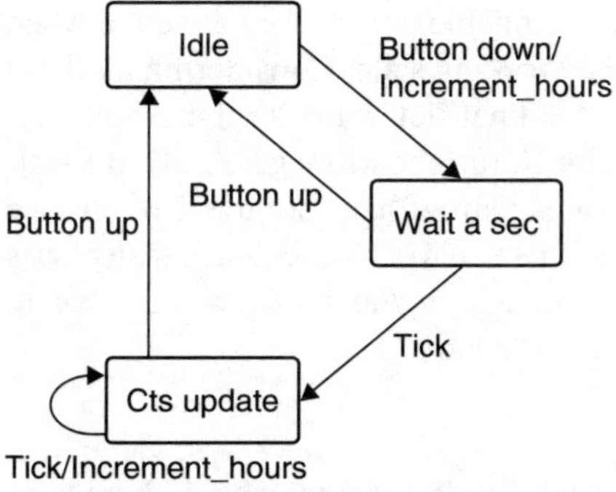

Figure 11.10 *State diagram showing part of the clock user interface*

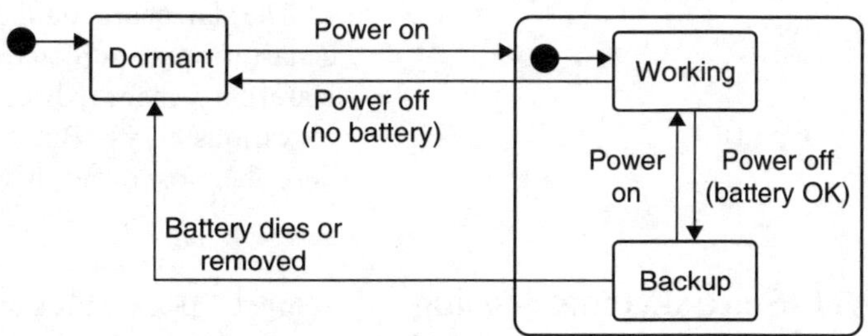

Figure 11.11 *Top-level statechart of the clock*

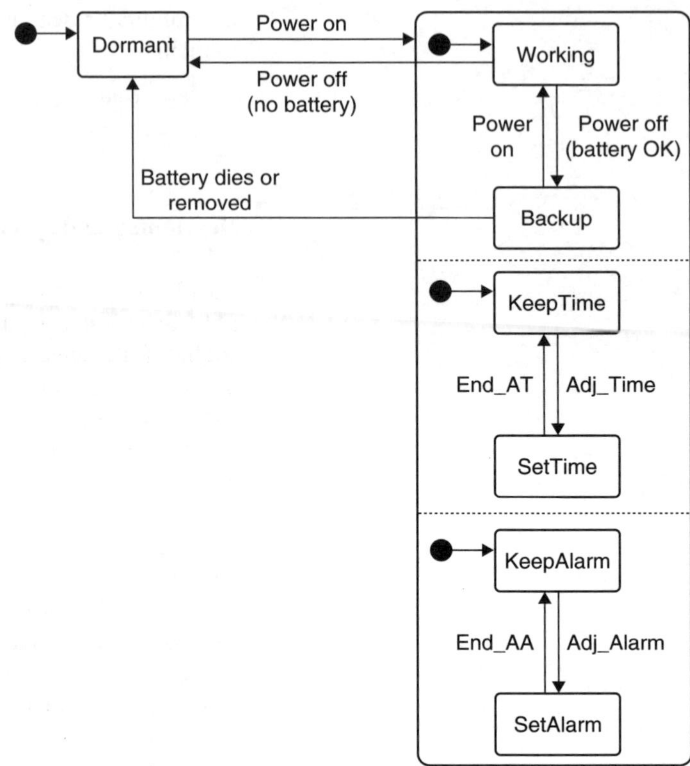

Figure 11.12 *Expanded top-level statechart for the clock*

Now the states Working and Backup are just associated with the special features of each. For example, in Working, the display will be enabled; in Backup it will not. But the time and the alarm are maintained either way.

We will fill in some more details of the time and alarm functions in the next diagram. But it seems that the Sleep function has been forgotten. Should this work in the Backup state as well as the Working state? It seems sensible to have it functioning (counting down) in both cases, although the radio will not be powered in Backup. We need another set of states for the alarm ringing/playing too; the 'snooze' facility can be incorporated there. And what of the user interface? It is to be decoupled from the rest of the system, but it needs to fit into the overall picture somewhere. Clearly, it is only working when the clock is switched on, and it would be reasonable only to have it active when the display is powered, i.e. in the Working state. Considering all these issues leads us to Figure 11.13, the final statechart for the example.

The statechart could be developed further. More generally, the statechart formalism includes facilities that we have not touched on. See Harel's paper for a discussion of these. UML also extends statecharts in various ways. But there is enough in the notation we have to describe most event-driven systems clearly.

11.11 State sketches – using state diagrams to understand algorithms

While UML extends statecharts beyond the coverage here, there is an alternative, simplifying, approach that gets extra mileage from state diagrams. I call this approach 'state sketching'. It is not appropriate

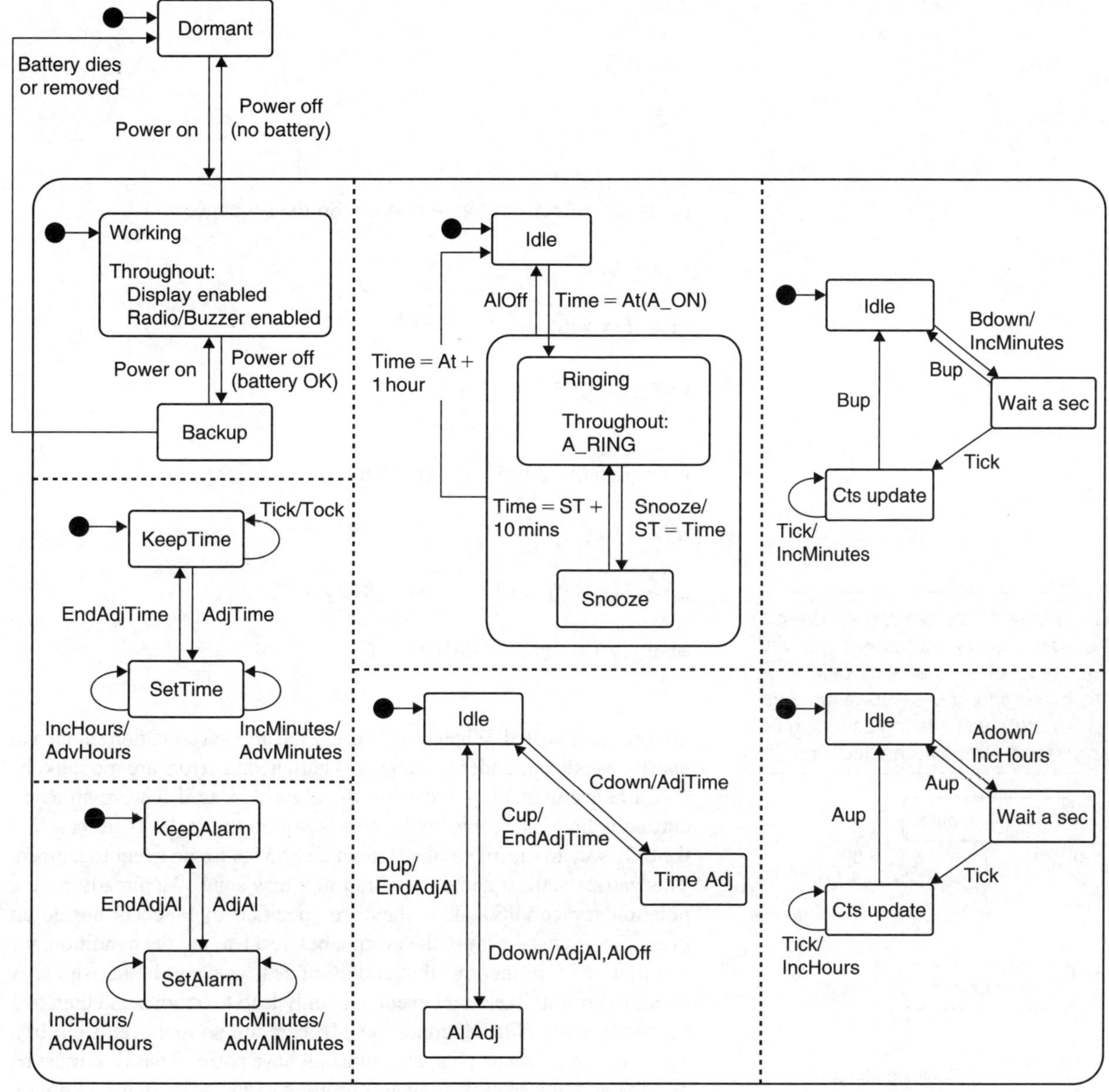

Figure 11.13 *Composite statechart for the clock*

for formal documentation – but to help you understand a method by re-representing it, it is invaluable.

When you first encounter an algorithm, it is likely it will be documented in one of three ways: first, it might be written out in a natural language (e.g. English); second, it might be diagrammed in a way that shows flow of control (e.g. a flowchart); third, it might be written in a computer language or pseudocode. If nicely structured, the third of these is often the best, because it explicitly shows for or while loops instead of using branching (i.e. gotos). But all three cases are likely to have multi-level conditionals (or 'tangled ifs'). This is partly because computers do test alternative cases sequentially.

```
if () {
    }
else if () {
    }
else {
    }
```

reads and runs from top to bottom. So the sequence

```
if (x < 0) {
    }
else if (x > 0) {
    }
else { // x == 0
    }
```

is doing a three-way branch, whereas

```
if (x < 0) {
    }
else if (y > 0) { // I.e. (x >= 0)&&(y > 0)
    }
else { // I.e. (x >= 0)&&(y <= 0)
    }
```

is more complicated. When ifs get nested, the implicit conditions hidden in else cases pile up, understanding gets harder, and errors are more likely.

Thinking in states – recasting the algorithm as if it were an asynchronous finite state machine – forces you to untangle ifs. Recall that the only way to move out of a state in an FSM is for an event to happen. This causes actions and a transition to a new state. We already have a notation for conditionals – they are specified in brackets beside an event. So, if we take away the event label, just leaving the condition, we can think of fulfilment of that condition as triggering the transition. A condition event, like a real event, can only lead to actions and then to a new state, *not* to further conditions. If there are no real events out of a state, just condition events, they must all have parity. That is, it must be possible to think of each as an asychronous event, so there is no implicit time or logical ordering of the different paths out of the state. To represent sequences of conditions, we must use sequences of states.

For the two cases of the three-way if above, the equivalent state sketches are shown in Figure 11.14.

State sketches explicate all the different alternatives for conditionals. Now you could enforce that with some other notation, but state sketches give you more. When a system, or an algorithm, is in a state, it is doing a well-defined set of things. The only way for a sequential computer 'to be doing' a set of things is for it to do them one after the other, but from the point of view of a state diagram, all the actions in a state are done at the same time. That is, it doesn't matter in what order they are done, so long as, at the time the state is left, all the actions in it are guaranteed complete. Just as conditionals out of a state have parity, so actions inside the state have parity.

> The programming language Ada has facilities for representing multiway branching code that can differentiate these two cases. However, Ada doesn't stop you nesting conditionals so you can still have tangled ifs.

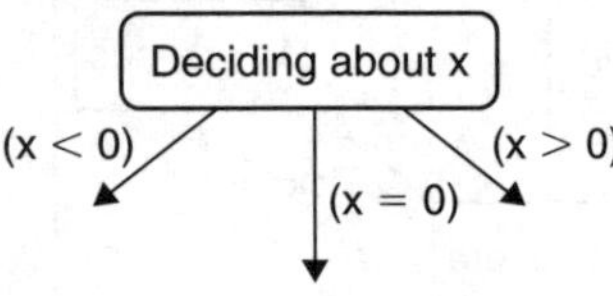

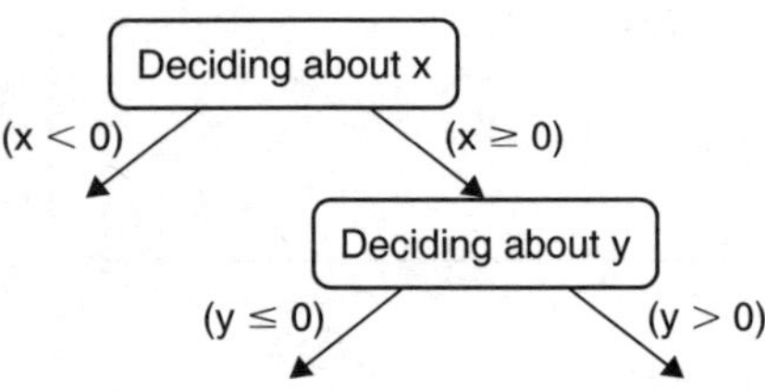

Figure 11.14 *State sketches for two types of three-way conditionals*

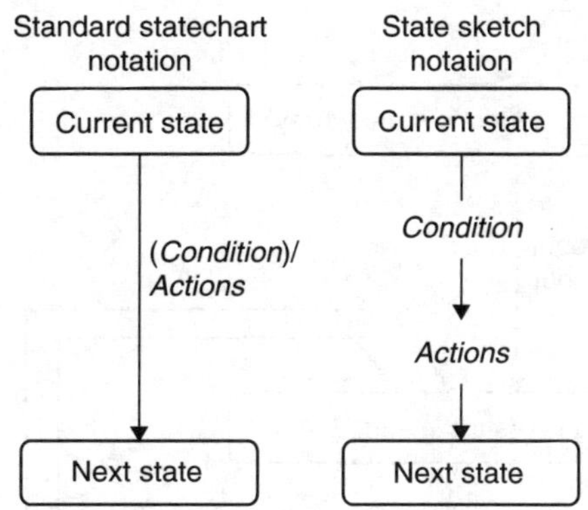

Figure 11.15 *Comparison of statechart and state sketch transition notations*

A state diagram therefore explicitly differentiates between tests (conditionals) and actions that have to be done in a particular sequence, and those where the sequencing is arbitrary.

A state sketch is just a conventional state diagram, with some minor notational short cuts. Because many states will be exited by condition events rather than real asynchronous events from outside, the notation for these and their associated actions is labels breaking an arrow into three pieces (see Figure 11.15).

Actions in a state don't need to be individual statements – they can represent whole nested functions, so long as, with respect to the state diagram as a whole, they are encapsulated in the state and have parity with all other actions in that state.

Example: binary search

So how do state sketches help for understanding algorithms? Let's consider an example: binary search. C++ code for binary search is given on page 223, and repeated here:

```
int search(int x, int array[], int num_in_array)
    {
    int middle;
    int start = 0;
    int end = num_in_array - 1;
    while(end >= start)
        {
        middle = (start + end)/2;
        if (array[middle] == x)
            return(middle);
        if (array[middle] < x)
            start = middle + 1;
        else
            end = middle - 1;
        }
    return(-1);
    }
```

Binary search is a short algorithm, but it's trickier than it looks (see sidebar on page 221). Suppose we just had an intuitive idea about how binary search works. Here's how we could map things out in a state sketch.

We might begin with two states: Searching, obviously, where we're doing the work, and Idle, where we're not. These are two states of some notional FSM that we might call the 'Binary Searcher'. Obviously this will be embedded in a longer sequential program, so when the Binary Searcher is idle, it's because other things are happening instead. But we can think of this little state machine as though it were a separate piece of kit that just does binary searching jobs for us. Presumably it will move from Idle to Searching when given a kick by another module after things have been initialized properly, that is, the sorted array has been prepared and a test probe is known. From Binary Searcher's point of view this is an event, so it gets written alongside the state transition

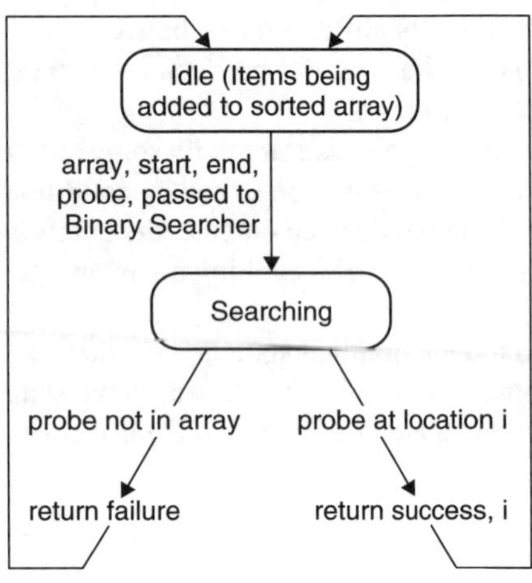

Figure 11.16 *Binary search state sketch: first stage*

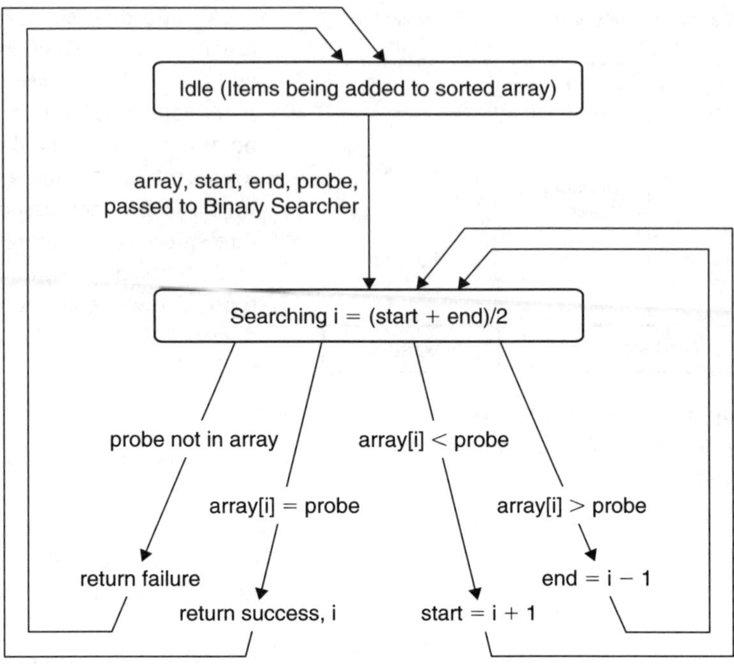

Figure 11.17 *Binary search state sketch: second stage*

Logically we don't *have* to introduce another state here. We can just make the conditionals more complicated: e.g. 'not known if probe is in array and array[i] < probe', but the whole point of state sketching is to make things clear.

arcs as ordinary events do. The state machine will get back from Searching to Idle either by finding the item, or not doing so. These will depend on internal condition tests of the Binary Searcher, so it can be represented as the state sketch in Figure 11.16.

Now we need to flesh out what happens in searching. Binary search's idea is that you look at the centre point of a range, then either find the item or limit the next search to a subrange on one side of it. Those steps are sketched in, and Figure 11.17 is the result.

But now it's clear that the conditions coming out of the searching state don't all have parity. The three comparisons between array[i] and probe do, so could be implemented in any order. One of them *must* be true so we have to take one of those three transitions out of the state. But there are circumstances when we might be able to take the 'probe not in array' route out too – indeed, we need to do so once we know it's true. That means we have to sequence things with another state. How will we detect that the probe is not in the array? When we can go no further with subdivision. So now we know what we have to test in the new state. The result is the modified sketch shown in Figure 11.18.

The sketch is complete, and coding can be done with reference to both the written algorithm and the state sketch. The main value of the sketch is in the process of its creation. It lets you look at an algorithm through a different syntactical lens, so helps you think through what happens when.

The state sketch does share with flowcharts and other diagrammatic notations for algorithms the shortcoming that gotos (i.e. transition arcs) can go anywhere. If you allow them to. But we can borrow notation from statecharts to encapsulate a group of states used in several contexts as an independent FSM. This will be triggered (i.e. moved out of its

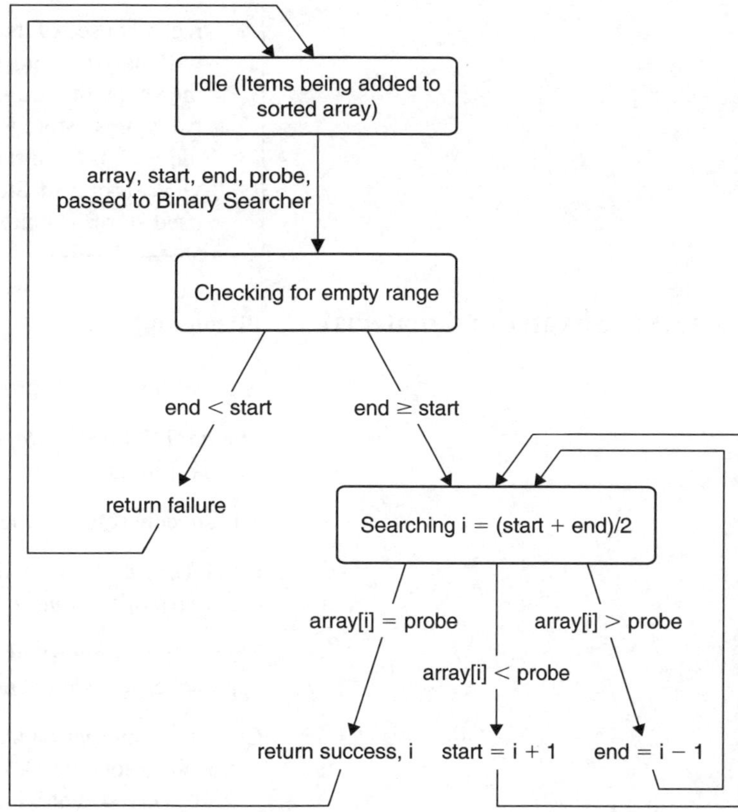

Figure 11.18 *Binary search state sketch: third stage*

idle state) by any of a number of states in the main FSM, then return, via the history notation, to the same place. This is directly analogous to a function call.

The exercises suggest algorithms suitable for sketching. In each case a deeper insight comes out of thinking in states. The simplex case study in Chapter 16 also shows the value of state sketching to clarify an ambiguity in English versions of an algorithm.

Develop a state sketch for

- insertion sort (page 227)
- all states in all access functions for an ordered list implemented with a linked list (page 210)
- a function that replaces repeated characters in a string with a single character: e.g. it maps aabbbbcdeffg to abcdefg.

The above exercises are all algorithms, but state sketching is especially useful when some parts of a program are procedural and some are event driven. Develop such a state sketch for:

- A module that double-buffers video input. On asynchronous request from another module, your module will signal acceptance as soon as it has an unread video frame in the less

> recently used of its two stores (Store A, say). Simultaneously it will start waiting for the start of a new video input frame, then copy that frame into the other (more recently used) of its two stores (Store B). Having signalled acceptance, your module will assume that the other module immediately uses the video frame in Store A, so that becomes the more recently used store. The next request will be serviced from Store B.

11.12 Chapter end material

Bibliography

A good treatment of prototyping is:

Maude, T. I. (1991). *Rapid Prototyping: The Management of Software Risk*, Pitman.

The statechart technique is introduced in:

Harel, D. (1987). 'Statecharts: a visual formalism for complex systems', *Science of Computer Programming*, **8**, pp. 231–274.

This paper contains many references back to other state machine approaches to specification, of which the most significant is probably:

CCITT (International Telecommunication Union), Functional Specification and Description Language (SDL), Recommendations Z.101-Z.104, Vol VI, Fasc. VI.7, Geneva, 1981.

The dominant set of notations for software design is UML, which is covered in many books. Perhaps the most prominent are:

Booch, G., Jacobson, I. and Rumbaugh, J. (1998). *Unified Modeling Language User Guide*, Addison-Wesley.
Jacobson, I., Booch, G. and Rumbaugh, J. (1999). *The Unified Software Development Process*, Addison-Wesley.
Rumbaugh, J., Jacobson, I. and Booch, G. (1999). *Unified Modeling Language Reference Manual*, Addison-Wesley.

Their common authorship is no accident: Booch, Jacobson and Rumbaugh are principal contributors to UML. For an independent view, try:

Fowler, M. (2003). *UML Distilled*, 2nd edn, Addison-Wesley.

Earlier notations, some of which are predecessors to UML, and some of which are alternatives, include:

Bohm, C. and Jacopini, G. (1966). 'Flow diagrams, Turing machines and languages with only two formation rules', *Communications of the ACM*, **9**, pp. 366–371.
Nassi, I. and Shneiderman, B. (1973). 'Flowchart techniques for structured programming', *ACM SIGPLAN Notices*, **8** (8), pp. 12–26.
Jackson, M. A. (1975). *Principles of Program Design*, Academic Press Inc.
DeMarco, T. (1978). *Structured Analysis and System Specification*, Yourdon Press.

Shneiderman, B. (1982). 'Control flow and data structure documentation', *Communications of the ACM*, **25**, pp. 55–63.

Cameron, J. R. (1983). *JSP & JSD: The Jackson Approach to Software Development*, IEEE Computer Society.

Martin, J. and McClure, C. (1985). *Diagramming Techniques for Analysts and Programmers*, Prentice-Hall.

Design of a flight reservation system through dataflow diagrams as in Figure 11.1 is explored as a case study in: Ince, D. (1991). *Object-Oriented Software Engineering*, McGraw-Hill.

12 Modularization

Breaking software into manageable parts is called modularization. A module (or component) is a subunit of the total system. Sometimes it will correspond to a work unit for one individual. Even when not, it's helpful to think in terms of modularization subdividing the development effort as well as the program. The aim is that modules are sufficiently decoupled that a different person could do each one. When the development team is small (say of size 1), modularization still helps to limit the amount of information to be held in mind at any time.

Modularization is perhaps more critical in software design than in any other field. Splitting a design project into units may be necessary in a standard engineering design project, but only for software does it assume paramount importance.

To see why this is, we return to a now familiar principle and rule:

Principle 5

The requirements will change.

Rule 9

**Design software to minimize the damage
caused by changes to requirements.**

This has a major effect on the design process model – instead of changes to the problem statement criteria or formal specification being seen as the thing-to-be-avoided-at-all-costs, they are accepted as inevitable, and the module decomposition, whose main emphasis is protecting the project against the inevitable mutations, becomes very important. The modularization process has the following goals:

- minimize the damage caused by changes to specifications
- maximize individual programmer productivity
- make designs comprehensible (and therefore maintainable and reusable).

Note that a module is not synonymous with a function or a class; it is usually bigger than one of these. It is also different from (and usually bigger than) a file. The fact that software tools allow a program to be split into many files, facilitates modularization, but there is no necessary equivalence between files and modules.

Namespaces

C++ provides the **namespace** facility to indicate the grouping of classes and non-member functions into modules. Each module may be enclosed in a **namespace** of a given name, and references to anything in that module must be explicitly marked as belonging to that **namespace**. This not only provides a wall around a module, it also allows the reuse of names for classes and functions in different **namespaces**, rather like the same street names used in different towns. Some people regard **namespaces** as *the* mechanism for modularization, and deprecate code that works around them, notoriously,

using namespace std;

which allows free use of **cout, string, istream**, etc., instead of the more 'modularity aware' **std::cout, std::string, std::istream**. Other people wonder if **namespaces** are really a wrapper too far, generating denser, less readable, code for little gain. In any event, the most important **namespace, std**, contains at least five modules, so is hardly an example of clean modularity. (The five modules are the STL (collections and iterators), strings, numerics, streams, and language support, the latter being things like the **exception** class hierarchy (e.g. **bad_alloc** as thrown by **new**).)

In this chapter we consider two modularization strategies, top-down design and information hiding.

12.2 Top-down design

The natural place to start a modularization is by thinking in terms of top-down design, that is, to address a task at successively lower levels of detail. Top-down design works by progressing through a hierarchy of What? How? steps as shown in Figure 12.1.

For example, to design a program that reads in a list of marks, finds the top ten and writes them out, we might use the decomposition in Figure 12.2.

There is one very good thing about this example: the use of a general-purpose step for sorting. It is always worthwhile to look for a way of using general-purpose modules. But there is a danger with it too: it suggests that the procedure, or sequence of steps to be taken, is the thing that drives the decomposition. This does nothing to ensure that the modules are independent. In this case, many modules have to share access to the array. In a small example like this, it doesn't matter too much. But you should recognize that this array sharing is a type of intermodule communication (via a shared pool). What is more, the illustration looks very neat in that a lower module always has a unique chain up to the top of the hierarchy. In practice this is unlikely to be the case. There are very likely to be low-level functions that are

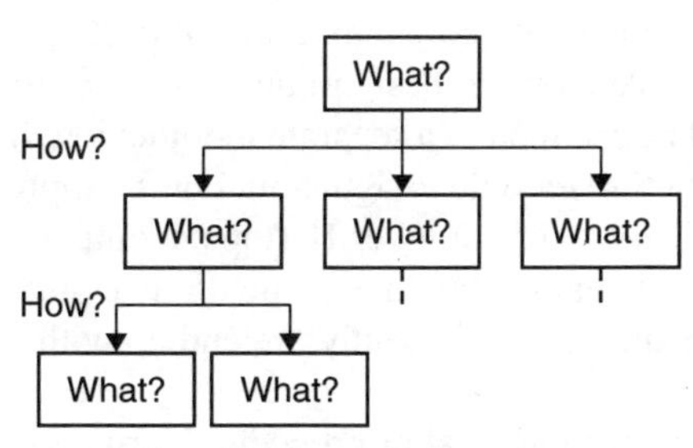

Figure 12.1 *Top-down design questions*

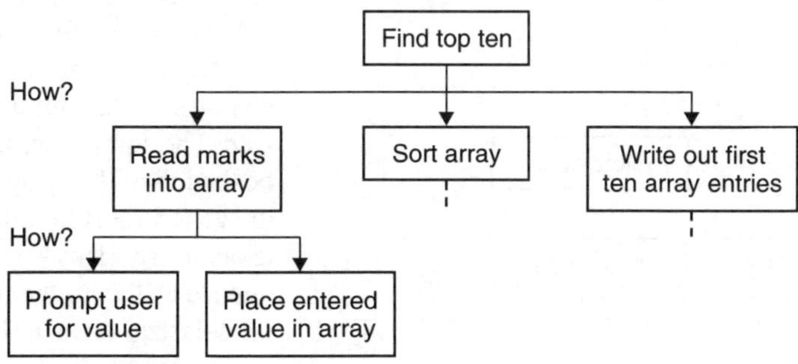

Figure 12.2 *Top-down design example*

useful to other modules from all over the tree. So the web of interconnections and dependencies will be much more tangled than it looks.

In general then, although top-down design is a natural and an important way to consider a problem, it may not give good modularity. By contrast, applying the principle of information hiding usually ensures that modules are as self-contained as possible.

12.3 Information hiding

Rule 6

Localize information.

Information hiding is the modularization of a program on the basis that each module must hide something. The more that is hidden within modules, and the less they reveal in their external interfaces, the better.

Information hiding is based on localizing risk, design flexibility, and consequently data. It is applied in the following way:

1. Begin with the success criteria (or specification, if you have one).
2. Identify the 'high risk' parts of the criteria or specification:
   * what is ambiguously stated
   * what is likely to change.
3. Generate an initial decomposition aimed at localizing code that deals with high risk portions of the specification.
4. If there are many plausible design alternatives, partition them into groups of dependent alternatives. For example, if you have dependent (or mutually exclusive) alternatives A, B, C and mutually exclusive alternatives 1, 2, such that any of A1 A2 B1 B2 C1 C2 are solutions, then group A, B, C together and group 1, 2 together.
5. Generate a further decomposition aimed at keeping dependent alternatives within a single module.

Now each module hides information (including design decisions) from the rest of the system. In consequence a module will consist of data structures and access functions. The data structures are not shared.

Now consider finding the top ten marks in a set. Using information hiding the program would be partitioned on the basis of risks inherent in the specification, and localization of design decisions. Consequently, all functions which communicate with the user would be put into a single module. This would allow a non-interactive version of the program, which reads and writes files, to be generated just by modifying the user communication module. The data structure in which the array is held should be the subject of another module, because the specification says nothing about the number of scores, and any impact that changes in this number would have on the design ought to be localized. The sort process can still be retained as a separate module; it will be the job of the array data structure module to figure out how to apply the generic sort facilities to the array. The number 10 (top ten outputs) is obviously risky – it might well change, but there is no direct consequence of this to the modularization. Consequently, we end up with a modularization as in Figure 12.3.

The system no longer has a chronological or procedural structure. Nor is it hierarchical, except in so far as there is a high-level control


```
            ┌────────────────┐
            │ System control │
            └────────────────┘

┌─────────────┐ ┌─────────────┐ ┌──────┐
│    User     │ │ Array data  │ │ Sort │
│ interaction │ │  structure  │ │      │
└─────────────┘ └─────────────┘ └──────┘
```

Figure 12.3 *Information hiding modularization*

module. If we were looking for a tag to describe modularization based on the criterion of information hiding, we might say 'object based' simply because the structure that drops out is one that emphasizes objects rather than actions.

12.4 A modularization example

Here is a modularization problem originally proposed by Parnas in his discussion of information hiding. It is about the right size to justify care in modularization, without being unwieldy for illustrative purposes.
The specification is as follows:

Input: Text file consisting of several lines of several words each.
Output: Text file consisting of all circular shifts of all lines of the input, in alphabetical order.
Definition: A circular shift is done by removing the first word from a line and appending it to the end of the line.

Example:

Input	(All circular shifts)	Output
the quick brown fox	the quick brown fox	brown fox the quick
jumps over	quick brown fox the	dog the lazy
the lazy dog	brown fox the quick	fox the quick brown
	fox the quick brown	jumps over
	jumps over	lazy dog the
	over jumps	over jumps
	the lazy dog	quick brown fox the
	lazy dog the	the lazy dog
	dog the lazy	the quick brown fox

(Note that giving an example clears some potential ambiguity: the alphabetization goes on to subsequent words after the first.)
Modularization A is based on a top-down decomposition.

1. **Input**
   Read lines into memory
   Set up pointer to start of each line
2. **Circular shift**
   Add pointer to start of every word
3. **Alphabetizing**
   Sort pointers (Sort function looks at words being pointed to and checks alpha order)
4. **Output**
   Go through pointer list sequentially.
   For any entry, print out until <CR>
   Go back to previous <CR> then print out from there to pointer position.
5. **Master control**
   Sequencing, error messages, etc.

This decomposition into modules has implicitly defined the data structures that will be used. It has also introduced design decisions which have global effects. For example, the decision has been made to index circular shifts and generate them at output time, rather than to store them all. This is probably a wise decision, but the important thing to realize is that the decision has been made. It means that both the Alphabetizing and the Output modules have to know how to generate circular shifts. For the Alphabetizer, this is because a decision about alphabetical order may involve wraparound of a line.

Modularization B is based on information hiding.

1. **Line storage**

   char *word(int r, int w)   returns wth word on rth line as a null-terminated string

   int setword(int r, int w, char *p)   sets wth word on rth line to word pointed to by **p**

   int numwords(int r)   returns number of words on rth line

   int numlines()

   int delline(int r)

   int delword(int r, int w)

2. **Input**

   Reads lines from input and calls module 1. to store.

3. **Circular shifter**

   int cssetup()   Sets up circular shifts

   char *csword(int l, int w)   Returns wth word of lth circular shift.

   int csnumwords(int l)   Number of words in lth circular shift.

(At this point, an addition to the definition of circular shift is needed:
   (a) if i < j, then shifts of line i precede shifts of line j
   (b) for each line the first shift is original line, the second is one-word rotation to original, etc.)

4. **Alphabetizer**

   int alphasetup()   Sets up alphabetical list of circular shifts

   int ith(int i)   Gives index of circular shift that comes ith in alphabetical ordering.

5. **Output**

   Uses 3. and 4. to produce output

6. **Master control**

   Sequencing, error messages, etc.

Table 12.1 compares the two modularizations in terms of programmer productivity, comprehensibility and the effects of specification changes.

What about efficiency? Modularization B may prove less efficient than A. But experiments with large systems suggest that a modularization based on information hiding is usually more efficient, because the code is less tangled. There have been no reported instances of an information hiding implementation thrown out because of inefficiency.

12.5 Modularizing from a statechart

If you have specified your system using a statechart (or a state transition diagram, or state sketch), top-down modularization would suggest partitioning according to major superstates. This is almost

Table 12.1

Modularization A	Modularization B
Programmer productivity and individual development Interface between modules is organized tables – must be developed jointly. Issues of memory usage must be worked out jointly.	Interfaces are function names and parameters – essentially messages. Programmers can concentrate on optimizing their own code given these interfaces. Programmers are given authority and ownership – good for morale.
Comprehensibility To understand the output module, you have to understand the jobs of circular shifter and alphabetizer (and their tables). On reflection, we seem to have put a lot of complexity into the output module – but if we were to split it, surely the first part should be incorporated into the circular shifter?	Information hiding has forced rigorous specification of the interfaces early. Now that is done, the modules are effectively self-contained and will therefore be comprehensible.

	Modularization A	Modularization B
Effect of changes Input format.	Input module only.	Input module only.
Decision to store all input lines in memory at once. (Suppose we ran out of memory.)	All modules.	Line storage only.
Decision to use 1D array for line storage, then index lines with pointers.	All modules.	Line storage only.
Decision to do circular shifts on the fly as required during alphabetizing and output.	Circular shift, alphabetizing, output.	Circular shift only (**cssetup** may or may not do anything).

certainly a mistake from the information hiding point of view. Rather, modularization of a statechart should be done as follows:

- If at all possible, the entire hierarchy of states, transition arcs, etc. should be encapsulated in one module, appropriately called the 'state sequencer'. The data structures in this module will contain all the information about states, events, conditions and actions. The algorithmic, procedural part of the module will simply be the mechanism for stepping through the data tables. An illustration of this is given on page 289.
- If the state sequencer *has* to be partitioned, minimize the number of modules in this area and enforce consistency in state sequencer design between programmers.
- The action methods should now be partitioned. Interfaces with file systems, data structures, etc. should be considered as in the examples above, designing to minimize risk. It might be appropriate to partition actions according to superstates, but there is no necessary guarantee that this is best. If there is no obviously better way, then superstates may be used to define modules.
- All input events should be dealt with in a single module. If possible this module should be the single layer between the user (and the rest of the asynchronous outside world) and the state sequencer.

The rule then for statechart modularization is to apply the principles of information hiding, while making every effort not to split the state sequencer.

12.6 Object-oriented modularization

Rule 8

Hide information in public; reveal it in private.

So far we've talked about modules or components as work-sized units, but in the previous section we saw that information hiding inside modules leads naturally to the units being abstractions of data and functionality, rather than procedures. The natural extension of this is to object-oriented design, where modules are actually classes that might be instantiated one or more times. In the case of one instance, the abstraction is similar to the simple information hiding modules we've just discussed. In the case of many instances, we have abstract data types. Once we have the ability to treat modules as types, the whole process of design is affected, for we might have many more modules than work units. But the important criterion of information hiding can still be used to guide design. Here's a recipe for object-oriented design that does so:

1. Identify the basic classes of objects that the system/program has to deal with. These need not be disjoint. For example, in a word processor, you may identify document, paragraph and page as three types of objects.
2. Identify the attributes of each class. These are things like an object's current settings, parameters and contents. For the word processor example, they could include font, alignment and so on.
3. Identify how classes are related to each other. What are the inheritance relationships? Should abstract base classes be defined?
4. For any instance of a class, identify the things it must be able to do (including the things it must do to itself). These are actions. For example, add character, change font. Which of these are inherited? Where will polymorphism be used?
5. Is the object modeless, or does it implement a state machine? If the latter, what are the states and actions associated with each transition?
6. What events are going to cause objects to take actions? Develop their FSMs.

Note that in steps 4, 5 and 6, possible states and actions are considered before events. At step 6, the FSM table can be developed, ensuring attention is given to the consequences of events in unexpected states.

Example

Consider the design of a theatre booking system (this example is due to Ince, although the way it is treated is different). We might go through the six steps as follows:

1. Classes of object will certainly include: Customer, Seat, Performance, Invoice, etc. At this stage you might well reserve judgement about whether seat should mean a physical seat with no time

Table 12.2

	Customer pays	**Customer cancels**	**Theatre cancels**	**Performance day**	**Performance day + two years**
Awaiting payment	Paid	Released	*Advise customer* Complete	*Stop customer orders* *Payment demand* Still waiting	*Error* Still waiting
Still waiting	Complete	*Payment demand* Still waiting	*Error* Still waiting	*Error* Still waiting	*Writeoff* Released
Paid	*Error* Paid	*Partial refund* Pending	*Full refund* Complete	Complete	*Error* Released
Pending	*Error* Pending	*Error* Pending	*Residual refund* Complete	Complete	*Error* Released
Complete	*Error* Complete	*Error* Complete	*Error* Complete	Complete	Released

limitation, or Seat-at-Performance, i.e. a physical seat at a particular time. Both of these might need to be thought of as objects.

2. For the customer class, the attributes of an object will include Name, Address, Telephone Number, and probably pointers to lists of Seat, Performance, and/or Invoice objects.

3. The classes have no commonalities that would allow us to derive one from another or two from a common abstract base class.

4. An action appropriate to Customer would be Associate-with-Seat. Note that if Seat does not mean Seat-at-Performance, then the action must associate Performance as well.

5. An invoice object certainly has modes. It can be awaiting payment, paid, cancelled, etc. The actions taken when moving between these modes are minimal – invoices rarely act on other objects.

6. Table 12.2 is a partial FSM table to deal with an invoice. It shows Actions in *italics* and Next State information in regular type. Note that the invoice is initially created in response to a customer order and starts in state Awaiting payment. When released, the invoice object is effectively destroyed. This FSM captures the theatre's policies about deposits, cancellations, etc. If these change later on, only this table need be revised.

> Consider a customer object in the theatre booking system. Develop an object-oriented specification for this type of object.

This example is really object based rather than object oriented. Step 3 did not apply. But in larger examples, classes (types/modules) do relate to each other. Some classes will contain others; some will use others. It's sometimes useful to show these and other associative relationships in diagrammatic form and the UML class diagram provides for this. However, the most important relationship between classes is inheritance ('isa'), and for this we can use a simple diagram like Figure 12.4.

Remember that arrows point up the tree, so a swallow *isa* Flying Bird, etc.

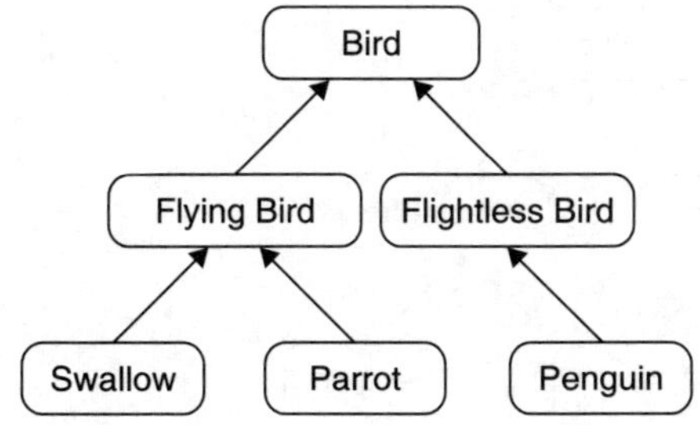

Figure 12.4 *An example class diagram*

In UML, *isa* arrows are shown with an open arrowhead. Other associations can be displayed. Figure 12.5 shows three of the most common.

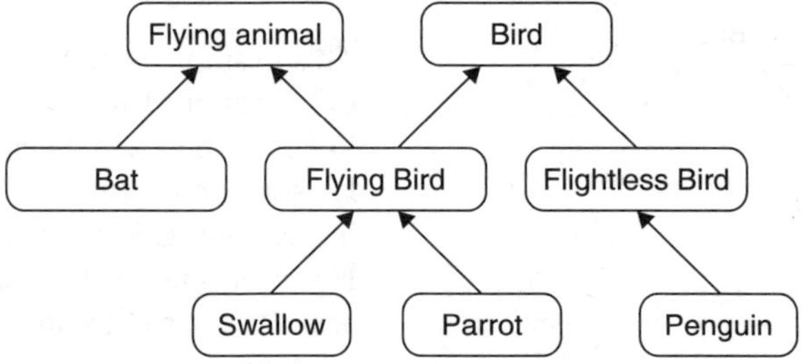

Figure 12.5 *Informal interpretations of UML inheritance, association and composition*

Sometimes, in developing a hierarchy, we want a little more information about which functions and data are inherited, which are redefined, and which are unique to derived types. Using a subset of UML notation, we can write within a class, the attributes and operations that are relevant in showing those differences. See Figure 12.6.

The symbol preceding an attribute or operation denotes its visibility: (+) public, (v) public and virtual, (#) protected, (−) private, with meanings exactly as in C++. This extra information can help in deciding where functionality belongs. For example, Flying Bird and Flightless Bird will both inherit the interface to walk() but may redefine it, perhaps by calling their private operations hop() and waddle() respectively. Although wingspan is more relevant to flying birds, all birds have one, so the appropriate place is as protected data in the Bird class. (Maybe flightless birds walk() depends on their wingspan too.)

It is also possible to use class diagrams to show multiple inheritance. This is a concept that C++ supports, although there wasn't space in Chapter 5 to cover it. Figure 12.7 shows how it might be represented with a class diagram.

In an expanded version of the diagram, with attributes and operations, it would be clear that Flying Bird inherits fly() from Flying animal, but layEgg() from Bird.

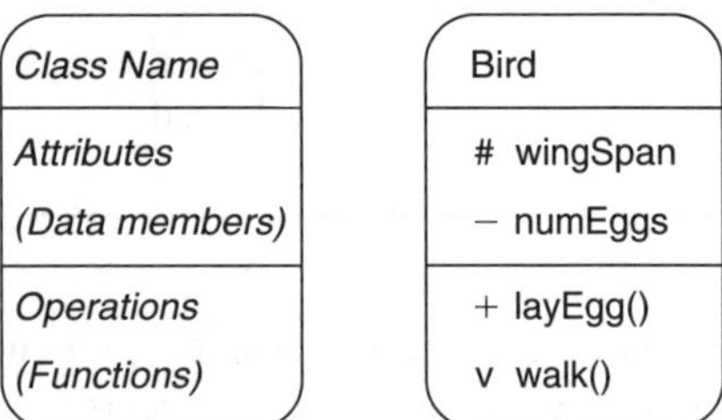

Class Name		Bird	
Attributes (Data members)		# wingSpan − numEggs	
Operations (Functions)		+ layEgg() v walk()	

Figure 12.6 *Putting more information into classes in class diagrams*

UML class diagrams additionally show the types of attributes and operations. They use the −, + and # symbols, but don't explicitly show an operation as being virtual. I use the v symbol for that purpose while working out hierarchies with class diagrams, simply because it reminds me that there are redefinitions further down the tree. (It's handy that v for virtual looks like a downward pointer.)

12.7 Documenting the modularization

As always, documentation depends on the size and structure of the project. For smaller programs, modularization is likely to be implicitly documented in the arrangement of files and classes and commentary

Figure 12.7 *Multiple inheritance in a class diagram*

in headers. For larger projects there may be a module description document.

For each module in the program, the module description contains the following information:

- the module's name
- its location in the code; i.e. the files it comprises
- the class hierarchies that it implements
- who is responsible for the module
- the services the module provides
- the secrets of the module; i.e. what the module hides. Here are stated the design decisions that are reserved to the implementer of the module, and which users of the module should make no assumptions about. This part of the module description is very malleable (unlike the rest). The developer adds secrets as design alternatives are realized, but does not reveal the actual design strategy chosen.
- the formal interface to the module's services:
  - black box picture of the module with just enough information for the programmer of another module to use its services.
  - includes list of access functions with their parameters, the externally visible effects of access functions, timing and accuracy constraints, details about error handling.

To illustrate this structure, the following two pages are a sample taken from the modularization description used in a multimedia communications system. These are the first part of the description of a module called 'Media Handling', which is one of seven modules in the system. You are not expected to understand the module from this extract, which comes from a 50-page document. It is included here simply to show how the modularization for a fairly large project is documented.

6   The Media Handling module

6.1   Name
Media Handling. MED_HANDLER

6.2   Files

meddisp.cpp	Dispatching of requests to this module. Contains all access functions.
medinfo.cpp	Interface to media configuration file. Loads and maintains information on the media that are to be handled.
medqueue.cpp	All request queues. Monitors the current state of objects in transit.
medcom.cpp	Communications interface. Initiates an object transfer with the peer Media Handling module at other sites, using services of Data Transport module.
medcode.cpp	Interface to medium-specific coders (e.g. JPEG coder, text coder).

6.3   Author

John Robinson

6.4   Services

1   The Media Handling module controls the encoding and decoding of all kinds of visual data. The Application module's only access to visuals is via direct calls to Media Handling. Medium-specific and device-specific submodules implement the specific requests for compression, display and conversion of visuals. These are loaded, initialized and supervised by the Media Handling module.

2   The Media Handling module uses the Data Transfer module to achieve transmission of visuals. It acts as a pipe between medium-specific submodules and the Data Transfer module, so that the former need know nothing of the current communications session structure.

3   The receiving operation is transparent – that is the Application module does not have to make calls to Media Handling in order to receive a visual. Once the transfer has been validated by the Security module, it is arbitrated entirely by the peer Media Handling modules in the sites involved.

6.5   Secrets

1   Names and locations of the loadable medium-specific submodules.

2   Communication packet format for visual data.

3   Strategy for checking for preexistence of particular visuals at remote sites, and for early termination, if all remote sites already have the visual.

4   Sizes of queues determining number of requests that can be pending.

(Note that the medium-specific submodules hide the compression and display algorithms for each medium, and the device-specific submodules hide the details of the hardware interfaces.)

6.6   Access Functions

6.6.1   med_handler

Function Prototype

int med_handler(char far *pMediaType, char far *pDevice, int RequestType)

pMediaType   (Unique name of medium, e.g. JPEG)
pDevice   (Name of input device etc. or NULL)
RequestType   (LOAD, UNLOAD or TEST)

Description

This function loads, unloads or tests for the existence of a local media handler of the specified type. Separate media handlers are used for different input devices, and a pInput Device of NULL loads a handler that does decoding and display from files only. After loading or unloading a handler, appropriate information is passed to the Monitor module which distributes it across the session to all connected sites.

Return Codes
MED_HANDLER_LOADED,
MED_HANDLER_DOES_ NOT_EXIST,
MED_HANDLER_UNLOADED,
MED_HANDLER_ EXISTS

Function Prototype
int med_available (char *pMediaType, char *pRemoteSite)
pMediaType   (Unique name of medium, e.g. JPEG)
pRemoteSite   (Name of receiving site)

Description
This function checks whether it is possible to exchange media of the specified type with the specified site. This information is obtained from the Monitor module.
... (etc.)

12.8 Chapter end material

Bibliography

The classic reference for the material in this chapter is:

Parnas, D. L. (1972). 'On the criteria to be used in decomposing systems into modules', *Communications of the ACM*, **15**, pp. 1053–1058.

Many object-oriented design books are good on modularization, stressing the importance of information hiding. Among them are:

Booch, G. (1993). *Object-oriented Analysis and Design with Applications*, 2nd edn, Benjamin/Cummings Publishing Co. (3rd edition is due in 2003).
Peters, J. R. and Pedrycz, W. (2000). *Software Engineering: An Engineering Approach*, Wiley.

13 Detailed design and implementation

13.1 Introduction

In software, detailed design is the implementation of module internals, design of data structures and algorithms, and coding in a programming language. This chapter connects the treatment of these issues in the software technology part of the book (Chapters 4 to 8) to the practicalities of applied software design. First, we will show how good solution notation makes life easy in programming with a small example of FSM programming. Then we will discuss harder cases where the programmer must make data structure and algorithm choices based on the issues discussed in Chapters 7 and 8.

13.2 Implementing from a higher-level representation

Recall that in Chapter 11 we had a specification expressed in terms of an FSM and an FSM table. These are reproduced here as Figure 13.1 and Table 13.1.

To implement this design we keep in mind the principles and advice developed in Chapters 4–8. The natural data structure is one that fits the structure of the FSM table – a simple two-dimensional array, and the algorithm is something that responds to events.

Clearly this can be implemented as an object with a single method **event** which takes a parameter of type **card**. Here is code that implements the state machine. For the purposes of illustration, **card**s are

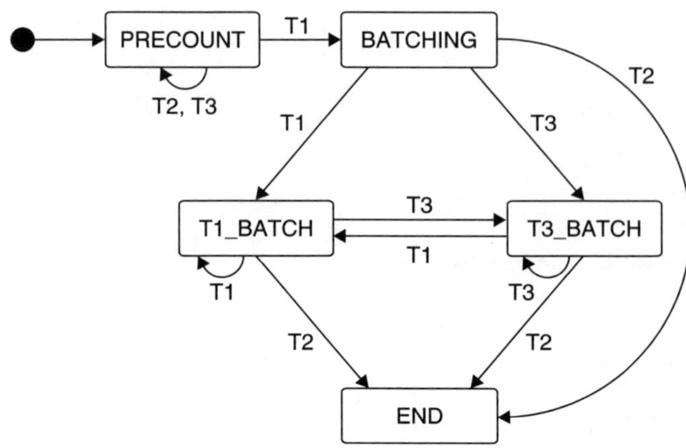

Figure 13.1 *A finite state machine*

Table 13.1

	T1	**T2**	**T3**
PRECOUNT	*Display T1* BATCHING	*Increment A* PRECOUNT	*Increment A* PRECOUNT
BATCHING	*Increment B and C* T1_BATCH	*Display T2, A, B, C, D* END	*Increment B and D* T3_BATCH
T1_BATCH	*Increment C* T1_BATCH	*Display T2, A, B, C, D* END	*Increment B and D* T3_BATCH
T3_BATCH	*Increment B and C* T1_BATCH	*Display T2, A, B, C, D* END	 T3_BATCH
END	*Signal error* END	*Signal error* END	*Signal error* END

supposed to contain the type number and another integer value; a **main** function is included that simulates the arrival of cards according to user input.

```cpp
#include <iostream.h>
int event_value;
// The "value" of the event; i.e. the integer data on the card
enum state {precount, batching, t1_batch, t3_batch, end};
state current_state = precount;
int a, b, c, d;        // Counters
// Now follow the action functions
void dis_t1()      { cout << "The current event's value\
    is " << event_value << "\n"; }
void inc_a()       { a++ ;}
void inc_b_c()   { b++; c++;}
void disp()
   {
   dis_t1();
   cout << a << ' ' << b << ' '<< c << ' '<< d << '\n';
   }
void inc_b_d()   { b++; d++;}
void inc_c()       { c++ ;}
void inc_b()       { b++ ;}
void error()        { cerr << "Received event after reaching end\
    state\n"; }
void do_nothing() {}
struct table_entry {
   void (*action)();  // Pointer to the action to do
   state next_state;
   } table[5][3] = {
      dis_t1, batching, inc_a, precount, inc_a, precount,
      inc_b_c, t1_batch, disp, end, inc_b_d, t3_batch,
      inc_c, t1_batch, disp, end, inc_b_d, t3_batch,
      inc_b, t1_batch, disp, end, do_nothing, t3_batch,
      error, end, error, end, error, end
   }; // So all the FSM is contained here
```

```cpp
int main()
  {
  int type;
  while(1)
    {
    cin >> type >> event_value; // User enters a card type
                                //  from 1 to 3 and a value
    if (type < 1 || type > 3)
       break;    // Enter illegal card type to quit
    (*table[current_state][type-1].action)();      // Do action
    current_state = table[current_state][type-1].next_state;
        // Change state
    }
  return 0;
  }
```

I originally wrote the above program in a more object-oriented style, making a class called **state_machine** that included all the actions as private member functions, and having a single public method called **event**. The ADT was then instantiated in a single FSM. The code for doing this is more complicated than that shown above, because it is necessary to take addresses of member functions for a particular object. Can you find a way to rewrite the code in a more object-oriented style that avoids this difficulty? Hint: Stick with a static, preinitialized table of static action functions and data. Access it using normal functions that are friends of a **state_machine** class. The class will contain all variable data for instantiated objects (e.g. **current_state**).

Can you write a program that reads in a textual description of a state/event table, and writes out C++ code like that above to implement the state machine? For example, you might allow the above table to be specified:

```
3 events T1 T2 T3
5 states PRECOUNT BATCHING T1_BATCH T3_BATCH END
PRECOUNT:
T1  { dis_t1();}  BATCHING
T2  { a++;}       PRECOUNT
T3  { a++;}       PRECOUNT
```

Your program would then be a translator of this miniature language into C++. Your translator involves simple parsing which you should represent with a finite state machine. Indeed, once your translator is working, you could try rewriting it in its input language, then using it as its own compiler!

13.3 Implementing with data structures and algorithms: rules of representation selection

Turning higher-level representations into code means deciding how objects and procedures are to be implemented in terms of data structures and algorithms. Assuming you know what is available, a choice has to be made to find the correct representation at the programming level.

The general rules of representation selection are as follows:

- Choice of data structure and choice of algorithm have to be considered together.
- The algorithm is (probably) in a book, and the general form of the data structure is too.
- The particular constraints and criteria of the problem to be solved have to be considered in choosing data structures and algorithms.
- The consequences of changes to those constraints and criteria should be considered in terms of the candidate data structures and algorithms.
- The constraints and criteria of the problem should be used, in conjuction with the known performance characteristics of the candidate algorithms to evaluate the performance consequences of the different choices. Use Big-O analysis to evaluate and compare algorithms. Calculate the memory requirements for data structures. Remember you can use the **sizeof** test program introduced in Chapter 4 (page 61) to tell you how big the built-in types are on your platform. You can then calculate the memory requirements of structures and collections built from those types.
- If N is known, then work out the running times in numbers of primitive operations: an $O(N)$ algorithm might be better than an $O(\log N)$ algorithm, simply because the first requires $N/100$ operations, the second requires $\log_2 N$ operations, and N is small enough. If N is not known, at least work out the running times for the largest possible N. Again, you may find that even for the worst case, the $O(N)$ algorithm beats the $O(\log N)$ algorithm.
- A program should be correct, reliable, maintainable, efficient. In choosing a representation it is easy to get absorbed in the last of these, forgetting that the others are important too. Never underestimate the virtues of a simple solution – it is likely to be easy to program correctly, reliable and maintainable. (If this sounds familiar, it is. Don't forget it!)

These rules are general guidelines. To see how they work in practice, we return to the problem from Chapter 7 – how to design a table of pairs.

13.4 The ADT table (again)

To recap from Chapter 7, the abstract data type table of pairs is to be implemented with the following access functions:

```
add_to_table(int, int)
remove_from_table(int)
get_other_stuff(int, int&)
number_in()
```

It was stated there that the choice of data structure and algorithm would be contingent on the particular constraints and criteria of the situation. Let us now consider a particular situation and develop a design.

A specific implementation

Suppose the ADT table of pairs must be able to store up to 5000 items with integer keys, any of which may take a value between $-1\,000\,000\,000$ and $+1\,000\,000\,000$. The other integer in the pair is from the same range. Typically the table will contain at least 3000 entries. There is a space constraint of 256 kbytes of memory, and the single evaluation criterion is that all access functions should be as fast as possible. The typical usage of the table will be instantiation, followed by many calls to **add_to_table** followed by interleaved calls to all four access functions.

How should the table be organised? The specification gives several pointers for the design:

- The range $[-1\,000\,000\,000,\ 1\,000\,000\,000]$ means that long integers are required. Each of these will occupy 32 bits or 4 bytes which means that 8 bytes are required for each pair (not counting any pointers that might be needed if the table is implemented with linked lists, a tree, or a hash table).
- A maximum of 5000 entries means there is plenty of room for extra pointers in the pair data structures. 256 kbytes are available, and there is no requirement that the memory used varies with number of entries. In a case like this, an array is clearly the simplest implementation.
- If a sorted array is used, the initial run of **add_to_table** calls can be speeded up by just appending new items to the end of the array, then when the first **remove_from_table** or **get_other_stuff** occurs, the array would be sorted and duplicate keys removed, retaining only the most recently added.

Let us consider the performance implications of the three most likely implementations:

- If an unordered array is used, each **add_to_table, remove_from_table** and **get_other_stuff** will require, on average, about 1500 operations. (In the case of **add_to_table** these operations are in the call to **remove_from_table** to remove any earlier items with the same key.)
- If an ordered array is used, an **add_to_table** requires $\log_2(3000) = 12$ operations to find the appropriate position, followed by a move of 1500 * 8 bytes (on average). **remove_from_table** requires the same number of operations, while **get_other_stuff** needs just 12 operations. In comparing the ordered and unordered array implementations, much depends on the speed of memory moves.
- The third alternative is a hash table. Because nothing has been said about the distribution of key values it is conceivable (but *very* unlikely) that they will all hash to the same table index, and that 5000 items will have to be chained into a single linked list. This highly inefficient situation is unlikely, but it is well to confirm that

there is no perverse pattern in the data before proceeding with a hash table. If the keys are well distributed (the best case), then every search, delete or add request will require a hash function step, followed by an average of $(3000/M)/2$ steps along a linked list (where M is the size of the hash table). If M were 101, as in the example in Chapter 7 (page 204), this would mean an average of 1 modulo operation and 15 steps per access. Most hash tables are larger than this; if M were 997, then the average would be 1 modulo operation and 2 steps.

So for the problem as specified, provided the keys are reasonably distributed, a hash table seems like the appropriate implementation.

But before committing this to code, step back and remember:

Rule 9

Design software to minimize the damage caused by changes to requirements.

What are the consequences of choosing a hash table? There are some things a hash table cannot do that an ordered array can. These are not in the specification – that is why we think a hash table is the right answer. But, what happens next?

The 'customer' (or the other programmer who is using your ADT) comes along and says 'Can you add an access function to print out the contents of the table in order?'

You ask with trepidation, 'What order?'

Unless the customer replies 'Any order will do', you are in trouble. If they want it in order of entry, then an unordered array is indicated, if they want it in order of key size (most likely), an ordered array or binary search tree is indicated. A hash table is hugely inefficient for either of these.

Surely it is unreasonable to spend your time second-guessing the specification? Yes, you can go too far. But you do need to know the properties of the data structures well enough to identify what changes to the specification will make your design collapse, and therefore to probe the customer beforehand about definitive specification change possibilities.

Assuming, then, that the ADT table of pairs as defined is not going to change, and assuming that the keys are well distributed, here is the recommended implementation.

```
//
//
// ADT table of pairs
// using hash table with separate chaining
//
//
struct pair {
  int key;
  int other_stuff;
  pair *next;
  };
```

```
class table_of_pairs {
  int table_size;
  pair hasht[997]; // Zeroth list entries - only the pointers
                   // matter here
  int num_in_table;
  int hashf(int key) const
      { return(key%997); };
public:
  table_of_pairs();
    ~table_of_pairs();
  int add_to_table(const int, const int);
  int remove_from_table(const int);
  int get_other_stuff(const int, int&) const;
  int number_in() const
      {return(num_in_table); };
  };

//
// Methods for the class table_of_pairs follow
//
table_of_pairs::table_of_pairs()
  {
  for (int i = 0; i < 997; i++)    // Initialize everything to empty
    {
    hasht[i].key = hasht[i].other_stuff = 0;
    hasht[i].next = 0;
    }
  };
table_of_pairs::~table_of_pairs() // Destructor has to
                                  // deallocate memory
    {
    for (int i = 0; i < 997; i++)
      {
      if (hasht[i].next)
        {
        pair *p = hasht[i].next;
        pair *q;
        while(q = p->next)
          {
          delete p;
          p = q;
          }
        delete p;
        }
      }
    }

int table_of_pairs::add_to_table(const int key, const int other)
    {
    remove_from_table(key);        // To prevent double entries
                                   // with this key
    pair *p = &hasht[hashf(key)];
    pair *q = new pair;     // Set up new item
```

```cpp
        q->key = key;
        q->other_stuff = other;
        q->next = p->next;     // Hook new item at
                               // beginning of list
        p->next = q;
        num_in_table++;
        return(0);
        }
int table_of_pairs::remove_from_table(const int key)
   {
   pair *p = &hasht[hashf(key)];
   while(p->next)
     {
     if (p->next->key == key)
       {
       pair *q = p->next;          // Unhook this item from the
                                   // linked list
       p->next = q->next;
       delete q;
       num_in_table--;
       return(1);
       }
     p = p->next;
     }
   return(0);       // Key not in table
   }
int table_of_pairs::get_other_stuff(const int key, int &other) const
   {
   const pair *p = &hasht[hashf(key)];
   while(p->next)
     {
     p = p->next;
     if (key == p->key)
       {
       other = p->other_stuff;
       return(1);
       }
     }
   return(0);       // Not found
   }


#include <iostream.h>
int main()
   {
   cout << "This program exercises a table of pairs.\n";
   cout << "In the instructions below, m and n are integers\n";
   cout << "Enter a m n to store integers m/n as a key/value\
        pair\n";
   cout << "Enter f m to retrieve a value associated with key\
        m\n";
   cout << "Enter r m to remove a pair with key m\n";
```

```cpp
        cout << "Enter q to quit\n";
        table_of_pairs table;
        while(1)
          {
          char c;
          int m, n;
          cout << "\n > ";
          cin >> c;
          switch (c)
            {
            case 'q':
              return 0;
            case 'a':
              cin >> m >> n;
              table.add_to_table(m, n);
              break;
            case 'f':
              cin >> m;
              if (table.get_other_stuff(m, n))
                cout << "value associated with key " << m
                << " is " << n;
              else
                cout << "Cannot find key " << m << " in the\
                table";
              break;
            case 'r':
              cin >> m;
              if (table.remove_from_table(m))
                cout << "Item with key " << m << " removed";
              else
                cout << "Cannot find key " << m << " in the\
                    table";
              break;
            default:
              cout << "Invalid input\n";
              break;
            }
          }
        }
```

> Read and understand the above code. Test it out using the simple
> interactive test shell.

Other scenarios and their implications

In the discussion above we have seen the analysis used for a particular situation. Clearly there are many other possible scenarios for an ADT table. Table 13.2 shows where similar analysis would lead if the ADT table of pairs has different constraints and criteria.

Table 13.2 *Alternatives for ADT table of pairs design*

Table size		Range of key values	Other access functions required or possible	Probable design (all other things being equal)
Max.	**Typical**			
5000	3000	$(-1B, +1B)$	None	Hash table
5000	3000	$(-1B, +1B)$	Iterate through table in order of update time	Unordered array (ordered by time)
5000	3000	$(-1B, +1B)$	Iterate through table in ascending key order	Ordered array
5000	30	$(-1B, +1B)$	Any	Linked list or hybrid because of large difference between typical and max. set size. (Might use resizable array – when full, allocate bigger array and copy)
50	30	$(-1B, +1B)$	Any	Unordered array (keep simple when overhead small)
5000	3000	$(-10k, +10k)$	Fairly even	Array of 20 000 structures, each with an **in_use** field and an **other_stuff** field. Index in array corresponds to key

Of course, a completely new set of solutions would be generated if the memory available were different, or the relative frequencies of access function calls were different.

In Chapters 15–17, we will see several more examples of representation selection based on analyses like those in this chapter.

The bottom line is that the analysis of alternatives must drive detailed design. Even in coding, analysis is essential.

14 Testing

14.1 Introduction

Testing has four roles in software design:

- finding faults
- assessing performance
- discovering behaviour
- driving code development.

This chapter is about all of these, but it concentrates on the first.

14.2 Finding faults

Rule 14

Test to break.

Testing is the process of exercising a system with the intention of finding errors. This definition carries the implicit belief that the system has errors, and it puts testing in the guise of detection – teasing out the secrets hidden in the program. It is derived from Myers, whose classic book *The Art of Software Testing* remains the standard reference. The definition is not without detractors who say that testing should be concerned with proving a system is correct. But software experience suggests that errors are lurking in every substantial program. By making a virtue of finding errors, Myers' definition prevents the human tendency to be a lenient judge of one's own, or someone else's, software creation.

In the minimalist design methodology, testing is shown as the last item. In its role of validation – checking whether a system meets customer requirements – it is prominent late in a project. But testing permeates the whole design process. To see how it fits at every stage, we'll consider the situation in a medium-sized project with a testing group independent of the developers. In a smaller project there may be no independent testers, but the jobs still have to be done.

Testing begins at the problem statement. When criteria are developed, they must not only be achievable, they must also be testable. That is why criteria should be quantified – things that can be measured, and measured at a reasonable cost. It is no use requiring that a piece of software work on 20 different platforms, if the project budget will not allow those platforms to be bought or rented. If a criterion has no hope of being tested, it should not be there.

If the problem statement is fleshed out into a specification, the issue of testing becomes more critical. Every statement of the specification should be testable so that it can ultimately be validated. Moreover, it should become clear at this stage what the equivalence classes of inputs and outputs are (see the discussion of black box testing below). This information will be used for calculating resources for the later system testing phase.

Test driven development

In the late 1990s a new and radical approach to testing became prominent: Test-Driven Development (TDD). TDD is one of the 12 core principles of Extreme Programming, and is also advocated by some people who are lukewarm about accepting the rest of the Extreme Programming agenda.

Coding is the central activity in TDD, and it is split into three: test case coding, functionality coding, and refactoring. All are done within the framework of an existing, working program. The first step in introducing a particular new piece of functionality is to write a test case for it. The existing program fails the test, so you next write just enough code to pass it. When the new modified program compiles and passes the test, you contemplate whether the code can be refactored – that is, revised so that it's more readable, simpler and more flexible, without altering its functionality. If it can, you do the change, then check that all tests still pass. Now you go round the loop again, starting with the test case for the next piece of functionality.

In test-driven development the conventional slow cycle of program, test, debug is replaced by a much faster rhythm: write a failing test case, code just enough to pass the test, refactor. The code grows in small increments, so small that writers on TDD say that you can often get round the loop several times an hour. At each iteration you have a working program, and, as it grows, it is constantly adjusted to keep it clean and under control.

Debate continues about the relative merits of conventional development and test-driven development. Both sides have tales of benefits and disasters. But, as you might have guessed from the difficulty of doing scientific studies of programming practice, there have been few controlled empirical comparisons. You might also have guessed, because this is a sidebar, and conventional testing is discussed in the main column, that this book is equivocal about TDD. That's true and I'm going to explain why, but you should regard my arguments with the usual scepticism and make your own judgement.

In scientific and engineering programming we are often concerned with difficult-to-understand methods. Numerical analysis is heavily used, with extended and subtle algorithms. We may be analysing uncertain data or using statistical tools. Most important, our program (or a major part of it) is an integrated whole and it is impossible to develop it bit by bit. If we tried, we might grow a candidate solution that solves the first 99% of test cases, but has no way of solving the remainder. To handle all 100% we have to develop holistically, then test holistically.

A more fundamental problem with TDD is that it embraces testing as a constructive activity. In the long run, testing *is* constructive. But at the program face, where you're subjecting your code to the hardest stresses that you can, your attitude should be that testing is *destructive*.

Rule 14

Test to break.

Real testing is like real science when it tests a hypothesis – throwing as much as it can against the hypothesis, so that, by survival, the hypothesis gains credibility. The test cases developed in TDD do not make this kind of demand: they simply give a supportive push to the development of the next piece of code. TDD testing is too shallow. In fairness, TDD programmers may realize that their unit testing is by no means complete when their final round of test case/code/refactor is done. But the embedding of testing in the coding process makes it harder to step back and come back with the big guns.

Having said all that, TDD is very, very good in at least one respect: it makes you think about re-representation.

Rule 2

**Don't expect to get the representation right first time. Look for opportunities
to improve code by *re*-representation.**

By putting refactoring into the tight, fast, loop of incremental development, it ensures that it doesn't get postponed.

TDD may be appropriate for broad, shallow, programs. That is, programs where lots of small pieces of functionality are collected, rather than building into a complicated process. The stable case study in Chapter 17 is an example, and the later iterations in its development get close to TDD.

By this stage, the testers know the scope of the project, and the detailed requirements. They can now proceed independent of the developers to generate a test plan.

Within the detailed design phase, programmers practise debugging. They may also do white box and black box tests of their own modules. As parts are assembled, the design team verifies that subsystems work correctly. This is integration testing, and checks not only the modules, but the modularization. Although the modularization is the most critical part of the design, an error in the way an access function was specified may not be too critical. It will be uncovered at this stage, if not earlier. When the developers are satisfied, they hand the whole system over to the testers for 'arm's length' system testing.

System tests are those developed by the testers from the specification. A subset of the system tests, called the acceptance tests, are designed to check conformance to specification, step by step. If a product fails one of these, it has failed to meet specification and can be rejected by the customer. The remaining system tests give more confidence about the robustness of the product.

Through development, another stream of testing takes place in reviews or inspections – formal structured meetings where participants read and check a document or program.

In the rest of this chapter, we will look at testing techniques, applicable on any sized project. First, we will consider static analysis – the reading of code without executing it on a computer. When done in a group, static analysis becomes an informal code review or a formal code inspection. When done in solitary, it is a frequent adjunct to machine debugging. Then we will consider dynamic analysis – running the code on the computer. Black box and white box testing will be described and illustrated.

14.3 Static analysis

There are three ways of approaching static analysis in testing, shown in Table 14.1.

Table 14.1 *Static analysis methods*

	Method 1: Solitary testing	Method 2: Informal review	Method 3: Code inspection
Purpose	Part of debugging	Error detection, consideration of alternative solutions	Error detection only
Organization	None	Little	Small group with defined roles
Leadership	N/A	Author	Moderator
Procedure	Read code, 'hand execution' = playing computer	Read code	Standardized protocol for reading and commenting. Documented
Training	None	None	For all participants
Error recording	Code annotations	Described	Formal method and tracking

We will consider code inspection only because this is the most formal technique. The others are less structured but similar approaches. It is important to note that error correction is not part of a code inspection: the object is to identify errors; their correction is the responsibility of the programmer concerned.

Code inspection

The success of code inspection in practical development environments has been widely reported. For example, American Express's first use of code inspections led to a product with a very low dynamic analysis error delivered on time and 10% under budget. The cost of inspections was 18% of the total project cost. A more usual figure is 10%. In any case, code inspections do consume a significant fraction of project resources.

A code inspection is conducted in a formal process, and participants have well-defined roles. Tables 14.2 and 14.3 give typical examples.

Table 14.2 *Roles in a code inspection*

Producer	the writer of the program under inspection; usually a quiet participant.
Moderator	the organizer, and leader of inspection meetings.
Reader	who presents the product, paraphrasing line by line.
Scribe	who records the errors detected.
Inspector	who inspects the product; all participants are inspectors.

Table 14.3 *Phases in a code inspection*

Planning	The moderator has to schedule meetings based on an inspection rate of: 300 lines/hour for overview; 150 lines/hour for preparation; 150 lines/hour for inspection meeting.
Overview	A meeting where the producer explains the product, and distributes it to the inspectors.
Preparation	Solitary examination of the code by each inspector.
Inspection	The reader reads through the code at an appropriate pace, paraphrasing and explaining. Inspectors comment as the reading progresses. The scribe classifies errors (see below) as they are detected. At the end the scribe presents errors for review.
Rework	The producer addresses all the errors.
Follow-up	The moderator checks the revisions and certifies the inspection. (A second inspection meeting may be used for checking.)

When errors are detected in a code inspection, they are classified according to a formal scheme along several 'dimensions'. For example:

Error severity:
    Major   (Observable product failure will result)
    Minor   (No failure of product)

Error classes:
    Missing
    Wrong
    Extra

Error types:
    Logic
    Performance
    Standards
    Documentation

A successful inspection is usually the result of careful preparation on the part of team members, and slow, methodical coverage during the inspection meeting (keeping to the 150 lines/hour rate). The team size (3, 4 or 5) seems to have little effect on the success of the inspection. The crucial thing is to inspect slowly and thoroughly.

14.4 Dynamic testing: (deterministic) black box test

This section and the next two describe black box and white box testing. Black box testing tests function, treating the internals of a program or module as hidden. Deterministic black box testing uses problem constraints and success criteria to derive test cases that probe the extreme cases of a module's performance. Statistical testing uses a large, perhaps random, set of inputs and systematically checks the outputs looking for faults. Both deterministic and statistical black box tests might be used in module testing, integration testing (i.e. as modules are added to a system) and acceptance testing (done to validate an entire product). White box, or 'glass box' testing uses the code of modules to decide what tests to run.

The following procedure may be used for developing black box test cases:

1. Divide the input conditions into equivalence classes. All items in an equivalence class are similar in a particular aspect.
2. Choose representative input samples from each equivalence class.
3. Choose boundary samples too, if the boundaries are not an equivalence class of their own.
4. Write down a test table with expected results.
5. Run test cases and compare results with the test table.

A useful heuristic for testing collections can be applied in step 1. Many algorithms on collections are sensitive to the special cases:

- zero items in the collection
- one item in the collection
- number of items in the collection is a power of 2
- number of items in the collection is odd.

The last two of these exercise different types of recursion. Therefore, for collections, suitable equivalence classes can be chosen to represent these four cases.

At steps 2 and 3 it is important to see whether some samples can be used for testing different equivalence classes. Typically two classes may be defined by relations which are independent of each other, and a single test case can exercise both classes.

Note the importance of writing down expected results before running the tests.

Example 1

Suppose we wish to test a binary search algorithm. We assume the precondition that the array is ordered. If this is untrue, the behaviour of the algorithm is undefined.

Following the procedure above, at step 1 we can identify the following equivalence classes:

Element is in the array/Element is not in the array.

Zero items in the array/One item in the array/2^n items in the array/Odd number items in the array/Other even number of items in the array.

At step 2, we can write down the following test cases:

1. Array size 0
2. Array size 1, item in array
3. Array size 1, item not in array
4. Array size 4, item in array
5. Array size 4, item not in array
6. Array size 5, item in array
7. Array size 5, item not in array
8. Array size 6, item in array
9. Array size 6, item not in array

At step 3, we can add:

10. Array size 4, item first element in array
11. Array size 4, item last element in array
…

The test table is created at step 4 and shown in Table 14.4.

Example 2

Consider the following program description: **triarea** is a program that takes three integer arguments which give the lengths of the sides of a triangle. It prints out the area of the triangle rounded to the nearest integer, or an error message if that is not possible.

Table 14.5 shows some equivalence classes for this problem, grouped according to the aspect that defines the equivalence relation.

From these equivalence classes, a test table can be drawn up. The first part of it might look like Table 14.6.

Table 14.4 *Black box test table for example 1*

Array size	Array	Item	Found?	Position
0	[]	17	false	– (don't care)
1	[17]	17	true	0
1	[17]	16	false	–
4	[17, 18, 23, 30]	23	true	2
4	[17, 18, 23, 30]	25	false	–
(etc.)				

Table 14.5 *Equivalence classes for example 2*

Relation	Equivalence classes		
Number of arguments	$=3$	<3	$\geqslant 3$
Type of arguments	All positive integer	At least one floating point	At least one negative or zero
Valid triangle	No argument is greater than the sum of the other two	(Converse)	
Order of length of sizes	$A > B > C$	$C > B > A$	(etc.)
Rounding	Result $<$ x.5	Result $=$ x.5	Result $>$ x.5
Area expressible as a short integer	Area $<$ 32 768	Area $\geqslant$ 32 768	
Shape of triangle	Scalene	Isosceles	Equilateral

Table 14.6 *First part of black box test table for example 2*

Input line	Expected output
Triarea 3 4 5	6
Triarea 1 2 3 4	Too many arguments (error message)
Triarea 1 2	Too few arguments (error message)
Triarea 3 4 5.1	Invalid argument (error message)
(etc.)	

14.5 Statistical black box testing

Many specifications are not as clear cut as the examples above. It is often hard to list all the equivalence classes. It could be that some special combination of input parameters causes an error. Then you have to fall back on statistical testing where you present to the system a large number of sample input sets, hoping that they probe the different possible failure situations. The output of the program must be compared with ground truth results independently derived from the input. Any discrepancies expose an error, and the task is then to home in on the cause.

Three different ways to generate the sample inputs are:

- Using random inputs might be appropriate when the program has to handle arbitrary data. You could, for example, write a program that generates random text files of arbitrary size to use in testing a find-string program.

- There may be a body of existing data that can be used as representative inputs. The good side of this is that the data will exhibit the characteristic of inputs the program has to process in practice and so focuses more on what matters for the application. The bad side is that the data may not be sufficiently diverse to uncover extreme-case faults. If the program is a new version, or a new solution to an old problem, there may be particular sets of data that were problematic to earlier programs. These can be specially included.
- The sample inputs can be specially designed based on the programmer's belief about possible different kinds of behaviour. This is an informal version of deterministic black box testing discussed above. Instead of being able to identify different equivalence classes explicitly, the programmer may simply have an intuition such as 'things might get unstable when the inputs are very small'. The test set is therefore organized to probe such cases, and to see what happens on the boundaries between different types of behaviour.

Similarly here are three ways to generate the ground truth results:

- For some applications the generation of results can be combined with the generation of test cases. In the case of the find-string example, the test-case generator could randomly choose a location to insert a particular random string. That string would then be a test probe for the program, with the known location the expected result.
- Have a completely different implementation that solves the same problem; feed the inputs into both programs and compare the outputs. In this case discrepencies could result from either program being wrong, but if they are relatively few, it should be possible to debug both simultaneously. You have to be aware that in some cases neither program may be correct, though this is only a serious problem when both are wrong in the same way.
- Hand compute some of the cases based on the program's output. Often it is possible to identify from outputs where the program is making mistakes. For these cases, the programmer can look at the corresponding input and work out what the computer should do.

Having a large representative test set is an important part of *regression testing*. Whenever a major change to a program is implemented, regression testing goes back over a large suit of cases that worked previously, making sure that nothing has been broken.

14.6 White box testing

White box testing aims to exercise the program based on a knowledge of its internals. The following procedure may be followed:

1. Trace through the code. Identify all the branches and loops.
2. List possible conditions for all branches and three cases for each loop: Skip (i.e. do not execute the body of the loop), Execute once, Execute more than once.
3. Devise a test case to exercise each item on the list prepared under step 2.
4. Determine what the expected results should be.
5. Run tests.

Example

Here is some code that finds the mean of the non-zero entries in an array:

```c
int mean(int *array, int num, float *avg)
{
int accumulator = 0;
int cnt = 0;
int i;
for (i = 0; i < num; i++)
  {
  if (*(array+i) != 0)
    {
    accumulator += *(array+i);
    cnt++;
    }
  }
if (cnt > 0)
  *avg = accumulator/cnt; //BUG HERE
else
  *avg = 0;
return(cnt > 0);
}
```

For this code, the branches and loops, together with their possible outcomes, are shown in Table 14.7.

Note that columns have been included for cross checking which conditions are satisfied as tests are developed. Four possible test cases are shown in Table 14.8, and the conditions list is repeated in Table 14.9, now with test case checks included.

Difficulties with white box testing

In white box testing, each test case exercises some combination of the decision and loop cases. It is sometimes hard to decide which combinations to test because of the interdependencies in the code. The limitation

Table 14.7 *White box test table for example program*

Decision or loop	Possible outcomes	Test cases			
		1	2	3	4
i < num	skip once >1				
*(array+i) != 0	True False				
cnt > 0	True False				

Table 14.8 *Four test cases for white box testing by case*

Case	Input values		Expected outcome		Actual outcome
	array	num	*avg	return value	
1	[89]	1	89.0	1	
2	[]	0	0	0	
3	[53, −77]	2	−12	1	
4	[27,0]	2	27	1	

Table 14.9 *White box test table with test case checks*

Decision or loop	Possible outcomes	Test cases			
		1	2	3	4
i < num	skip		√		
	once	√			
	>1			√	√
*(array+i) != 0	True	√	−	√	√
	False		−		√
cnt > 0	True	√		√	√
	False		√		

to three cases for loops is another potential deficiency of white box testing, though it is a practical compromise. More importantly, white box testing does nothing to test if something has been left out. In the above example, no test expects a fractional average, yet the bug in the code is that it always does an integer division to work out the mean. For this reason alone, it is often appropriate to concentrate on black box testing.

14.7 Final words on testing for finding faults

Developing test cases is an art. There are books that give good advice, particularly Myers (1979), which also has the none-too-encouraging graph shown in Figure 14.1.

In words, the graph says the more errors you have found so far, the more errors there are still to find. The moral of this is that if testing reveals a highly error prone module, don't fix it, scrap it.

14.8 Assessing performance

Your program will be designed to meet measurable criteria.

Rule 10

Steer design by clear criteria.

Some criteria may effectively be constraints – if you don't meet them, you've failed – while others give a scale on which you strive to do as well as possible. The fault-finding approach to testing ensures that you meet constraints. But you will also test to see how well you've done,

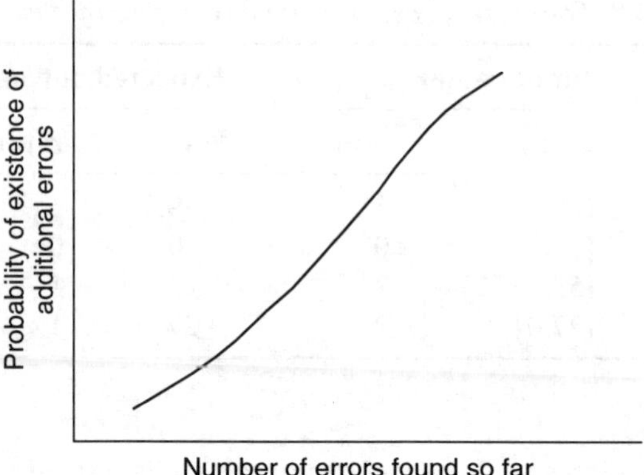

Figure 14.1 *Myers' graph of bug probability as a function of discovered bugs*

usually by comparison with some earlier approach. Even when you're assessing performance against a secondary criterion (execution speed, perhaps), you will compare your current implementation at least against its earlier versions.

In assessing performance it is crucial to be fair, particularly when comparing against other people's methods and implementations. This means testing with a large and diverse set of inputs that are representative of the range of inputs seen in practice. The regression test set that you use in statistical black box testing is probably acceptable for this purpose, though it's also important, if your program is in some way trained by a stock set of inputs, that the same inputs should not be used in assessing performance. Just as in fault finding testing, where you test to break, you must adopt an antagonistic stance towards your own method, making sure that any benefit-of-the-doubt is given to competitors.

14.9 Testing to discover

Rule 7

Test to discover information.

Rule 8

Hide information in public; reveal it in private.

Chapter 11 discussed the tools available for researching possible solutions including experiments, mock-ups and simulations. These are useful not just early in the design process, but during development too. When your initial program appears to be working, if it is at all complex, it's worth trying to write at least one test program that helps you visualize what's going on. This is illustrated in the case study in Chapter 16. As the program develops don't remove the ability to link it with visualization tools. Remember,

Principle 4

Every program has surprises.

Even the one you've spent your recent life developing.

14.10 Release

A product is released to the customer or market when it meets acceptance tests, fulfilling the problem statement success criteria. The performance of the product is monitored and feedback from the market gives pointers to future development and maintenance.

Many programs that you write as an apprentice programmer are not released at all – they are for your development, or stepping stones towards a solution. There is only one thing to be done with such programs:

FINISH THEM

14.11 Chapter end material

Bibliography

The classic reference on software testing is:

Myers, G. J. (1979). *The Art of Software Testing*, John Wiley & Sons, Inc. A second edition has been advertised for 2004.

Hetzel, W. (1993). *The Complete Guide to Software Testing*, John Wiley & Sons, Inc. is the only more recent book wholly devoted to software testing.

Most software engineering books include coverage on testing. See the references in Chapter 2. Books on style and debugging are listed in Chapter 6.

Beck, K. (2002). *Test Driven Development*, Addison-Wesley, gives the Extreme Programming take on TDD.

15 Case study: Median filtering

15.1 Introduction to the case studies

The case studies in Chapters 15 to 17 discuss the development of three real programs. Each illustrates multiple facets of the design processes, with different emphases. Each of the problems being solved is a sub-problem of scientific and engineering R&D, and each represents the sort of task an individual scientist or engineer might face in practice. Because the programs are not long, the requirements for specification, modularization and detailed design are modest compared to large software systems. But the principles and practices are constant, as we will see.

15.2 Introduction to this chapter

In this chapter we will develop an n-point median filter for signal processing. Median filtering is widely applied in signal conditioning (immediately after collecting measurements from a sensor) and later manipulation of signals for enhancement, coding or analysis. The particular application that we'll focus on is in video signal processing, and we'll see how this presents some important constraints on the design. On the other hand, we'll always keep in mind that we'd like to develop software that is as widely useful as possible.

First, we'll look at the context: why you might want to filter signals with a non-linear operator like a median rather than a conventional filter. Then we turn to the problem: the particular video analysis application of median filters that motivated this case study. Next, the steps of the design process will illustrate:

- Rapid prototyping

Rule 5

**Make the first version durable not
functional, and get it running early.**

- The search for relevant prior art

Rule 11

Exploit existing solutions.

- Use of a doubly-linked list data structure
- Implementation and testing strategy

Rule 14

Test to break.

- Finishing: polishing the final version.

Several versions of the code will be presented. Beware: the second one contains a subtle bug that does not get resolved until the statistical testing stage. The final version implements a robust and efficient median filter class that solves the problem.

15.3 Background

The median of a set of numbers is the middle one when they are listed in ascending or descending order. So 3 is the median of $\{5, 4, 3, 2, 1\}$, of $\{1, 2, 3, 5, 8\}$, and of $\{3, 1, 4, 1, 5, 9, 2\}$ (which, like most real input sets, isn't in order yet). Those examples all involved an odd number of numbers; when you have an even number, the median is the average of the two middle items in a sorted list. So the median of $\{2, 0, 0, 4\}$ is 1. On the face of it, finding the median of $m = 2k + 1$ numbers means sorting the set, then choosing the kth, while finding the median of $m = 2k$ numbers means sorting the set, then averaging the $k - 1$ and kth values (assuming the sorted list is indexed from 0). There are faster ways to calculate a median than doing a full sort, but we need not consider them here, because our task is to design a median filter.

In *filtering* we have an input signal or sequence $S = (x_1, x_2, \ldots, x_n)$ of numbers from a limited range, and apply to them a *sliding window function f* of length m so that the output y_i is given by

$$y_i = f(x_j | j = i - o, \ldots, i + m - o)$$

where f could be any function of the samples in the window and o is an offset. Often filters are centred on the output index, and may even be symmetric or antisymmetric. In these cases, the window length is odd, and, as above, we can write $m = 2k + 1$ and

$$y_i = f(x_j | j = i - k, \ldots, i + k)$$

Some important filter functions are linear, and in this case the filtering operation is known as convolution. But non-linear filters can be used too, and one of the most popular is the median filter:

$$y_i = median(x_j | j = i - k, \ldots, i + k)$$

Of course, median filtering could use an even window length, but in all practical cases, an odd-length window is appropriate. We therefore confine our discussion to this case.

15.4 Why use median filtering?

Signals are often filtered to remove noise. This can be done with a linear filter, such as a simple average of the samples in the window, but that often causes unwanted side effects. For example, suppose you had a signal with values $\{0, 0, 0, 0, 0, 100, 100, 100, 100, 100\}$ that was corrupted by random noise to become $\{3, 15, -7, 2, -10, 94, 102, 100, 112, 91\}$. You could apply an averaging filter with window size 3 to obtain $\{?, 3.667, 3.333, -5, 30.667, 62, 96.667, 104.667, 101, ?\}$ where the '?'s indicate that the filter overlaps the end of the signal. This certainly smoothes out some of the noise, but it also blurs the strong step that is in the original. If, instead, you use a length 3 median filter, the result is $\{?, 3, 2, -7, 2, 94, 100, 102, 100, ?\}$ which preserves the step discontinuity much better. As a result median filters are commonly used for signal enhancement. If applied to an image, for example, a median filter will remove graininess and point noise but preserve the sharp edges between objects.

Linear filters are fairly easy to implement in software. Averaging is particularly easy and efficient: simply keep a running sum of the m samples in the window; as it is slid to the next position, subtract the sample just leaving the window and add the sample just entering; then divide the sum by the window size. A median filter is more complicated – perhaps an ordered list can be maintained with items removed and added as in heaps (see page 214) – but before considering alternatives let us turn to a particular application and look at its demands.

15.5 The application

One important module in a particular kind of video analysis is a temporal median filter. Each output video frame is the median of m consecutive input frames. That is, each individual output pixel is the median of the m values the pixel at that position had in m consecutive input frames. In a frame of $r \times c$ pixels (r for rows, c for columns), there will be $r \times c$ independent, simultaneously running median filters. By employing such a temporal median filter, a system can distinguish short-term temporal 'noise' (such as an object whizzing across the foreground of a picture over a couple of frames and occluding the scene behind) from a sequence of fast cuts between different scenes. The problem that faces us is how to implement a median filter that can do this accurately and efficiently.

15.6 Approaching the problem

Several features of the temporal median filter problem are immediately evident:

- The obvious decomposition of the total problem is to develop a single median filter and then replicate it many times (equal to the number of pixels in a picture – maybe quarter of a million). Object-based coding is therefore ideal: we will define an abstract data type that encapsulates median filter functionality, then instantiate a 2D array of objects of that type.
- There is a simple naïve algorithm for the median filter: read the m values in the current window into an array; sort the array; return the middle element. Such an implementation could be prototyped very rapidly.
- Median filtering is a well-known operation so presumably there are established, tested and true algorithms.

Bearing these three in mind, our best approach is to begin coding a class and a test program with the naïve algorithm; use this to evolve our public class interface and identify important design and test issues, then turn to prior art and do the 'real' implementation based on a study of the literature. Why is this approach best? Why is postponing the search for a good algorithm until *after* the first implementation good policy? Partly the reason is the primacy of first version as framework: we want to concentrate on getting the overall structure right first. But a second and probably more important reason is to do with testing. The key problem faced by this design is algorithmic. The naïve method will work, but it won't be good enough because it will be slow and clunky. Yet a high-efficiency method is likely to be tricky to code correctly. Because we are fairly sure that our naïve initial method will have little in common with the final version, it will make an excellent reference version for statistical black box testing. At this stage we can already see that once we have basic functionality, we will want to generate many test cases automatically. Our initial naïve implementation will provide a set of results for these. We can't regard the naïve implementation as the gold standard – it may have bugs – but it provides a cross check.

15.7 Rapid prototyping

We can use the framework generator on page 190 to create median-filter.h, medianfilter.cpp, testmedianfilter.cpp and a Makefile. Then opening medianfilter.h and medianfilter.cpp in side-by-side windows, we begin by adding expected data items and functions to median-filter.h.

```
// Header for medianfilter class, created Thu Aug 28
// 12:42:56 2003
// John Robinson

#ifndef medianfilter_H
#define medianfilter_H

class medianfilter {
   double *buf;
   int length;
   int bufptr;   // Index into circular buffer
public:
   medianfilter (const int len);
   ~medianfilter ();
   double addnext(const double newval);
       // Returns value displaced from buffer
   double getmedian() const ;
   void print() const ;// Debug function to show current state of
                       // buffer
};

#endif
```

As usual the choice of data structures affects the choice of algorithms and vice versa. We have to store the members in the set somewhere, so our first decision is to have a data item **buf** that will point to an array of length **length**. But how will items be added and removed? Assuming we want our data type to function *as a filter*, we should keep the current window in the object and each time we slide the window by one position of the input, we load the new item in and displace the item that's just left the window. So we need to store the window contents in a circular buffer. Therefore we need **bufptr**, an index into the **buf** array showing where the next item will be placed.

We have also defined the public interface for the median filter: a constructor in which the length of the window is specified, a member function **addnext()** which adds a new value, causing the displacement of the oldest value (which it returns, which may be helpful in debugging), and a member function **getmedian()** which simply returns the current median. Finally there is a **print()** function to show the current state of the buffer, which, again, will be useful in testing. What seems like a big decision has also been made – that the samples will be **doubles**. Indeed, the type of the samples is an important issue: since we're dealing with video coding, our samples will actually all be integers between 0 and 255, so could be represented as **int**s, **unsigned int**s or even **unsigned char**s. At this stage, we're fine choosing a type (**double**) and developing with that. Later we can make the code into a template version to handle diverse types, or perhaps have a single **type-def** declaration that can be changed depending on the application.

The next thing we do is think about **const**s. Already the class declaration includes some **const**s, where values are passed into functions. But we should also decide at this stage which member functions leave the object unchanged and preface their declaration with **const**. **print()** should just output current state information and should therefore be **const**. In our naïve implementation, we would expect to code the sorting as part of **getmedian()** but that need not alter the object itself. Therefore, we also provisionally declare **getmedian()** as **const**.

We now move on to write our initial implementation file median-filter.cpp:

```
// Implementation for medianfilter class, created Thu Aug 28
// 12:42:56 2003
// John Robinson

#include "medianfilter.h"
#include <cstdlib>
#include <iostream>
using namespace std;

// Private function for qsort. Will remove when
// we improve median finding
static int comparedoubles(const void *a, const void *b) {
  const double *da((const double *)a);
  const double *db((const double *)b);
  const double diff(*da-*db);
  if (diff < 0)
    return -1;
  else if (diff > 0)
    return 1;
  return 0;
  }

medianfilter::medianfilter(const int len) {
  length = len;
  buf = new double[length];
  for (int i = 0; i < length; i++)
    buf[i] = 0;
  bufptr = 0;
}

medianfilter::~medianfilter() {
}
```

```cpp
double medianfilter::addnext(const double newval) {
    double retval = buf[bufptr];
    buf[bufptr] = newval;
    if (++bufptr == length)
      bufptr = 0;
    return retval;
}

double medianfilter::getmedian() const {
    // Initial implementation: do sort each time. Must improve!
    double tempbuf[length];
    for (int i = 0; i < length; i++)
      tempbuf[i] = buf[i];
    qsort(tempbuf, length, sizeof(double),comparedoubles);
    if (length & 1) // length is odd
      return tempbuf[length/2];
    return (tempbuf[length/2] + tempbuf[length/2-1])/2;
}
void medianfilter::print() const {
    for (int i = 0; i < length; i++)
      cout << "buf[" << i << "] = "<< buf[i] << endl;
    cout << "Current buffer index is " << bufptr << endl;
}
```

The code here is straightforward. Only two things to note. First, our window is a circular buffer, and therefore we have the test on **bufptr** in **addnext()** which cycles it back to the start. Second, we have used **qsort()** as discussed in Chapter 4 to sort a copy of the buffer. (If we were using the standard template library, we would have accomplished the same thing with the function **sort()** on a **vector**.) Obviously, as the code itself says, this is a clunky implementation, but it is easy to code quickly, and thus it forms our first attempt.

We now develop the test framework for our new class. Initially we want to check that the circular buffer is working correctly, the true medians are being calculated, and the implementation does support arbitrary window sizes. We therefore make our initial test program an interactive loop where we enter values by hand and at each stage check that both the output and the internal state are correct:

```cpp
// Test program for medianfilter class, created Thu Aug 28
// 12:42:56 2003

#include "medianfilter.h"
#include <iostream>
using namespace std;

int main(int argc, char *argv[]) {
    int filterlengh;
    int inputlength;
    double val;
    cout << "Enter filter length: ";
    cin >> filterlengh;
    cout << "Enter input length (number of iterations): ";
    cin >> inputlength;
```

```
        medianfilter testmedianfilter(filterlengh);
        cout << "Buffer state:\n";
        testmedianfilter.print();
        for (int i = 0; i < inputlength; i++) {
          cout << "Next input value: ";
          cin >> val;
          val = testmedianfilter.addnext(val);
          cout << "Value displaced was " << val << endl;
          val = testmedianfilter.getmedian();
          cout << "Current median is " << val << endl;
          cout << "Buffer state:\n";
          testmedianfilter.print();
          }
        return 0;
}
```

The test program uses all the debugging features provided by the class itself to monitor what is going on.

When we compile and test the program it seems to work – except for negative filter lengths. Adding numbers one by one, we keep a running hand calculation of the median, and the program agrees. We therefore have an initial framework which can serve as reference comparison for our final version.

15.8 Exploit existing solutions

Our search for existing solutions begins with a search for 'median filtering' on Citeseer, Yahoo! and Google (see Chapter 11). Variations like 'median filter' + algorithm also are worth trying. Very quickly we find a pointer to http://iris.usc.edu/Vision-Notes/bibliography/twod266.html which is a whole list of publications on median filtering. Digging deeper draws a blank on online resources but some good leads on papers. One that sounds particularly promising is: M. Juhola, J. Katajainen and T. Raita, 'Comparison of algorithms for standard median filtering', *IEEE Transactions on Signal Processing*, Vol. 39, No. 1, January 1991, pp. 204–208. It's a little old, but most of the more recent stuff seems to be about hardware, and it has two attractive features: (1) it is a comparison paper, so it will be concise in describing algorithms and will give advice about how to choose an appropriate one, (2) it's only five pages long. So we head for the library and get a copy.

With permission of the authors and the IEEE, that copy is reproduced in the Appendix, page 398.

This paper seems like just what we want. After one read-through, it looks likely that we will want to concentrate on method D or E, since these are clearly the fastest. See Figure 4 on page 401 and Table 1 on page 402. Our next step is therefore to rough out an implementation based on one of these. When I originally attempted the design, I chose method E, because it's fastest overall, and its restriction to integers doesn't matter for the end application of video processing. I began by ensuring I understood the method, sketching out a histogram and running some numbers through the algorithm by hand, 'playing computer'. To be sure, I wrote the code for the key central section of the algorithm that involves stepping a counter up (or down) through the number of samples that are equal to the current median as samples are added to each side. This was important because the authors' description of the algorithm didn't explicitly say what happens when a new item has the same value as the current median, and I needed to check the assumptions I was making about that. Next I rewrote the header file and implementation just to use integers from lowest to highest. This introduced a lot of extra range checks so I decided to restrict the range to [0, 255], since these would be the values in the video frames. And then ... at last ... I realized what was wrong with method E.

Although the histogram method is fast, each median filter requires a histogram of bins (i.e. an array of counters) in the range of input values. Even restricting that range to [0, 255], and the window size to 255 (so that each histogram bin can be an **unsigned char**), means that each filter object must maintain 256 bytes of histogram. If we have a quarter million median filters running at once (for a typical-sized picture), that's 64 Mbytes of memory for what is just one component of the system. Sixty-four Mbytes is perhaps not prohibitive, but it is still a huge overhead on what method D requires.

Given that D can cope with unrestricted ranges, it surely is the better choice. How long did I waste on method E? About an hour. Could I have avoided that? Certainly – right there on page 207 of the paper (page 402 of this book), the authors state that the space requirement of method E is $O(U + m)$ where U is the number of integers in the range and m is the window size. All their other tested methods required space $O(m)$. I missed that, or perhaps I didn't register it until I'd started programming. But the cost was not too great, because developing incrementally exposed the problem more quickly than if I'd been occupied trying to write the whole of method E in one go.

Method D is a doubly linked list method, and as we know, linked list manipulation can be tricky. We always have to keep in mind what happens at the ends of the lists. So here is a carefully considered implementation that treats the ends specially. First, the header file which includes the whole class definition for a linked list node called **bufitem**.

```
// Header for medianfilter class, created Thu Aug 28
// 12:42:56 2003
// Revised Aug 28 15:15
// John Robinson

#ifndef medianfilter_H
#define medianfilter_H

class bufitem {
// All items private. Just used by friends
  double value;
  bufitem *smaller;// Using pointers rather than indices
    // means bufitems can be collected into other DSs than just
    // arrays.
  bufitem *greater;
  bufitem () {
    value = 0;
    smaller = 0;
```

```cpp
      greater = 0;
    }
    ~bufitem() {
    }
  void set(const double v) {
    value = v;
  }
  double get() const {
    return value;
  }
  void initpointers(bufitem *const predecessor,bufitem *const
      successor) {
    smaller = predecessor;
    greater = successor;
  }
  void putafter(bufitem *predecessor) {
    // Break existing link from predecessor to its successor
    // and insert this one
    bufitem *successor = predecessor->greater;
    predecessor->greater = this;
    if (successor)
       successor->smaller = this;
    smaller = predecessor;
    greater = successor;
  }
  void putbefore(bufitem *successor) {
    // Can't just use putafter(successor->before) because
    // this may be 0 on list
    bufitem *predecessor = successor->smaller;
    successor->smaller = this;
    if (predecessor)
       predecessor->greater = this;
    smaller = predecessor;
    greater = successor;
  }
  void removefromlists() {
    if (smaller)
       smaller->greater = greater;
    if (greater)
       greater->smaller = smaller;
  }
  bufitem *next() const {
    return greater;
  }
  bufitem *prev() const {
    return smaller;
  }
  friend class medianfilter;
  };
class medianfilter {
  bufitem *buf;
  int length;
```

```cpp
    int nextinsertion;   // Index into circular buffer
    bufitem *medianloc;   // Not index but pointer
public:
    medianfilter (const int len);
    ~medianfilter ();
    double addnext(const double newval);
        // Returns value displaced from buffer
    double getmedian() const;
    void print() const;   // Debug function to show current
                          // state of buffer
};

#endif
```

Here is the implementation file. Note that most of the functionality is in **addnext()** as implied by the paper. Although care has been taken with pointer manipulation, *this version does have a bug*. See if you can track how the data structures and algorithms on page 401 have been implemented in the code, and perhaps spot the bug.

```cpp
// Implementation for medianfilter class, created Thu Aug 28
// 12:42:56 2003
// Revised Aug 28 15:15
// John Robinson

#include "medianfilter.h"
#include <cstdlib>
#include <iostream>
using namespace std;

medianfilter::medianfilter(const int len) {
    length = len;
    buf = new bufitem[length];
    // Have to treat the items at each end differently to the others
    // because they are at the ends of the linked lists
    buf[0].set(0);
    buf[0].initpointers(0, &buf[1]);
    for (int i = 1; i < length-1; i++) {
        buf[i].set(0);
        buf[i].initpointers(&buf[i-1],&buf[i+1]);
    }
    buf[length-1].set(0);
    buf[length-1].initpointers(&buf[length-2],0);
    nextinsertion = 0;
    medianloc = &buf[length/2];// Actually doesn't matter
        // because initial buf values are all equal to 0
}
medianfilter::~medianfilter() {
}

double medianfilter::addnext(const double newval) {
    int thisinsertion = nextinsertion++;
    if (nextinsertion == length)
```

```cpp
          nextinsertion = 0;
      double oldval = buf[thisinsertion].get();
      if (newval == oldval)   // Everything fine. Nothing to do
        return oldval;
      bufitem *p = &buf[thisinsertion];
      p->set(newval);
      int movedmedian = 0;
      if (newval > oldval) { // May need to move this node upwards
        if (!p->next() || p->next()->get() >= newval)
        // Again, don't need to change
          return oldval;
        // Otherwise have to move upwards
        if (p == medianloc) {
          medianloc = p->next();
          movedmedian = 1;
          }
        p = p->next();
        while(p->next()) {
          if ((p == medianloc)&&!movedmedian) {
            medianloc = medianloc->next();
            movedmedian = 1;
            }
          if (p->next()->get() >= newval) {
            buf[thisinsertion].removefromlists();
            buf[thisinsertion].putafter(p);
            break;
            }
          p = p->next();
          }
        if (!p->next()) {
          buf[thisinsertion].removefromlists();
          buf[thisinsertion].putafter(p);
          }
        }
      else {
        if (!p->prev() || p->prev()->get() <= newval)
        // Again, don't need to change
          return oldval;
        // Otherwise have to move downwards
        if (p == medianloc) {
          medianloc = p->prev();
          movedmedian = 1;
          }
        p = p->prev();
        while(p->prev()) {
          if ((p == medianloc)&&!movedmedian) {
            medianloc = medianloc->prev();
            movedmedian = 1;
            }
          if (p->prev()->get() <= newval) {
            buf[thisinsertion].removefromlists();
            buf[thisinsertion].putbefore(p);
```

```
            break;
        }
        p = p->prev();
    }
    if (!p->prev()) {
        buf[thisinsertion].removefromlists();
        buf[thisinsertion].putbefore(p);
    }
}
return oldval;
}
double medianfilter::getmedian() const {
    // Initial implementation: do sort each time. Must improve!
    return medianloc->get();
}
void medianfilter::print() const {
    for (int i = 0; i < length; i++) {
        cout << "buf[" << i << "] = " << buf[i].get();
        cout << " has address " << (void *) &buf[i];
        cout << ", prev pointer " << (void *) buf[i].prev();
        cout << ", next pointer " << (void *) buf[i].next() << endl;
    }
    cout << "Current buffer index is " << nextinsertion << endl;
}
```

After compiling and quick hand testing, this version seems to work. But
here is where we need to do rigorous, extensive testing. Because the
code is a little tricky to understand, white box testing can't be guaranteed to uncover situations that have not been properly catered for in the
code. Black box testing by hand is also susceptible to missed cases:
we could enumerate possibilities like add-value-greater-than-current-
median-when-current-median-equals-value-to-remove-from-the-
window-but-more-than-one-other-value-still-in-window-equals-
current-median, but it's going to be tedious to devise test sets that cover
all combinations. This is a case when statistical testing is justified: we
can generate long streams of random numbers, filter with both our
initial naïve method and our 'provisionally final' method and look for
differences.

Generating long random sequences is easy and we can even use the
stable data type discussed in Chapter 17 to help analyse outputs. When
we find a difference, we can check by hand which of the two systems
is wrong, then work step by step using **print()** (or other debugging tools)
to find the problem.

When we do such testing on the above implementation we discover
its problem: **addnext()** does not properly reset the pointer to the median
when the new item becomes the median. This can be fixed as demonstrated in the version of **addnext()** listed below. The location of the
fix is flagged by comments.

```
double medianfilter::addnext(const double newval) {
    int thisinsertion = nextinsertion++;
    if (nextinsertion == length)
```

```
                      nextinsertion = 0;
        double oldval = buf[thisinsertion].get();
        if (newval == oldval)   // Everything fine. Nothing to do
          return oldval;
        bufitem *p = &buf[thisinsertion];
        p->set(newval);
        int movedmedian = 0;
        if (newval > oldval) {   // May need to move this node
                                 // upwards
          if (!p->next() || p->next()->get() >= newval)
          // Again, don't need to change
            return oldval;
          // Otherwise have to move upwards
          if (p == medianloc) {
            medianloc = p->next();
            movedmedian = 1;
            }
          p = p->next();
          while(p->next()) {
            // THE NEXT 11 LINES FIX THE BUG BY REORDERING
            // TESTS
            // A CORRESPONDING 11-LINE FIX IS FLAGGED
            // LOWER DOWN
            if (p->next()->get() >= newval) {
              buf[thisinsertion].removefromlists();
              buf[thisinsertion].putafter(p);
              if ((p == medianloc)&&!movedmedian)
                medianloc = medianloc->next();
              break;
              }
            if ((p == medianloc)&&!movedmedian) {
              medianloc = medianloc->next();
              movedmedian = 1;
              }
            p = p->next();
            }
          if (!p->next()) {
            buf[thisinsertion].removefromlists();
            buf[thisinsertion].putafter(p);
            }
          }
        else {
          if (!p->prev() || p->prev()->get() <= newval)
          // Again, don't need to change
            return oldval;
          // Otherwise have to move downwards
          if (p == medianloc) {
            medianloc = p->prev();
            movedmedian = 1;
            }
          p = p->prev();
          while(p->prev()) {
```

```
                    // THE NEXT 11 LINES FIX THE BUG BY REORDERING
                    // TESTS
                    // A CORRESPONDING 11-LINE FIX IS FLAGGED
                    // HIGHER UP
                    if (p->prev()->get() <= newval) {
                       buf[thisinsertion].removefromlists();
                       buf[thisinsertion].putbefore(p);
                       if ((p == medianloc)&&!movedmedian)
                          medianloc = medianloc->prev();
                       break;
                    }
                    if ((p == medianloc)&&!movedmedian) {
                       medianloc = medianloc->prev();
                       movedmedian = 1;
                    }
                    p = p->prev();
                 }
              if (!p->prev()) {
                 buf[thisinsertion].removefromlists();
                 buf[thisinsertion].putbefore(p);
              }
           }
        return oldval;
        }
```

15.9 Finishing

Now extensive random number tests succeed for the implementation.
We therefore consider what should be done to finish the program. The
helper class **bufitem** is only used inside **medianfilter**. It could be an
embedded class, but at the very least it can be moved out of the header
file and into the implementation. How shall we make our median filter work on other types than **double**s (we're particularly interested in
ints for video processing)? One possibility is to use templates, but
because **export** is not supported by many compilers, that would mean
relocating a large part of the implementation to the header file.
Instead we use a **typedef** in the header file with a comment to remind
us that this has to be changed and things recompiled for different
types. We also need to fix the problem identified in the very first version: a negative argument to the constructor causes errors. There are a
number of ways of dealing with this, but the best option is to signal a
problem but proceed anyway with a default value replacing what the
user asked for. We need a default value for the constructor too, so that
arrays of **medianfilter**s can be instantiated. Having made all those
changes, here is the final version. First the header file:

```
// Header for medianfilter class, created Thu Aug 28
// 12:42:56 2003
// Revised Aug 28
// Version 0.1
// John Robinson

#ifndef medianfilter_H
#define medianfilter_H
```

```
typedef double itemtype; // Change and recompile for other
    // input sample types

class medianfilter {
  class bufitem *buf;
  int length;
  int nextinsertion; // Index into circular buffer
  class bufitem *medianloc;    // Not index but pointer
public:
  medianfilter (const int len = 15);
  ~medianfilter ();
  itemtype addnext(const itemtype newval);
      // Returns value displaced from buffer
  itemtype getmedian() const;
  void print() const;    // Debug function to show current state
                            // of buffer
};

#endif
```

Note the addition of the **class** keyword before **bufitem**'s first mention, because now we have taken the declaration of **bufitem** out of the
header. Here is the final implementation file:

```
// Implementation for medianfilter class, created Thu Aug 28
// 12:42:56 2003
// Revised Aug 28 ... Corrected same day
// Version 0.1
// John Robinson

#include "medianfilter.h"
#include <cstdlib>
#include <iostream>
using namespace std;

// bufitem now defined in implementation file because not part
// of interface
class bufitem {
// All items private. Just used by friend medianfilter
  itemtype value;
  bufitem *smaller;// Using pointers rather than indices means
        // bufitems can be collected into other DSs
        // than just arrays.
  bufitem *greater;
  bufitem () {
    value = 0;
    smaller = 0;
    greater = 0;
  }
  ~bufitem() {
  }
  void set(const itemtype v) {
    value = v;
```

```cpp
        }
    itemtype get() const {
        return value;
        }
    void initpointers(bufitem *const predecessor,bufitem *const
            successor) {
        smaller = predecessor;
        greater = successor;
        }
    void putafter(bufitem *predecessor) {
        // Break existing link from predecessor to its successor
        // and insert this one
        bufitem *successor = predecessor->greater;
        predecessor->greater = this;
        if (successor)
            successor->smaller = this;
        smaller = predecessor;
        greater = successor;
        }
    void putbefore(bufitem *successor) {
        // Can't just use putafter(successor->before) because
        // this may be 0 on list
        bufitem *predecessor = successor->smaller;
        successor->smaller = this;
        if (predecessor)
            predecessor->greater = this;
        smaller = predecessor;
        greater = successor;
        }
    void removefromlists() {
        if (smaller)
            smaller->greater = greater;
        if (greater)
            greater->smaller = smaller;
        }
    bufitem *next() const {
        return greater;
        }
    bufitem *prev() const {
        return smaller;
        }
    friend class medianfilter;
    };

medianfilter::medianfilter(const int len) {
    if (len < 0) {
    // Can add an upper limit later if necessary
        cout << "Specified length " << len << " is out of bounds\n";
        cout << "Using length = 15 instead\n";
        length = 15;
        }
    else
```

```cpp
        length = len;
    buf = new bufitem[length];
    // Have to treat the items at each end differently to the others
    // because they are at the ends of the linked lists
    buf[0].set(0);
    buf[0].initpointers(0, &buf[1]);
    for (int i = 1; i < length-1; i++) {
        buf[i].set(0);
        buf[i].initpointers(&buf[i-1],&buf[i+1]);
    }
    buf[length-1].set(0);
    buf[length-1].initpointers(&buf[length-2],0);
    nextinsertion = 0;
    medianloc = &buf[length/2];
            // Actually doesn't matter because initial
            // buf values are all equal to 0
}

medianfilter::~medianfilter() {
}

itemtype medianfilter::addnext(const itemtype newval) {
    int thisinsertion = nextinsertion++;
    if (nextinsertion == length)
        nextinsertion = 0;
    itemtype oldval = buf[thisinsertion].get();
    if (newval == oldval)   // Everything fine. Nothing to do
        return oldval;
    bufitem *p = &buf[thisinsertion];
    p->set(newval);
    int movedmedian = 0;
    if (newval > oldval) { // May need to move this node upwards
        if (!p->next() || p->next()->get() >= newval)
        // Again, don't need to change
            return oldval;
        // Otherwise have to move upwards
        if (p == medianloc) {
            medianloc = p->next();
            movedmedian = 1;
        }
        p = p->next();
        while(p->next()) {
            if (p->next()->get() >= newval) {
                buf[thisinsertion].removefromlists();
                buf[thisinsertion].putafter(p);
                if ((p == medianloc)&&!movedmedian)
                    medianloc = medianloc->next();
                break;
            }
            if ((p == medianloc)&&!movedmedian) {
                medianloc = medianloc->next();
                movedmedian = 1;
            }
```

```cpp
        p = p->next();
      }
      if (!p->next()) {
        buf[thisinsertion].removefromlists();
        buf[thisinsertion].putafter(p);
      }
    }
    else {
      if (!p->prev() || p->prev()->get() <= newval)
      // Again, don't need to change
        return oldval;
      // Otherwise have to move downwards
      if (p == medianloc) {
        medianloc = p->prev();
        movedmedian = 1;
      }
      p = p->prev();
      while(p->prev()) {
        if (p->prev()->get() <= newval) {
          buf[thisinsertion].removefromlists();
          buf[thisinsertion].putbefore(p);
          if ((p == medianloc)&&!movedmedian)
            medianloc = medianloc->prev();
          break;
        }
        if ((p == medianloc)&&!movedmedian) {
          medianloc = medianloc->prev();
          movedmedian = 1;
        }
        p = p->prev();
      }
      if (!p->prev()) {
        buf[thisinsertion].removefromlists();
        buf[thisinsertion].putbefore(p);
      }
    }
  return oldval;
  }

itemtype medianfilter::getmedian() const {
  return medianloc->get();
  }

void medianfilter::print() const {
  for (int i = 0; i < length; i++) {
    cout << "buf[" << i << "] = " << buf[i].get();
    cout << " has address " << (void *) &buf[i];
    cout << ", prev pointer " << (void *) buf[i].prev();
    cout << ", next pointer " << (void *) buf[i].next() << endl;
    }
  cout << "Current buffer index is " << nextinsertion << endl;
  }
```

Is this the end of the story? Not quite. **addnext()** can be rewritten to avoid repetition. The two cases (**newval < oldval** and **newval > oldval**) can be combined by adding extra functions to **bufitem** that switch on a parameter to say whether the search is going upwards or downwards. Making the change certainly tightens up the code, but the clarity of shortness and saying something only once is traded for more variables being maintained and repeatedly checked. The upshot is an execution time increase of about 25%, so such a change is probably best avoided.

Finally we install the module in the system where it will solve the problem. The median filter developed here was integrated in a video processing system, where it achieved an order of magnitude speedup over a previous median filter that sorted the window each time.

16 Multidimensional minimization – a case study in numerical methods

16.1 Numerical methods

Many scientists and engineers write programs to solve numerical problems such as

- finding roots of non-linear equations
- solving systems of equations
- interpolating data
- fitting models to data
- numerical integration
- solving differential equations.

Numerical methods have a long history. Until the invention of mechanical calculating machines, they were used by unfortunate human computers to produce numerical tables. (Computer = machine *or person* who computes.) For example, someone generating a table of square roots could use fixed-point interation, Newton–Raphson, or some other general-purpose (or polynomial) iterative improvement scheme to refine a numerical approximation to the required level of accuracy. The sidebar explains how. For one-shot calculations, such methods might still be appropriate, but the speed of calculators and computers today often means that suboptimal methods are acceptable. In the case of root finding, as the sidebar explains, the first step is to find an approximation by plotting. In many cases, this first step can be generalized for the whole job: with a graphing calculator, it is likely to be easy and quick enough to zoom in and replot successively until the root is located to sufficient accuracy. Plotting many values unnecessarily is computationally inefficient, but practically it doesn't matter.

On the other hand, scientists often want to analyse large amounts of data quickly or in real time, so efficient numerical methods are required. In this case study we discuss the minimization of an eight-dimensional function. Even as a one-shot problem, this is difficult enough to merit use of appropriate efficient numerical techniques. But when we want to solve the problem many times per second (as the R&D project from which this problem is taken does), computational efficiency is very important.

Calculating the square root of 2

The numerical techniques for solving single non-linear equations have two steps:

- find an approximate solution
- improve it iteratively.

If we have a non-linear equation in one variable (x), we reorganize it so that all the terms are on one side, and put this equal to $f(x)$. Solving the original equation is equivalent to finding the roots of $f(x)$ (i.e. the values of x for which $f(x)$ is 0). So we begin by plotting $f(x)$ and look for zero crossings.

To find the square root of 2, we seek to solve the equation

$$x^2 = 2$$

So we want to find the roots of

$$f(x) = x^2 - 2$$

To plot this function, we could consider some values of x:

X	0	1	2	3
$F(x)$	−2	−1	2	7

Clearly there is a zero-crossing between 1 and 2, so we could choose one of these as our initial root estimate. The Newton–Raphson method updates estimates according to the following formula:

$$s_{n+1} = s_n - \frac{f(s_n)}{f'(s_n)}$$

For our equation,

$$f'(x) = 2x$$

So, choosing 2 as our initial root estimate, we have:

$$s_1 = 2$$
$$f(s_1) = 2$$
$$f'(s_1) = 4$$

Therefore, using the Newton–Raphson formula,

$$s_2 = 2 - \frac{2}{4} = 1.5$$

So

$$s_2 = 1.5$$
$$f(s_2) = 0.25$$
$$f'(s_2) = 3$$

Applying the Newton–Raphson formula again gives

$$s_2 = 1.5 - \frac{0.25}{3} = 1.4167$$

And so on.

The human computers doing this calculation by hand would also have to be familiar with the limitations of the iterative improvement method. (In the case of Newton–Raphson, this means ensuring that the derivative doesn't change sign between the current estimate and the root.)

16.2 The problem

Because this case study deals with a real R&D problem, its background needs explanation. The issues are not subtle, but it will take a moment to describe what they are and why they're important.

Finding minima in 1D

In school mathematics, we learn how to find the minima of a function $y = f(x)$: we differentiate the function and find out where the derivative (which is the slope of the original function) equals 0. That's possible if $f(x)$ has an analytic form and is differentiable, but what about the $f(x)$ in Figure 16.1?

Ugly isn't it? Where would a function like this come from? Well, x might be the control parameter for a complicated system and y might be the measurable output we're using to evaluate how stable the system is. y depends on x but not in an analytic way. What we have to do is put in xs and see what ys come out, then pick the x with the lowest y. But how many xs do we try?

Like the root-finding problem in the sidebar, finding a minimum of a non-analytic function has two parts:

1. Find an x that is close enough to the global minimum so that step 2 will descend to it rather than to some other local minimum.
2. Descend step by step to the minimum.

For the function in Figure 16.1, we hope that step 1 gives an x in the range shown in Figure 16.2. That range might be called the 'catchment basin' of the global minimum.

For many practical problems, y, the criterion function to be minimized, is guaranteed to have just one minimum, so optimization is simply descent from an arbitrary starting point. Just as for root finding (see sidebar) there are numerical techniques for descent so that the number of evaluations of the function is minimized. In one dimension, golden section search is the appropriate method.

For other problems, it is guaranteed, or very likely, that minima are well separated. For example, in Figures 16.1 and 16.2, it might be certain that each local minimum has a 'catchment basin' of at least d. Then global optimization would amount to sampling x at intervals of d, descending from each of these start points, then selecting the lowest as the global minimum.

For yet other problems, there might be no guarantees about local minima. Then descent would be useless. We would be forced to do an exhaustive search (at some fine spacing) of xs to find the global minimum.

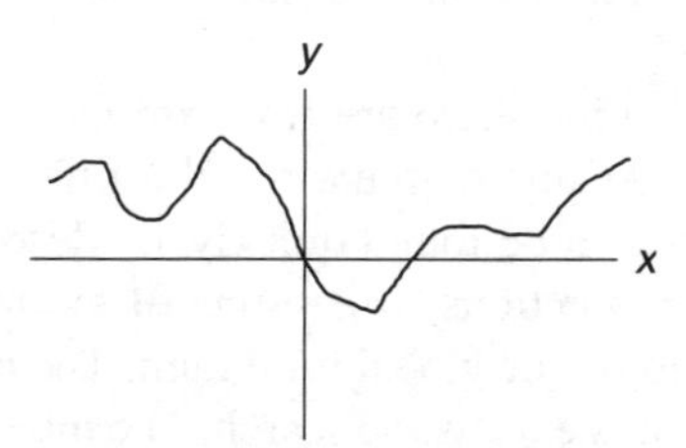

Figure 16.1 *A non-analytic function*

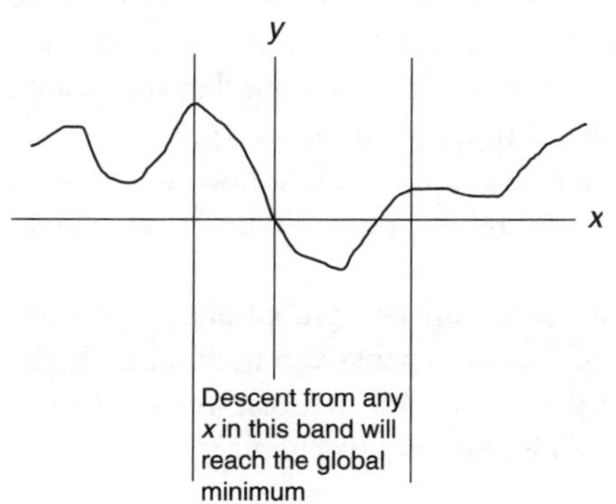

Figure 16.2 *The 'catchment' basin of the global minimum for the function in Figure 16.1*

Finding minima in multidimensions

For more than one dimension, the story is the same. Descent algorithms are available and fast, but they only find the local minimum of the catchment area that you start from. From its current point, a descent algorithm works out the steepest direction downhill. It then makes its next function evaluation in that direction (or in some direction related

Figure 16.3 *A two-dimensional function with a single minimum at the origin*

Figure 16.4 *A two-dimensional function with many minima*

Figure 16.5 *A noisy two-dimensional function*

Figure 16.6 *A noisy one-dimensional function*

to the steepest but taking account of previous directions). For example, Figure 16.3 shows a function in 2D, where lightness represents the function value.

This has a single minimum at the origin, shown at the centre of the diagram. A descent algorithm started at any point would just walk towards the minimum, adjusting its step size as it converged.

But Figure 16.4 shows a different kind of 2D function.

This time there are many minima and descent is practically useless. The problem is like trying to find the highest tuft in the pile of a carpet.

So in multidimensional optimization there are two extremes, and some interesting places between. At one extreme are the functions with one minimum. The minimum can be found quickly, by descent. At the other extreme are noiselike functions, where one of the huge number of local minima happens to be the global minimum. The only hope of finding the global minimum is exhaustive search. The interesting cases in between are where there are a significant number of local minima that are good enough – they may not be the global minimum, but they are close in value. Finding one of these will take a lot longer than a single descent, but much shorter than searching every possible value. Methods for doing just this include simulated annealing and various kinds of evolutionary algorithms. There is another interesting case though, exemplified by the 2D function in Figure 16.5.

This is the sum of a simple, single minimum, function and noise. If we were to take a cross-section through it, we would get a 1D equivalent something like Figure 16.6.

This is another intermediate case between global and local optimization. But we don't want to use as expensive a method as simulated annealing here. We really want a kind of noise-resilient local optimizer. *That* is the problem addressed in this chapter.

16.3 Researching possible solutions

Multidimensional optimization is a standard numerical method, so the place to start is the contents and index pages of standard textbooks. Well, not quite the first. A Google search takes only a few seconds, and there might be appropriate introductory but comprehensive learning material available on the web. Here's what happened when I started researching possible solutions with a Google search in August 2003. I began with a search for ' "multidimensional optimization" "noisy function" ' yielding one hit that didn't look promising. Next I tried 'multidimensional minimize "noisy function" ' giving several hits including a Citeseer directory. Better. After some more experimentation with search terms I had a few pages of possibilities. But nothing that appeared to be teaching materials or general references. Resisting the temptation to jump straight into specialized articles, I headed for the library. I looked for terms like 'noisy function', 'noisy criterion function' and 'noise resilience' in every promising-looking text in the numerical methods section of my university's library. There were no references. This search took about half an hour. So why was that seemingly futile effort worthwhile? The point is, in trying to

Rule 11

Exploit existing solutions.

it's important to know about solutions that are established, standard practice first. You may well end up using a standard solution, and even if you use something newer instead, or more esoteric, you need to understand the context. Finding a lack of promising references is good information: it gives the impetus to go back to the Google hits lists to look at the research papers and more obscure hits. Meanwhile, I checked out three of the most used texts from the library. It's likely that specialists are going to refer to things I've not heard of and it will be handy to have some textbook help while I'm reading what they have to say.

The next step – digging deeper into the Google hits list – took a little time and turned up some false leads, but also the information that several people say the Nelder–Mead simplex is appropriate for noisy functions. That's good news, because although they don't mention the noise issue explicitly, two of the textbooks I've borrowed, plus my copy of *Numerical Recipes*, talk about the Nelder–Mead method. It seems that it is regarded as slow and clunky compared to gradient-based local optimizers, but there are a huge number of people who have used it, and it is even reasonably respectable among mathematicians. Perhaps this is what I need.

I downloaded some particularly promising papers to use with the standard textbooks and set to work understanding the method. Remember,

Principle 4

Every program has surprises.

So, although *Numerical Recipes* plus three online sites offered code to do simplex minimization (and you'll find an alternative at the end of this chapter), I needed – and we now need – to be clear on what it was doing.

16.4 Nelder–Mead Simplex Optimization

Here is a description of Nelder–Mead Simplex Optimization, similar to what is given in many textbooks.

A *simplex* is the non-degenerate convex hull of $n + 1$ vertices in n-dimensional space. In 2D, the simplex is a triangle, in 3D, it is a tetrahedron, and so on. The Nelder–Mead algorithm iteratively moves a simplex through n-space by attempting to replace its worst vertex (the one at which the function evaluation is highest) with a better point. This is done through operations of reflection, expansion, contraction and shrinkage, as specified in the sequence below and illustrated, for $n = 2$, in Figure 16.7.

1. At the beginning of iteration k, the current simplex is defined by its $n + 1$ vertices $x_1^{(k)}, \ldots, x_{n+1}^{(k)}$, and their function values, ordered so that $f(x_1^{(k)}) \leqslant f(x_2^{(k)}) \leqslant \cdots \leqslant f(x_{n+1}^{(k)})$. So $x_1^{(k)}$ is the *best* vertex on iteration k and $x_{n+1}^{(k)}$ is the *worst*. A tie-breaking rule can be used for determinacy.

2. The *reflection point* x_r is computed using $x_r = x_m + \rho(x_m - x_{n+1})$ where the (k) superscript is assumed, x_m is the mean of all the vertices except x_{n+1} ($x_m = \sum_{i=1}^{x} x_i/n$) and ρ is a reflection coefficient greater than 0. So the worst point is reflected through the centroid of the hyperplane joining all the other points. Usually ρ is set equal to 1, making for a true reflection. If $f(x_1) \leqslant f(x_r) < f(x_n)$, the reflected point is accepted and the iteration ends.

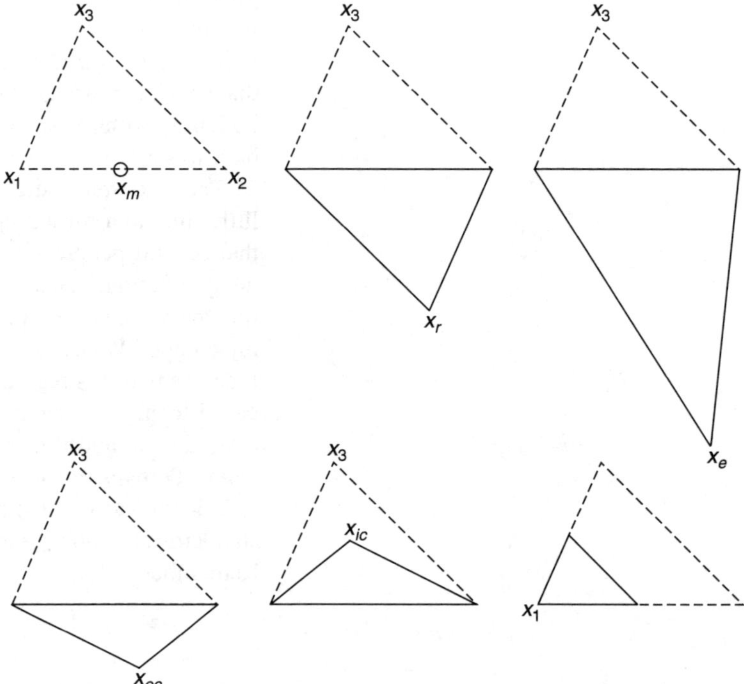

Figure 16.7 *Nelder–Mead operations for a 2D simplex. For top left: original simplex with* x_3 *the worst point and* x_m *the centroid of the hyperplane (in 2D, line) opposite* x_3*; a reflection; an expansion; an outside contraction; an inside contraction; a shrink (about the best point* x_1*)*

3. If $f(x_r) < f(x_1)$ the reflected point is better than the previous best point, so the algorithm tries to go further in the same direction by calculating the expansion point $x_e = x_m + \chi(x_r - x_m)$. χ is greater than 1 and typically has the value 2. If $f(x_e) < f(x_r)$ the expansion point is accepted; otherwise the reflected point is accepted. In either case the iteration ends.

4. If $f(x_n) \leqslant f(x_r) < f(x_{n+1})$, the reflected point was an improvement over the worst point, but is still no better than the second worst. To avoid the same point being moved on the next iteration, the algorithm now tries an *outside contraction*: $x_{oc} = x_m + \gamma(x_r - x_m)$. γ must be between 0 and 1, and is usually set to 1/2. If $f(x_{oc}) \leqslant f(x_r)$, x_{oc} is accepted and the iteration ends.

5. If $f(x_r) < f(x_{n+1})$ the reflected point was worse than the worst point, so an *inside contraction* is now attempted. $x_{ic} = x_m - \gamma (x_r - x_{n+1})$. If $f(x_{ic}) \leqslant f(x_{n+1})$, x_{ic} is accepted and the iteration ends.

6. If this point is reached, none of the attempts to improve the worst point have succeeded, so the algorithm makes do with a *shrink* operation. N new vertices are defined by $x_i^{(k+1)} = x_i^{(k)} + \sigma(x_i^{(k)} - x_1^{(k)})$ for $i = 2, \ldots, n + 1$. σ must be between 0 and 1 and is usually set to 1/2. The best point is retained, so $x_1^{(k+1)} = x_1^{(k)}$.

16.5 Understanding the method with state sketches

The description of the Nelder–Mead simplex in the previous section is similar to those found in textbooks. A single iteration of the algorithm is described in a linear way, and there are many 'if's which interact. That particular description is in my words and uses my notation, so it differs subtly from other people's descriptions, and they differ subtly from each other. Some authors give a flowchart where 'if's are put into chains.

Because we have an algorithm with many interacting 'if's, it is tricky to understand. The approach advocated in Chapter 11 is to put a procedural algorithm like this into a statechart-like representation. That way the interactions of the conditionals become clear. That is what we can do and the resulting state sketch is shown in Figure 16.8.

The sketching of the state diagram helps in two ways. The first can only be experienced by going through the process yourself. Converting the do-this, do-that algorithm description to the doing-this, doing-that state diagram is a way of understanding through re-representation; by thinking in two modes about what is going on, you are able to appreciate the mechanism more deeply. The second way in which state sketching helps is that, as usual, all alternative conditions are lined up, showing their parity (as opposed to being cascaded with if-else if- … -else) and the sequencing of actions at the ends of states removes ambiguities that other kinds of description have. In the case of the simplex, doing the state sketch revealed that two different versions of the Nelder–Mead algorithm have been widely documented. In the second line of actions two-thirds down the diagram, you see the two alternatives, 'Shrink $x_2, \ldots, x_n, x_r$' and 'Shrink $x_2, \ldots, x_{n+1}$'. So the role of x_r in the first case is played by x_{n+1} in the second. In some descriptions of the algorithm, *including the one in the previous section*, both of these actions are 'Shrink $x_2, \ldots, x_{n+1}$'. Doing the state sketch gives a good idea how this mixup has happened. In many written and flowchart

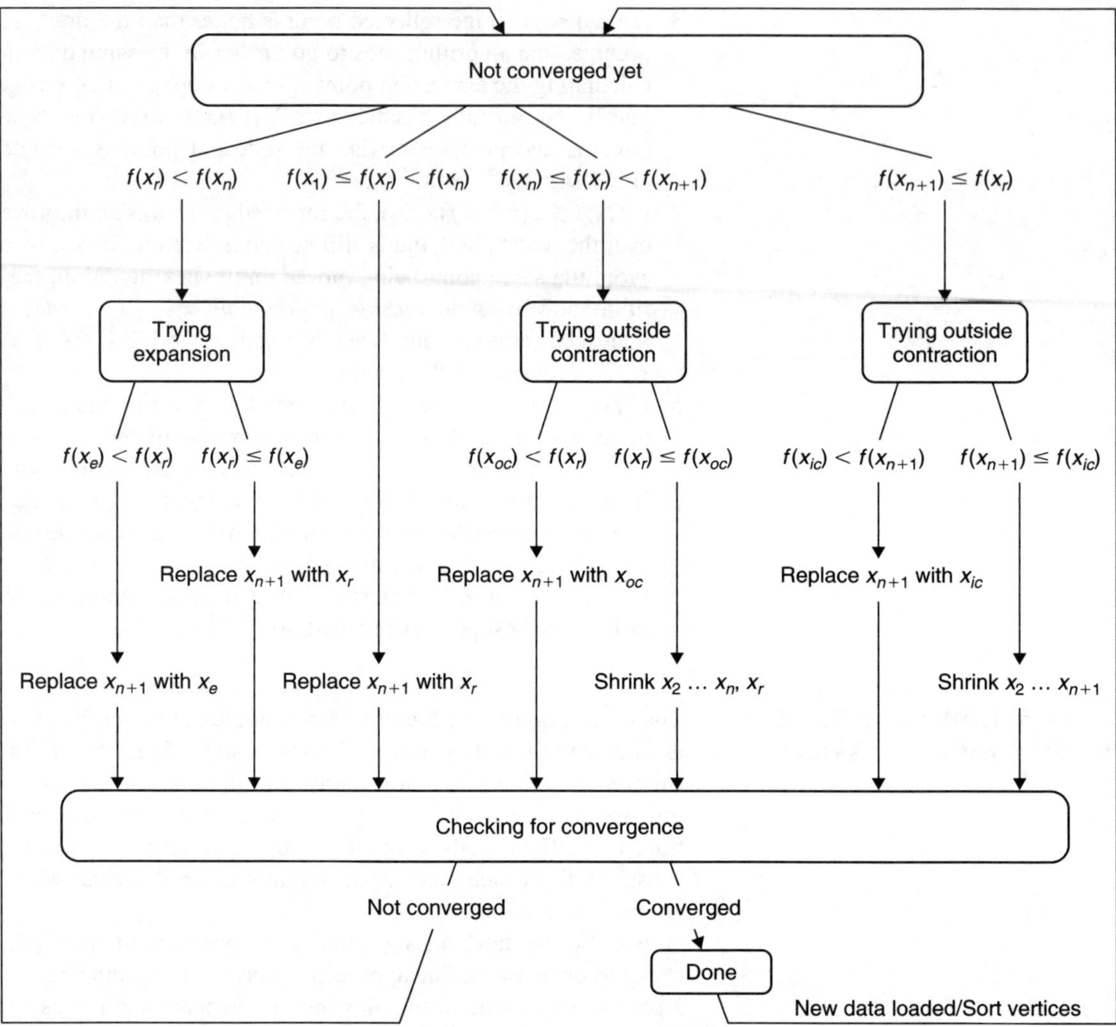

Figure 16.8 *State sketch of simplex*

descriptions, x_{n+1} is replaced by x_r at an intermediate stage on the 'Trying outer contraction' branch. Some authors have tried to structure the algorithm better by deferring this replacement until after the second set of tests, but have forgotten to put it back in all the right places. And not surprising. When you're using a linear description with interacting 'if's, it's difficult to keep track of everything. Using a state sketch helps to identify and avoid these errors.

Doing the statesketch has shown us that the simplex algorithm is tricky. The *Numerical Recipes* implementation is probably correct, but its structuring of the 'if's is tangled. We therefore are going to generate our own solution. In this, we can take on board any advice that other implementers have given.

Before starting to program the simplex it's worth noting any nuances of the method that particular authors have emphasized. For example, one author comments that although the simplex was originally designed for unbounded variables, in many practical situations there will be limits on the values variables can take. They dealt with this by ensuring that if the simplex tried to move outside the bound in any dimension, the value for that dimension would be set at the bound. Another possibility might be to set criterion function values for points outside the bounds to very high levels so the simplex would never have an out-of-bound vertex. This would make it harder to move towards the boundaries, which might be a heuristic advantage. Clearly there are two alternatives to experiment with. Writing the simplex code ourselves also makes it easy to experiment with different possibilities for the shrink step. After all, perhaps someone found that the single type of shrink actually works better and deliberately described that version in their writeup.

16.6 Experiment-driven development

In 'test-driven development' the rhythm of programming is write a test case that will fail, fix the code to pass the test, then refactor. Chapter 14 discussed the pros and cons and came out in favour of conventional development and testing except for broad, shallow modules. In the simplex, we are not dealing with a shallow algorithm (although, of course, there are many that are much deeper), so we write the code carefully, embedding debugging facilities, but concentrating on capturing the whole algorithm at once. However, once the first version is running, the simplex is a very good example of a module that needs heavy exploratory testing to see if it truly is working correctly and to uncover hidden issues. Like test-driven development, experiment-driven development sets up multiple test cases, but they are at a higher level, exercising the whole functionality of the program in different ways.

We therefore must consider how to design test programs and cases to uncover surprises, and how to embed within the simplex, code that reveals what it is doing.

In my work on the simplex class, I simultaneously developed:

- a vizualization tool that would allow me to watch the simplex solving 2D optimization problems
- an *n*-dimensional analytic function test tool where a complicated analytic function could be specified as the criterion function (so the true minimum could be compared with that found by the simplex)
- a simple pattern recognition program that would use the simplex to find boundaries between classes of objects
- the main program for the eight-dimensional problem I was working on.

Testing the simplex on these different problems not only helped to find bugs, it also revealed initialization to be a key issue. Rather than presenting the code developed, the following sections review the experiments at each stage, showing how they fed into new ideas. At the end, the header and implementation of the final simplex class are provided.

Rule 7

Test to discover information.

Rule 8

Hide information in public; reveal it in private.

Figure 16.9 *Minimizing with the simplex. For the function, lightness represents value. For the simplex, lightness represents iteration number, so the start position is black, and successive positions are lighter*

Basic working

Figure 16.9 illustrates the output from the simplex visualization program. The program animates a triangular simplex as it tumbles towards a minimum. In this static visualization, only the first ten locations of the simplex are shown and each successive position is shown in lighter grey.

The function to minimize is the simple one-minimum function of Figure 16.3. Again lightness (for the function) corresponds to value.

The simplex starts as an equilateral triangle in the top left corner. Its first move is an expansion, then a reflection, then an outer contraction. Starting the simplex from a variety of positions verifies that the decisions it makes about which vertex to move where corresponds to those in the state sketch. Every path in the state sketch can (and should) be exercised by starting the simplex in a different initial configurations.

During the initial visualization process I discovered and fixed a number of bugs. For example, when shrinking, my initial version moved towards x_1 either x_{n+1} or x_r as specified by the state sketch, plus the other vertices from x_2 to x_{n-1}. But it missed out x_n – a classic example of an off-by-1 error, and one that wouldn't have been caught if I hadn't tested carefully. The visualization also uncovered higher-level issues.

Learning from experiments

The initial stopping criterion was based on the distance between the function values of the best and worst vertices in the simplex. This is often used. But some of my test criterion functions were coarsely quantized, meaning that widely separated x_is could have the same $f(x_i)$. The result was a tendency to report convergence for a level simplex whose vertices had the same quantized value. This led me to consider a different stopping condition: terminating when the separation between the x_is is small enough. But that takes significant computation time, so it should probably only be done every few iterations. In any event, when the criterion function is of high enough resolution (i.e. the $f()$ variation is finely quantized enough), there is less likelihood of a non-converged level simplex, especially with a higher number of dimensions.

Initially I provided ways to initialize the simplex with random vertices and as a regular hypersolid (i.e. equal distance between all the vertices). But if the starting point is reasonable, then why not initialize the other vertices from it by one-dimensional searches along each of the individual dimensions? This produced different behaviour from the other cases – sometimes better, sometimes worse – and so led on to further experimentation.

Initial testing also gave the opportunity to try different methods for dealing with bounds. As expected, with problems where the actual minimum is more likely to be closer to the middle of ranges than their endpoints, it was better to prevent the simplex crossing the bounds by returning very high criterion values for those points rather than letting it move up to the bound.

Minimizing noisy functions

Well, the proof of the pudding is, can the Nelder–Mead simplex work as advertised and solve the problem of minimization of noisy functions? Its magic is that it always tries to improve the worst point rather than

Figure 16.10 *The simplex striding over a local minimum to find the global minimum*

Figure 16.11 *Which minimum the simplex finds depends on initialization*

Figure 16.12 *Convergence towards a noisy minimum from an initial equilateral triangle*

Figure 16.13 *Convergence towards a noisy minimum by a simplex initialized with 1D line searches*

the best, so if the best happens to be stuck in a local minimum, there are many chances for the simplex just to stride over and find a different catchment area. Figure 16.10 shows an illustration of it doing just that for a function with a local minimum at the origin and the global minimum at $(0, 100)$.

Although the simplex actually starts closer to the false minimum, it manages to step over it and descend to the true global minimum. Of course, its ability to do that depends on how the vertices actually fall. In this example the catchment area of the minimum at $(0, 0)$ is significantly larger than that of the global minimum, so sometimes the simplex never sets foot in the region of the global one as in the case shown in Figure 16.11. The first simplex in this example is initialized by 1D line searches instead of as an equilateral triangle and that contributes to the problem. The choice about initialization interacts with the type of data.

The example in Figure 16.11 shows that the simplex can't always find a global minimum embedded in the catchment basin of a local minimum. But how does it fare with noisy functions? Figures 16.12, 16.13

Figure 16.14 *Convergence towards the noisy global minimum from different starting points*

and 16.4 show some examples. Although the time to converge varies, the simplex consistently finds a minimum very close to the true global minimum.

The simplex successfully minimized higher-dimensional analytic functions too. It was incorporated into the eight-dimensional estimator and performs fast and accurate minimization of noisy functions.

16.7 Program code

The final header file for the simplex is:

```
// simplex class
// John Robinson
// Version 0.1
// 26 August 2003
#include <cmath>
using namespace std;
class simplex {
    double **vertices;
    double *fvals;
    double *sums;
    double *lowbounds;
    double *highbounds;
    double (*function)(const double *vals);
    int highest, secondhighest, lowest;
    int num_dims;
    int num_vertices;
    double initfrom(const double source[], double dest[], int
    dim, double startval);
    // Helper function for initat. Initialize vertex dest from vertex
    // source by a 1D search along dimension dim. See
    // comments on initialization functions below
    int init(const int dim);
    // Initialize dimension with random values in range
    // previously set by call to bound()
```

```cpp
    int replace_highest(double *rep, double repval);
    // Replace highest vertex with rep. Invariant: lowest. sums,
    // highest, secondhighest all maintained here
    int findkeyvertices();
    // Finish initializing the simplex by working out the
    // sums, highest, secondhighest, lowest vertices
    // (which are all used internally to calculate simplex moves)
public:
    simplex(const int ndims, double (*f)(const double *));
    // Instantiate the simplex with the number of dimensions
    // and the criterion function to minimize
    ~simplex ();
    int bound(const int dim, const double lowval, const double
    highval);
    // Bound a dimension between specified low and high limits
    // Can initialize the simplex in various ways:
    // initrand() provides random initialization values.
    // initat() functions take either
    // a given start vertex or choose one randomly, then initialize
    // the others from it by 1D line searches.
    // initequ initializes a regular simplex with one vertex at a
    // given startpoint
    int initrand();
    // Initialize all dimensions randomly within ranges previously
    // set by calls to bound()
    int initat(const double startpoint[]);
    // Initialize first dimension, then rest using line searches
    int initat();
    // Initialize one dimension randomly, then rest using line
    // searches
    int initequ(const double startpoint[], const double size);
    // Initialize regular simplex of size size.
    // based on Haftka and Guerdal's rule (1992)
    int move(double tolerance);
    // Does one simplex move: Keeps testing alternative possible
    // steps until one makes things better. I.e. one call
    // corresponds to one iteration of the simplex algorithm.
    // Returns 0 if the difference between highest and lowest
    // function values in the simplex is less than specified
    // tolerance.
    // Otherwise returns an integer signalling what kind of move
    // has been made:
    // return value 1: simple reflection
    // return value 2: reflection followed by extension
    // return value 3: contraction (may have been preceded by a
    // simple reflection)
    // return value 4: contract about lowest
    double *getv(const int vertex) const;
    // Get pointer to list of coords for the specified vertex.
    // Returns 0 if vertex is outside range of dimensions.
    double getfval(const int vertex) const;
    // Get function value for specified vertex
```

```cpp
    double *getmin() const;
    // Get pointer to list of coords for current minimum vertex.
    double getminval() const;
    // Get function value for current minimum vertex
    double getmaxval() const;
    // Get function value for current maximum vertex
    void print() const;
    // Debugging function that prints out the current coords and
    // function values. The vertices are always printed in the
    // same order: the highest and lowest are marked with 'H'
    // and 'L' respectively.
    // Prints out whole simplex on a single line.
    };
```

And the implementation:

```cpp
// Implementation of simplex class
// John Robinson
// 26 August 2003
// Version 0.1
#include <cstdlib>
#include <ctime>
#include <iostream>
#include <iomanip>
#include "simplex.h"
using namespace std;
simplex::simplex(const int ndims, double (*f)(const double *)) {
  num_dims = ndims;
  num_vertices = ndims+1;
  vertices = new double *[num_vertices];
  for (int i = 0; i < num_vertices; i++)
      vertices[i] = new double[num_dims];
  fvals = new double[num_vertices];
  sums = new double[num_dims]; // Stores sums of all vertices
  lowbounds = new double[num_dims];
  highbounds = new double[num_dims];
  for (int i = 0; i < num_dims; i++)
     lowbounds[i] = highbounds[i] = 0;
  function = f;
  srand((int) time(NULL));
  }
simplex::~simplex () {
  for (int i = 0; i < num_vertices; i++)
     delete [] vertices[i];
  delete [] fvals;
  delete [] sums;
  delete [] vertices;
  delete [] lowbounds;
  delete [] highbounds;
  }
int simplex::findkeyvertices() {
// Finish initializing the simplex by working out the
```

```cpp
// sums, highest, secondhighest, lowest
  for (int j = 0; j < num_dims; j++)
    sums[j] = 0;
  for (int i = 0; i < num_vertices; i++) {
    // fvals[i] = (function)(vertices[i]);
    for (int j = 0; j < num_dims; j++)
      sums[j] += vertices[i][j];
  }
  if (fvals[1] > fvals[0]){
    highest = 1;
    secondhighest = lowest = 0;
    }
  else {
    highest = 0;
    secondhighest = lowest = 1;
    }
  for (int i = 2; i < num_vertices; i++) {
    if (fvals[i] > fvals[highest]){
      secondhighest = highest;
      highest = i;
      }
    else if (fvals[i] > fvals[secondhighest])
      secondhighest = i;
    else if (fvals[i] < fvals[lowest])
      lowest = i;
    }
  return 0;
  }
int simplex::bound(const int dim, const double lowval, const
double highval) {
// Bound dimension
  if ((dim < 0) || (dim >= num_dims))
    return -1;
  if (lowval >= highval) {
    cout << "bound(" << dim << "," << lowval << "," <<
highval << ")";
    cout << " is not allowed: Have to have positive range\n";
    return -1;
    }
  lowbounds[dim] = lowval;
  highbounds[dim] = highval;
  return 0;
  }
int simplex::init(const int dim) {
// Helper function to initialize one dimension
// with random values between bounds
  if ((dim < 0) || (dim >= num_dims))
    return -1;
  double min = lowbounds[dim];
  double range = highbounds[dim] - min;
  if (range <= 0.0) {
    cout << "init(" << dim << "):";
```

```cpp
            cout << "range = "<< range <<". Can't initialize";
            cout << "without a positive range.\n";
            cout << "Did you remember to call bound()";
            cout << "for all dimensions before initat()";
            cout << "or initrand()?\n";
            return -1;
        }
    for (int i = 0; i < num_vertices; i++)
        vertices[i][dim] = (double)rand()/(RAND_MAX/range)+
            min;
    return 0;
    }
int simplex::initrand() {
// Initialize all dimensions randomly
    for (int i = 0; i < num_dims; i++)
        if (init(i) < 0)
            return -1;
    for (int i = 0; i < num_vertices; i++)
        fvals[i] = (function)(vertices[i]);
    findkeyvertices();
    return 0;
    }
double simplex::initfrom(const double source[], double dest[],
    int dim, const double startfval) {
// Helper function for initat. Initialize vertex dest from vertex
// source by a 1D search along dimension dim
// Returns function value for vertex
    double stepsize;
    double range = highbounds[dim] - lowbounds[dim];
    if (range == 0.0)
        // Bounds/ranges not set. Arbitrary choice for step
        stepsize = 0.1;
    else
        // May need to be able to set granularity
        stepsize = (range/32);
    // First copy vertex
    for (int j = 0; j < num_dims; j++)
        dest[j] = source[j];
    // Try to move in positive direction along dimension
    dest[dim] += stepsize;
    double fval1, fval2;
    // Got do deal with exceptional cases of start point being
    // near the dimensional limits
    if (dest[dim] > highbounds[dim]) {
        // Try going other way
        stepsize = -stepsize;
        dest[dim] += 2*stepsize;
        fval1 = (function)(dest);
        if (fval1 > startfval)
            // We have to accept this one anyway because
            // the other way is out of bounds
            return fval1;
```

```
        // else we can carry on looking in the same direction
      }
  else {
    fval1 = (function)(dest);
    if (fval1 > startfval) {
      // Try going other way
      stepsize = -stepsize;
      dest[dim] += 2*stepsize;
      if (dest[dim] < lowbounds[dim]) {
        // Go back to first choice: Even
        // though it wasn't as good as
        // startpoint, this direction is
        // out of bounds.
        dest[dim] -= 2*stepsize;
        return fval1;
      }
      // else
      fval2 = (function)(dest);
      if (fval2 > startfval) {
        if (fval2 > fval1) {
          // Go back to first choice: Even
          // though it wasn't as good as
          // startpoint, it was better
          // than this one.
          dest[dim] -= 2*stepsize;
          return fval1;
        }
        return fval2;
      }
      // If got to here fval1 was worse than startfval
      // but fval2 was better, so...
      fval1 = fval2;
    }
    // else fval1 was better than startfval.
  }
// However we've got to here dest[] is better than
// source, so we'll keep moving in the same direction
// so long as there is improvement
fval2 = fval1;
while (fval2 <= fval1) {
  fval1 = fval2;
  dest[dim] += stepsize;
  if ((dest[dim] > highbounds[dim])||
      (dest[dim] < lowbounds[dim]))
    fval2 = fval1 + 1.0; // To end loop
  else
    fval2 = (function)(dest);
}
// Have now gone one step too far, so...
dest[dim] - = stepsize;
return fval1;
}
```

```cpp
int simplex::initat(const double startpoint[]) {
   // Initialize simplex using simple 1D line searches about
   // startpoint
   for (int j = 0; j < num_dims; j++) {
     if ((startpoint[j] < lowbounds[j]) ||
          (startpoint[j] > highbounds[j])) {
       cout << "initat:startpoint[" << j;
       cout << "] =" << startpoint[j];
       cout << "is out of bounds for that dimension.\n";
       cout << "Did you remember to call bound()";
       cout << "for all dimensions before initat()?\n";
       return -1;
       }
     vertices[0][j] = startpoint[j];
     }
   fvals[0] = (function)(vertices[0]);
   for (int i = 1; i < num_vertices; i++)
     fvals[i] = initfrom(vertices[i-1],vertices[i],i-1, fvals[i-1]);
   findkeyvertices();
   return 0;
   }
int simplex::initat() {
   // Initialize one dimension randomly, then rest using line
   // searches.
   // First use init() to initial all dims, all vertices randomly
   for (int i = 0; i < num_dims; i++)
     if (init(i) < 0)
       return -1;
   // But now we only use first one
   fvals[0] = (function)(vertices[0]);
   for (int i = 1; i < num_vertices; i++)
     fvals[i] = initfrom(vertices[i-1],vertices[i],i-1, fvals[i-1]);
   findkeyvertices();
   return 0;
   }
int simplex::initequ(const double startpoint[],
   const double size) {
// Equilateral simplex
// Initialize based on Haftka and Guerdal's rule (1992)
   // First step as initat
   for (int j = 0; j < num_dims; j++) {
     if ((startpoint[j] < lowbounds[j]) ||
          (startpoint[j] > highbounds[j])) {
       cout << "initHG:startpoint[" << j;
       cout << "] =" << startpoint[j];
       cout << "is out of bounds for that dimension.\n";
       cout << "Did you remember to call bound()";
       cout << "for all dimensions before initHG()?\n";
       return -1;
       }
     vertices[0][j] = startpoint[j];
     }
```

```cpp
      fvals[0] = (function)(vertices[0]);
      double sqrt2 = sqrt(2.0);
      double sqrtnplus1 = sqrt((double)num_dims+1.0);
      double p = (size/(num_dims*sqrt2))*(sqrtnplus1+
      num_dims-1);
      double q = (size/(num_dims*sqrt2))*(sqrtnplus1-1);
      // cout << "p =" << p << endl;
      // cout << "q =" << q << endl;
      for (int i = 1; i < num_vertices; i++) {
        // cout << "vertex" << i << endl;
        for (int j = 0; j < num_dims; j++) {
          vertices[i][j] = vertices[0][j];
          if (j == i - 1)
            vertices[i][j] += p;
          else
            vertices[i][j] += q;
        }
        fvals[i] = (function)(vertices[i]);
      }
    findkeyvertices();
    // cout << "out" << endl;
    }
int simplex::replace_highest(double *rep, double repval) {
// Replace highest vertex with rep. Invariant: lowest. sums,
// highest, secondhighest all maintained here
    for (int i = 0; i < num_dims; i++) {
      sums[i] -= vertices[highest][i];
      vertices[highest][i] = rep[i];
      sums[i] += vertices[highest][i];
    }
    // cout << "Replacing" << fvals[highest] << "with"
    // << repval << endl;
    fvals[highest] = repval;
    if (fvals[highest] >= fvals[secondhighest]) {
    // cout << "Replaced highest still worst than second
    // highest\n";
      return 0;
    }
    int temp = highest;
    highest = secondhighest;
    if (fvals[temp] < fvals[lowest])
      lowest = temp;
    secondhighest = temp;
    for (int i = 0; i < num_vertices; i++) {
      if (i == highest)
        continue;
      if (fvals[i] > fvals[secondhighest])
        secondhighest = i;
    }
    return 1;
    }
int simplex::move(double tolerance) {
```

```cpp
// Invariant: highest, secondhighest, lowest and sums, all
// maintained by replace_highest
  // cout << "move" << endl;
  int i;
  if (fvals[highest] - fvals[lowest] < tolerance)
    return 0;
  double highval = fvals[highest];
  double tryextendval, tryreflectval, trycontractval;
  // Try reflection
  double average[num_dims];
  double tryreflect[num_dims];
  int outofbounds = 0;
  for (i = 0; i < num_dims; i++) {
    average[i] = (sums[i]-vertices[highest][i])/num_dims;
    tryreflect[i] = 2*average[i]-vertices[highest][i];
    if (tryreflect[i] < lowbounds[i])
      outofbounds = 1;
    else if (tryreflect[i] > highbounds[i])
      outofbounds = 1;
  }
  if (outofbounds) // Reject this point
    tryreflectval = highval;
  else // in bounds, so calculate its value
    tryreflectval = (function)(tryreflect);
  // cout << "sums(" <<sums[0] << "," << sums[1]
  // << ")\n";
  // cout << "avg(" << average[0] << "," << average[1]
  // << ")\n";
  // cout << "ref(" << tryreflect[0] << "," << tryreflect[1];
  // cout << ") =" << tryreflectval << endl;
  if (tryreflectval < fvals[lowest]) {
    // Try to go further
    double tryextend[num_dims];
    for (i = 0; i < num_dims; i++) {
      tryextend[i] = 3*average[i]-2*vertices[highest][i];
      if (tryextend[i] < lowbounds[i])
        break;
      else if    (tryextend[i] > highbounds[i])
        break;
    }
    if (i == num_dims)
      tryextendval = (function)(tryextend);
    else // Premature break: Reject this point
      tryextendval = highval;
    // cout << "ext(" << tryextend[0] << "," << tryextend[1];
    // cout << ") = " << tryextendval << endl;
    if (tryextendval < tryreflectval) {
      replace_highest(tryextend,tryextendval);
      return 2;  // Extended
    }
  }
  replace_highest(tryreflect,tryreflectval);
```

```cpp
      return 1;    // Reflected
    }
  // When compiled with g++ -O2, would sometimes pass next
  // test, fail replace_highest test, then drop though to the
  // following replace_highest. I don't understand it, but I
  // do know that when I inserted:
  // cout << "try";
  // here, it stopped it. Also, using g++ -O stopped it.
  if (tryreflectval < fvals[highest]) {
    if(replace_highest(tryreflect,tryreflectval))
      // Replacement better than secondhighest too
      return 1;
    // cout << "Taken ref, but also got to contract\n";
    }
  // If got to here we've got to try a contraction
  double trycontract[num_dims];
  for (i = 0; i < num_dims; i++)
    trycontract[i] = (average[i]+vertices[highest][i])/2;
  trycontractval = (function)(trycontract);
  // cout << "con(" << trycontract[0] << "," << trycontract[1];
  // cout << ") = " << trycontractval << endl;
  if (trycontractval < fvals[highest]) {
  // Accept any improvement over highest
    replace_highest(trycontract, trycontractval);
    return 3;
    }
  // Got to contract the whole thing
  for (i = 0; i < num_vertices; i++) {
    if (i == lowest)
      continue;
    for (int j = 0; j < num_dims; j++) {
      vertices[i][j] = (vertices[i][j] +
        vertices[lowest][j])/2;
      }
    fvals[i] = (function)(vertices[i]);
    }
  findkeyvertices();
  return 4;
  }

double *simplex::getv(const int vertex) const {
  if ((vertex < 0) || (vertex > num_dims))
    return 0;
  return vertices[vertex];
  }
double simplex::getfval(const int vertex) const {
  if ((vertex < 0) || (vertex > num_dims))
    return 0;
  return fvals[vertex];
  }
double *simplex::getmin() const {
```

```cpp
    return vertices[lowest];
  }
double simplex::getminval() const {
  return fvals[lowest];
  }
double simplex::getmaxval() const {
// Provided for debugging purposes
  return fvals[highest];
  }
void simplex::print() const {
  double *verts;
  double currval;
  double highval = getmaxval();
  double lowval = getminval();
  for (int i = 0; i < num_vertices; i++) {
    verts = getv(i);
    currval = getfval(i);
    if (currval == lowval)
      cout << 'L';
    else if (currval == highval)
      cout << 'H';
    else
      cout << ' ';
    cout << "f(";
    for (int j = 0; j < num_dims-1; j++)
      cout << setw(6) << setprecision(5) << verts[j]
      << ",";
    cout << setw(6) << setprecision(5)
<< verts[num_dims-1] << ") = ";
    cout << setw(8) << setprecision(7) << currval
    << " ";
  }
  cout << endl;
  }
```

The final listing is a simplified test program that runs the simplex on the Rosenbrock 'banana' function whose minimum is at $(1, 1)$, one end of a banana-shaped valley. The simplex is started round the bend of the valley at $(-1, 1)$. The function is: $f(x, y) = 100(y - x^2)^2 + (1 - x)^2$. Figure 16.15 shows Rosenbrock's function represented with lightness corresponding to value in the range $(-2, 2)$ in both x and y. Also shown is the progress of a simplex as it moves towards the minimum.

```cpp
// Test of simplex on 2D Rosenbrock banana function
#include "simplex.h"
#include < iostream >
double valueat(const double *dims) {
  double x1 = dims[0];
  double x2 = dims[1];
```

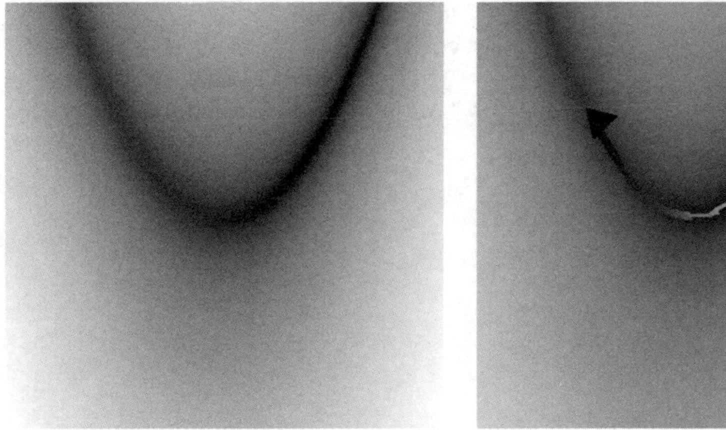

Figure 16.15 *The Rosenbrock 'banana' function (left) and simplex progress along the valley (right)*

```
    double f = 100*(x2 - x1*x1)*(x2 - x1*x1)+(1 - x1)*
      (1 - x1);
      return f;
  }
int main(int argc, char *argv[]) {
  simplex s(2, valueat);
  for (int i = 0; i < 2; i++)
    s.bound(i, -3, 3);
  if (argc == 4) {
    double startcoords[2];
    startcoords[0] = atof(argv[1]);
    startcoords[1] = atof(argv[2]);
    double size = atof(argv[3]);
    s.initequ(startcoords,size);
    }
  else {
    cerr << "Usage: rosenbrock x1start x2start initsize\n";
    return -1;
    }
  cout << "Start simplex:\n";
  s.print();
  int movetype;
  int cnt = 0;
  while(movetype = s.move(0.0001)) {
    // cout << "After move type "<< movetype << endl;
    // s.print();
    cnt++;
    }
  cout << "End simplex:\n";
  s.print();
  cout << "Found minimum at\n";
  double *verts = s.getmin();
```

```
        cout << "f(" << verts[0] << "," ;
        cout << verts[1];
        cout << ") = " << s.getminval();
        cout << " in " << cnt << " iterations\n";
}
```

17 stable – designing a string table class

<table>
<tr><td>

17.1 A perennial problem in data analysis

</td><td>

Scientific programmers working in a tool-based mode develop multiple programs to collect (or simulate), organize and process their data. In the work of many scientists and engineers, programs produce textual output that streams together a host of information about a particular experiment. Later they want to gather some of the information for new analyses. If the data are clearly structured into a table, then it might be appropriate to load them into a spreadsheet and use the spreadsheet's programming language to extract cells and combine them into a new table. But often, the data are only semi-structured: the textual outputs might embed debugging information, or formatting may vary according to particular run parameters. In these cases what's needed is a kind of semi-structured text processing facility – perhaps a language, but more simply a class library (in this case in C++) – that lets the required manipulation be done easily.

</td></tr>
</table>

Here are two particular examples that illustrate the sort of thing that happens often in processing of textual experimental output.

Collating one type of table into another type

We run programs that apply methods $y_1, y_2, \ldots, y_n$ to inputs $x_1, x_2, \ldots, x_m$ yielding $m \times n$ text files $x_i \cdot y_j$. Some of these contain the word 'Exact' preceded by a numerical parameter. We want to consolidate the 'Exact' outputs of the tests into a new table, for which we could write pseudocode of the following form:

```
For all files in format X.Y
    If the file contains "Exact"
        If the file contains more than one "Exact"
            Signal that there's an ambiguity in file X.Y
        Else
            Let variable val equal the numeric value preceding
            "Exact"
            Insert val at row labelled X and column labelled Y in
                a table
Save the table
```

Sifting and computing

We run a program that produces virtual reams of output, monitoring the state of a process. We want to go through that output and pull out and combine some semi-structured data. Every few thousand lines the output contains a number in the range 0.002 to 0.003 in the first column. When that happens, we want to go back a line and pick up some parameters separated by commas after an equals sign. Depending on how many parameters there are, we want to save them in different accumulators. Eventually we want to compile a table comparing the ranges of outputs for different numbers of parameters. (It sounds convoluted, but this is just the sort of thing that happens in practice.) What we need here is some convenient way of getting the computer to do this for us easily. Something like:

```
For all rows
    If row i column 0 is a number x and 0.002 <= x <= 0.003
    If row i-1 contains '='
        Find position of '=' in row i-1; call this j
        Let the number of commas on row i-1 from j to the
            line end be k
        Split line i-1 from j to the end of the line at commas
        For each item m in the split, add this value to
            accumulator[k][m]
Print out values of all accumulators in a 2D table
```

C++ does not have a strong following as a text processing language, but it has a string type and facilities to support regular expressions (which we won't be using in this case study). This means that construction of classes to handle messy manipulations is possible. We will therefore consider such a project by looking at the evolution of a class called 'stable' (short for string-table). In this case study we will see many different versions of the same class. The purpose is illustrative – to show how the specification and design developed together in what is essentially an open-ended problem.

17.2 Design approach

The first step in design is to define the problem. We have outlined a couple of use cases – examples of what we want to be able to do once the program is complete. Is this sufficient? After all, we have a rule:

Rule 10

Steer design by clear criteria.

But this particular project is one where we expect technology to some extent to drive use. That is, as the stable class is developed, we would expect to see new uses for it. (Indeed, this happens. See page 321.) Because the use cases are grounded in specific real needs, we can continue with them for now. We'll know we have succeeded when our stable class provides effective functions to do the work.

The project is one in the manipulation of strings.

Rule 11

Exploit existing solutions.

This turns us towards the C++ string type. Perhaps other standard library types would be useful too. A map, for example, is a kind of templated associative array. It can be used to index strings by strings, e.g.

```
capital["France"] = "Paris";
```

But we want something more like a two-dimensional associative array, for which maps will probably be a complication.

Because we are designing a data type that will provide many facilities, the design will be broad rather than deep. That means this will be a typical organic-growth project, with features being added incrementally.

Principle 3

A program grows along the framework of its first version.

So our approach should be to establish a robust framework right from the start, then fill in functionality piece by piece. This could be done by 'failing tests' – the extreme programming approach of writing a test case, watching it fail, then writing the code to fix it. But we will proceed more conventionally by writing the implementation and then trying to break it.

The **stable** class will hide all its methods. These will initially be simple. Because it is a collection, it will probably also need an item helper class to store individual entries.

With these preliminary thoughts, we can begin coding.

17.3 Rapid prototyping a framework

Here is the first version. Because this is the first, rapid, step in development, it is not only inefficient, it also does not contain much in the way of comments. But check the main function at the bottom. The few comments there are give a clue as to what is going on in this far-from-finished version.

```cpp
// string table class. JAR. v0.01 20/8/03
#include <string>
#include <iostream>
using namespace std;
class tcell {
// Structure of each cell in the table: basic unit of the collection
  bool cellused;
  int rowindex;
  int colindex;
  string content;
public:
  tcell() {
    cellused = 0;
    rowindex = colindex = -1;
    };
  ~tcell() {
    };
  void setindices(const int r, const int c) {
```

```cpp
        rowindex = r;
        colindex = c;
      };
    void setval(const string& cont) {
      cellused = 1;
      content = cont;
    };
    void delval() {
      cellused = 0;
      content = "";
    };
    const string getval() {
      return content;
    };
    const int row() {
      return rowindex;
    };
    const int col() {
      return colindex;
    };
    const bool used() {
      return cellused;
    };
  };
class stable {
// The 2D table of tcells plus row and column labels
  int max_rows;    // Set on instantiation
  int max_cols;    // Set on instantiation
  int lowest_available_row;   // Row numbers equal to or
                              // higher than this are not yet
                              // used.
  int lowest_available_col;   // No rows have entries in this or
                              // any higher numbered column
  tcell *rlabel;   // Row labels are outsize the table itself in
                   // a 1D array
  tcell *clabel;  // Column labels are outsize the table itself
                  // in a 1D array
  tcell **cell;  // The array of pointers to arrays (rows) of cells
public:
  stable(const int r, const int c) {
    max_rows = r;
    max_cols = c;
    lowest_available_row = 0;
    lowest_available_col = 0;
    rlabel = new tcell[max_rows]; // Row labels
    clabel = new tcell[max_cols]; // Column labels
    cell = new tcell *[max_rows]; // All the cells
    for (int i = 0; i < max_rows; i++)
       cell[i] = new tcell[max_cols];
    };
  ~stable() {
    for (int i = 0; i < max_rows; i++)
```

```cpp
      delete [] cell[i];
    delete [] cell;
    delete [] clabel;
    delete [] rlabel;
    };
  int set(const int row, const int col, const string& val) {
    if ((row < 0)||(col < 0)||(row >= max_rows)||
        (col >= max_cols))
      return -1;
    if (row >= lowest_available_row)
      lowest_available_row = row+1;
    if (col >= lowest_available_col)
      lowest_available_col = col+1;
    cell[row][col].setval(val);
    return 0;
    };
// May later need versions of set that have int in one of row
// and col and string in the other. But for now, just do both
// same.
  int set(const string& row, const string& col, const string&
        val) {
    int r;  // Integer value of row
    for (r = 0; r < lowest_available_row; r++) {
      if (rlabel[r].getval() == row)
        break;
      }
    if (r == lowest_available_row){
    // Not already in array. (Note inefficiency: usually will
    // call set(...) when labels not yet used, but every time do
    // futile linear search through all all labels)
      if (lowest_available_row == max_rows)
        return -1;
      // Otherwise, using r = lowest_available_row and
      // move lowest_available_row up
      rlabel[lowest_available_row].setval(row);
      lowest_available_row++;
      }
    for (int c = 0; c < lowest_available_col; c++) {
      if (clabel[c].getval() == col) {
        cell[r][c].setval(val);
        return 0;
        }
      }
    // Not already in array. (Note inefficiency: as for rows)
    if (lowest_available_col == max_cols)
        return -1;
    clabel[lowest_available_col].setval(col);
    cell[r][lowest_available_col].setval(val);
    lowest_available_col++;
    return 0;
    };
  int get(const int row, const int col, string& val) {
```

```cpp
          if ((row < 0) || (col < 0) || (row >= max_rows) ||
              (col >= max_cols)){
            val = "";
            return -1;
          }
          if (!cell[row][col].used()) {
            // Works for row, col greater than lowest_available too
            val = "";
            return 0;
          }
          val = cell[row][col].getval();
          return 1;
        };
      int get(const string& row, const string& col, string& val) {
        int r;  // Integer value of row
        for (r = 0; r < lowest_available_row; r++) {
          if (rlabel[r].getval() == row)
            break;
        }
        if (r == lowest_available_row) {
          val = "";
          return 0;
        }
        for (int c = 0; c < lowest_available_col; c++) {
          if (clabel[c].getval() == col) {
            val = cell[r][c].getval();
            return 1;
          }
        }
        return 0;
        };
      void print() {
        cout << "\t";  // Top left corner: No label
        int i;
        for (i = 0; i < lowest_available_col; i++)
          cout << clabel[i].getval() << "\t";
        cout << endl;
        for (i = 0; i < lowest_available_row; i++) {
          cout << rlabel[i].getval() << "\t";
          for (int j = 0; j < lowest_available_col; j++)
            cout << cell[i][j].getval() << "\t";
          cout << endl;
        }
      };
    };


// Test program
int main() {
  stable mytab(5,3); // Little test table of 5 rows, 3 cols
  mytab.set("w","y","wy"); // Set up a row labelled "w" and a
                           // col labelled "y" and at their
                           // intersection in the table, put "wy"
```

```
mytab.set("w","z","wz"); // Already got row "w", so set up
              // another col "z" and put "wz" in cell.
mytab.set("x","y","xy");
mytab.set("x","z","xz");
mytab.set(2,0,"Num");  // Check that we can index by
                          // numbers too
mytab.print();
string retval;
mytab.get(1,1,retval);  // Get a cell by numbers
cout << "Value at 1,1 is " << retval << endl;
mytab.get("w","w",retval); // Get a cell by row, col labels
        // This should fail because there is
        // no column "w". Return empty string
cout << "Value at w,w is " << retval << endl;
mytab.get("x","y",retval); // This should work though
cout << "Value at x,y is " << retval << endl;
}
```

The first version of **stable** can be written in one session. It is all in one file – class declaration, definitions and test program. There are lots of inefficiencies and repeated code. For example, to find a particular row or column label uses linear search. There is no syntactic sugar. Later, we would expect the [] operator to be overloaded to support indexing by **string**s, but so far everything is done through normal member functions.

Testing is very basic. It is really a kind of 'existence test' – to show that something works – rather than a serious attempt to find errors. **print()** is a debug method that will doubtless come into its own later.

So all of this is an outline, and a rather rushed one. But it is an outline that works. The key functions **set()** and **get()** indexed on strings make the **stable** usable like a 2D associative array, though an inefficient one. The implementation has already exposed issues that need design choices. One is the decision not to have a label in the top left corner of the table.

17.4 A quick fix

Even in the far-from-demanding **main** function included in stable's first version, testing reveals a bug. The fix was to add one line:

```
val = "";
```

before the final

```
return 0;
```

in the

```
int get(const string& row, const string& col, string& val)
```

function. Can you see how this error was exposed by the tests already done on the **stable**?

17.5 Reading and writing

Much of the practical usage of **stable** will be reading and writing text files, so in the third version we add functions for that:

```
void readwithoutlabels(ifstream& in, const char
        coldelim = '\t') {
```

```cpp
      // Read from file. Newline is row delimiter.
      lowest_available_row = 0;
      lowest_available_col = 0;
      string line;
      int r = 0;
      while(!in.eof()) {
        getline(in,line);
        if (line == "")  // Nothing after final '\n'
          break;
        int c = 0;
        string temp;
        string::size_type currpos, delimpos;
        currpos = 0;
        rlabel[r].setval("");
        while ((delimpos = line.find(coldelim, currpos))
                != string::npos){
          temp = line.substr(currpos,delimpos-currpos);
          set(r,c++,temp);
          currpos = delimpos+1;
          }
        temp = line.substr(currpos);
        set(r,c,temp);
        r++;
        }
      for (int c = 0; c < lowest_available_col; c++)
        clabel[c].setval("");
      };
    void readwithlabels(ifstream& in, const char coldelim = '\t') {
      // Read from file. Newline is row delimiter.
      lowest_available_row = 0;
      lowest_available_col = 0;
      string line;
      int r = -1;
      while(!in.eof()) {
        getline(in,line);
        if (line == "") // Nothing after final '\n'
          break;
        int c = -1;
        string temp;
        string::size_type currpos, delimpos;
        currpos = 0;
        while ((delimpos = line.find(coldelim, currpos))
                != string::npos){
          temp = line.substr(currpos,delimpos-currpos);
          if ((r == -1)&&(c == -1)) {
            if (currpos != delimpos) {
              cout <<
"Warning: readwithlabels from file that doesn't begin with \n\
        a column delimiter. The first entry will be lost.\n";
              }
            c++;
            }
```

```cpp
      else if (r == -1)
        clabel[c++].setval(temp);
      else if (c == -1) {
        rlabel[r].setval(temp);
        c++;
        }
      else
        set(r,c++,temp);
      currpos = delimpos+1;
      }
    temp = line.substr(currpos);
    if (r == -1)
      clabel[c].setval(temp);
    else if (c == -1)
      rlabel[r].setval(temp);
    else
      set(r,c,temp);
    r++;
    }
  };


  void writewithlabels(ofstream& out, const char
      coldelim = '\t') {
    out << coldelim;  // Top left corner: No label
    int i, j;
    for (i = 0; i < lowest_available_col-1; i++)
      out << clabel[i].getval() << coldelim;
    out << clabel[i].getval() << endl;
    for (i = 0; i < lowest_available_row; i++) {
      out << rlabel[i].getval() << coldelim;
      for (j = 0; j < lowest_available_col-1; j++)
        out << cell[i][j].getval() << coldelim;
      out << cell[i][j].getval() << endl;
      }
  };
  void writewithoutlabels(ofstream& out, const char
      coldelim = '\t') {
    int i, j;
    for (i = 0; i < lowest_available_row; i++) {
      for (j = 0; j < lowest_available_col-1; j++)
        out << cell[i][j].getval() << coldelim;
      out << cell[i][j].getval() << endl;
      }
  };
```

There are two ways to read and two ways to write depending on whether the row and column labels are included. We have duplicated some code here. That is something to deal with when we refactor later: probably we will seek to consolidate the common code to make things more concise and maintainable. For the time being, however, we are seeking to make progress as quickly as possible.

The main program changes too. Here it is:

```
int main(int argc, char *argv[]) {
    if (argc != 3) {
        cerr << "Usage: stable tablefile outfile\n";
        return -1;
    }
    ifstream i(argv[1]);
    stable mytab(100,20);
    mytab.readwithlabels(i);
    i.close();
    mytab.print();
    ofstream o(argv[2]);
    mytab.writewithlabels(o);
    o.close();
}
```

Clearly this needs a testfile to work with, and a simple one looks like:

```
a       b       c       d
00      11      22      33
000     111     222     333
odd line in the middle
0000    1111    2222    3333
```

What would you expect the output to look like? Here's a clue:

```
        b       c       d
00      11      22      33
000     111     222     333
odd line in the middle
0000    1111    2222    3333
```

Is that what you would expect? Why?

Note that in writing out columns in **writewithlabels()** and **writewithoutlabels()** we go up to the next-to-last column in a loop and then add the last. This is so the column delimiter is not repeated at the end of each line.

The actual output raises the issue of what to do about output lines that are not full length. Do we output the extra column delimiters (as is in fact done in this version) or should we maintain a **lowest_available_col** for each row, so that we only go up to it? But then perhaps we also need a **start_col** for each row?

17.6 Finding things

The next version of **stable** adds find functionality to get the coordinates (row and column numbers) of a specified string.

```
int find(const string& val, int& row, int& col) {
    int r, c;
    for (r = 0; r < lowest_available_row; r++)
        for (c = 0; c < lowest_available_col; c++) {
```

```
                if (cell[r][c].used()) {
                    if (cell[r][c].getval() == val) {
                        row = r;
                        col = c;
                        return 1;
                    }
                }
            }
        return 0;
    };
```

It provides similar facilities to find the number of a specified row or column label.

```
    int findrowlabel(const string& val, int& row) {
        for (int r = 0; r < lowest_available_row; r++) {
            if (rlabel[r].used()) {
                if (rlabel[r].getval() == val) {
                    row = r;
                    return 1;
                }
            }
        }
        return 0;
    };
    int findcollabel(const string& val, int& col) {
        for (int c = 0; c < lowest_available_col; c++) {
            if (clabel[c].used()) {
                if (rlabel[c].getval() == val) {
                    col = c;
                    return 1;
                }
            }
        }
        return 0;
    };
```

To balance this, methods to set labels and to get labels on the basis of row or column number are added.

```
    int setrowlabel(const int row, const string& val) {
        if ((row < 0) || (row >= max_rows))
            return -1;
        if (row >= lowest_available_row)
            lowest_available_row = row+1;
        rlabel[row].setval(val);
        return 0;
    };
    int setcollabel(const int col, const string& val) {
        if ((col < 0) || (col >= max_cols))
            return -1;
```

```
          if (col >= lowest_available_col)
            lowest_available_col = col+1;
          clabel[col].setval(val);
          return 0;
          };
      int getrowlabel(const int row, string& val) {
        if ((row < 0) || (row >= max_rows)) {
          val = "";
          return -1;
          }
        if (!rlabel[row].used()) {
          val = "";
          return 0;
          }
        val = rlabel[row].getval();
        return 1;
        };
      int getcollabel(const int col, string& val) {
        if ((col < 0) || (col >= max_cols)) {
          val = "";
          return -1;
          }
        if (!clabel[col].used()) {
          val = "";
          return 0;
          }
        val = clabel[col].getval();
        return 1;
        }
```

This version is developed and compiled, but clearly we are close to doing
something useful with **stable**, so rather than testing, we move straight
to version 5.

17.7 Matching the requirements

Version 5 is the first test of the requirements. It uses all the versions
before but changes the test program to actually do the kind of analysis
outlined on page 353 above. Here is the main function of the new test
program:

```
int main(int argc, char *argv[]) {
  if (argc < 3) {
    cerr << "Usage: stable files_containing_Infinity\n";
    return -1;
    }
  stable intab(40,4);
  stable outtab(80,6);
  for (int a = 1; a < argc; a++) {
    string fname(argv[a]);
    ifstream i(fname.c_str());
    if (i == 0) {
      cerr << "can't open " << fname.c_str() << endl;
      return -1;
```

```
            }
          intab.readwithoutlabels(i, ' ');
          i.close();
          string::size_type index = fname.rfind('/');
          string prefix = fname.substr(index+1);
          index = prefix.find('.');
          prefix = prefix.substr(0,index);
          int row, col;
          if (intab.find("Infinity",row, col)) {
            string value;
            intab.get(row,col-1,value);
            string suffix;
            intab.get(row-1,0,suffix);
            outtab.set(prefix,suffix,value);
          }
        }
      outtab.print();
      ofstream o("losslesstable");
      outtab.writewithlabels(o);
      o.close();
      }
```

All the additional functions are incorporated in this version. As well, in testing it became clear that we need the same bounds checking on the setting of row and column labels as on the table itself. This was fixed by adding **setval()** member functions and making every set of these use the member function. Also added are some **setw()** manipulators for improving the appearance of the printed table.

stable5 was tested using the command line

stable5 dataexamples/*

where **dataexamples** was a directory containing many files of results. Here is one example. Others followed a similar pattern in that the relevant number (the lossless data rate) could be obtained from the line where the word 'Infinity' appears.

```
XUnitText: bpp
YUnitText: PSNR
TitleText: airplane.ppm
"btpc5
12.634 Infinity
3.20627 41.3915
2.67166 40.5021
1.69696 38.2776
1.12238 36.4754
0.902802 35.5615
0.677032 34.3041
0.539368 33.2836
0.427673 32.2656
0.362701 31.5511
0.289612 30.4847
0.233429 29.3926
0.193726 28.561
```

Randomized checking of the results showed that every facility tested by the program was working correctly.

17.8 Generalizing stable to do more

stable5 uses **string** manipulation to break the filename apart, but could we parse it using the stable ADT we've invented?

Well this is really a 1D problem, so we're overengineered, but why not? To do this, we have to (a) accept other row delimiters than just '\n' and (b) be able to read strings into **stable** objects as well as whole files.

So we add some **stringstream** processing to use existing code with some small extra wrappers. That is, the **stream** reading is already implemented, so we change **strings** to **streams**, and our use of **ifstream** to **istream**. Clearly we could also extend the processing to C-strings. Here is an example of one of the new member functions:

```
void readwithoutlabels(string& instring, const char
      coldelim = '\t', const char rowdelim = '\n') {
   istringstream is(instring);
   readwithoutlabels(is, coldelim, rowdelim);
   };
```

Having done all that fiddly stuff, the only thing we actually use it for is to replace a bit of filename manipulation using string manipulation (.rfind, .substr, .find, .subr) with the **parsetable** towards the end of main(). This also seems to involve a retreat into using C-strings from the **argv[]** array directly for the filenames. Here is the revised **main()** function:

```
int main(int argc, char *argv[]) {
  if (argc < 3) {
    cerr << "Usage: stable files_containing_Infinity\n";
    return -1;
    }
  stable intab(40,4);
  stable outtab(80,6);
  stable parsetable(20,20);
  for (int a = 1; a < argc; a++) {
    string fname(argv[a]);
    parsetable.readwithoutlabels(fname, '.', '/');
    ifstream i(argv[a]);
    if (i == 0) {
      cerr << "can't open " << argv[a] << endl;
      return -1;
      }
    intab.readwithoutlabels(i, ' ');
    i.close();
    int row, col;
    if (intab.find("Infinity",row, col)) {
      string value;
      intab.get(row,col-1,value);
      string suffix;
      intab.get(row-1,0,suffix);
```

```
            string prefix;
            parsetable.get(parsetable.nrows()-1,0,prefix);
            outtab.set(prefix,suffix,value);
            }
        }
    outtab.print();
    ofstream o("losslesstable");
    outtab.writewithlabels(o);
    o.close();
    }
```

17.9 Size flexibility

Having to specify the row and column sizes is a limitation. We'd prefer some kind of size adaptivity. We therefore keep the number of columns fixed at a high value and allocate rows on an as-needed basis. This makes instantiation of **stable** objects easier.

It's also clear now that we need to clean up the table before reusing it, so **read** functions now call **reset()**.

17.10 Yet more generality: using templates to store other types in stable

stable8.cpp fixed a couple of bugs, such as in **findcollabel()**. (Can you see the problem in the version above?) Its main change though was use of templates to convert via **streams** to and from **strings** for storage in a **stable** object. Thus we can now store any types that can be represented as **strings** via stream processing. **stable8** also provides symmetry in write functions so that we can use any row delimiter and not just '\n' and it fixes the issue of writing column delimiters after the last entry on a row. **write()** gained functionality so that the caller could choose whether to include the last row delimiter.

All the cosmetic **iomanip** stuff from version 6 can be removed and replaced with substring processing.

In the main program, we reorganize the use of **parsetable**, renaming to **parsetab** for consistency with other tables, and add functionality to put the best of two candidate columns into another column. This provides a test on conversion between **doubles** and **strings** using the new template functions.

17.11 A final program before refactoring

stable9.cpp modifies the previous main program to give even more information about the various files collated. It displays how much the entries in one particular column differ from the best of the remaining columns. This sort of operation is typical of what could be done in a spreadsheet, but now we have the stable member functions it is as easy to use **stable** itself.

The full code of **stable9.cpp** is given below as this is the last version before major refactoring.

```
// string table class. JAR. v0.09 23/8/03
#include <string>
#include <iostream>
#include <fstream>
#include <sstream>
#include <iomanip>
```

```cpp
using namespace std;
class tcell {
// Structure of each cell in the table: basic unit of the collection
  bool cellused;
  int rowindex;
  int colindex;
  string content;
public:
  tcell() {
    cellused = 0;
    rowindex = colindex = -1;
    };
  ~tcell() {
    };
  void setindices(const int r, const int c) {
    rowindex = r;
    colindex = c;
    };
  void setval(const string& cont) {
    cellused = 1;
    content = cont;
    };
  void delval() {
    cellused = 0;
    content = "";
    };
  const string getval() {
    return content;
    };
  const int row() {
    return rowindex;
    };
  const int col() {
    return colindex;
    };
  const bool used() {
    return cellused;
    };
  };
class stable {
// The 2D table of tcells plus row and column labels
  int max_rows;  // Set on instantiation
  int max_cols;  // Set on instantiation
  int lowest_available_row;   // Row numbers equal to or
                              // higher than this are not
                              // yet used.
  int lowest_available_col;   // No rows have entries is this or
                              // any higher numbered column
  tcell *rlabel;   // Row labels are outsize the table itself in
                   // a 1D array
  tcell *clabel;   // Column labels are outsize the table itself
                   // in a 1D array
```

```cpp
    tcell **cell;  // The array of pointers to arrays (rows) of cells
public:
  stable(const int r = 2048, const int c = 256) {
    max_rows = r;
    max_cols = c;
    lowest_available_row = 0;
    lowest_available_col = 0;
    rlabel = new tcell[max_rows]; // Row labels
    clabel = new tcell[max_cols]; // Column labels
    cell = new tcell *[max_rows]; // All the cells
    for (int i = 0; i < max_rows; i++)
      cell[i] = 0;
  };
  ~stable() {
    for (int i = 0; i < max_rows; i++) {
      if (cell[i])
        delete [] cell[i];
    }
    delete [] cell;
    delete [] clabel;
    delete [] rlabel;
  };
  void reset() {  // Clean up the table for reuse
    for (int i = 0; i < max_rows; i++) {
      if (cell[i]) {
        for (int j = 0; j < max_cols; j++)
          cell[i][j].delval();
      }
    }
    lowest_available_row = 0;
    lowest_available_col = 0;
  }
  int nrows() {
    return lowest_available_row;
  }
  int ncols() {
    return lowest_available_col;
  }
  int set(const int row, const int col, const string& val) {
    if ((row < 0) || (col < 0) || (row >= max_rows) ||
        (col >= max_cols))
      return -1;
    if (row >= lowest_available_row) {
      while (lowest_available_row <= row) {
        if (!cell[lowest_available_row])
          cell[lowest_available_row] = new tcell[max_cols];
        lowest_available_row++;
      }
    }
    if (col >= lowest_available_col)
      lowest_available_col = col+1;
    if (!cell[row])
```

```cpp
        cell[row] = new tcell[max_cols];
      cell[row][col].setval(val);
      return 0;
      };
    template <class T>
    int set(const int row, const int col, const T& val) {
      ostringstream os;
      os << val;
      return set(row, col,os.str());
    }
    int set(const string& row, const string& col, const
        string& val) {
      int r;  // Integer value of row
      for (r = 0; r < lowest_available_row; r++) {
        if (rlabel[r].getval() == row)
          break;
      }
      if (r == lowest_available_row){
      // Not already in array. (Note inefficiency: usually will
      // call set(...) when labels not yet used, but every time do
      // futile linear search through all all labels)
        if (lowest_available_row == max_rows)
          return -1;
        // Otherwise, using r = lowest_available_row and
        // move lowest_available_row up
        rlabel[lowest_available_row].setval(row);
        if (!cell[lowest_available_row])
          cell[lowest_available_row] = new tcell[max_cols];
        lowest_available_row++;
      }
      for (int c = 0; c < lowest_available_col; c++) {
        if (clabel[c].getval() == col) {
          cell[r][c].setval(val);
          return 0;
        }
      }
      // Not already in array. (Note inefficiency: as for rows)
      if (lowest_available_col == max_cols)
        return -1;
      clabel[lowest_available_col].setval(col);
      cell[r][lowest_available_col].setval(val);
      lowest_available_col++;
      return 0;
      };
    template <class T>
    int set(const string& row, const string& col, const T& val) {
      ostringstream os;
      os << val;
      return set(row, col,os.str());
    }
    int setrowlabel(const int row, const string& val) {
      if ((row < 0)||(row >= max_rows))
```

```cpp
        return -1;
    if (row >= lowest_available_row) {
        while (lowest_available_row <= row) {
            if (!cell[lowest_available_row])
                cell[lowest_available_row] = new tcell[max_cols];
            lowest_available_row++;
            }
        }
    rlabel[row].setval(val);
    return 0;
    };
int setcollabel(const int col, const string& val) {
    if ((col < 0)||(col >= max_cols))
        return -1;
    if (col >= lowest_available_col)
        lowest_available_col = col+1;
    clabel[col].setval(val);
    return 0;
    };
void readwithoutlabels(istream& in, const char
        coldelim = '\t', const char rowdelim = '\n') {
    // Read from file.
    reset();
    string line;
    int r = 0;
    while(!in.eof()) {
        getline(in,line,rowdelim);
        if ((line == "")&&in.eof())  // Nothing after final
                                     // rowdelim
            break;
        int c = 0;
        string temp;
        string::size_type currpos, delimpos;
        currpos = 0;
        setrowlabel(r,"");
        while ((delimpos = line.find(coldelim, currpos))
                != string::npos){
            temp = line.substr(currpos,delimpos-currpos);
            set(r,c++,temp);
            currpos = delimpos+1;
            }
        temp = line.substr(currpos);
        set(r,c,temp);
        r++;
        }
    for (int c = 0; c < lowest_available_col; c++)
        clabel[c].setval("");
    };
void readwithoutlabels(const string& instring, const char
        coldelim = '\t', const char rowdelim = '\n') {
    istringstream is(instring);
    readwithoutlabels(is, coldelim, rowdelim);
```

```cpp
};
void readwithoutlabels(const char *p, const char
    coldelim = '\t', const char rowdelim = '\n') {
  string instring(p);
  istringstream is(instring);
  readwithoutlabels(is, coldelim, rowdelim);
  };
void readwithlabels(ifstream& in, const char coldelim = '\t',
    const char rowdelim = '\n') {
  // Read from file.
  reset();
  string line;
  int r = -1;
  while(!in.eof()) {
    getline(in,line, rowdelim);
    if ((line == "")&&in.eof())  // Nothing after final rowdelim
      break;
    int c = -1;
    string temp;
    string::size_type currpos, delimpos;
    currpos = 0;
    while ((delimpos = line.find(coldelim, currpos))
          != string::npos){
      temp = line.substr(currpos,delimpos-currpos);
      if ((r==-1)&&(c==-1)) {
        if (currpos != delimpos) {
          cout <<
"Warning: readwithlabels from file that doesn't begin with a \n\
    column delimiter. The first entry will be lost.\n";
        }
        c++;
        }
      else if (r == -1)
        setcollabel(c++,temp);
      else if (c == -1) {
        setrowlabel(r,temp);
        c++;
        }
      else
        set(r,c++,temp);
      currpos = delimpos+1;
      }
    temp = line.substr(currpos);
    if (r == -1)
      setcollabel(c,temp);
    else if (c == -1)
      setrowlabel(r,temp);
    else
      set(r,c,temp);
    r++;
    }
  };
```

```cpp
void readwithlabels(const string& instring, const char
    coldelim = '\t', const char rowdelim = '\n') {
  istringstream is(instring);
  readwithoutlabels(is, coldelim, rowdelim);
};
void readwithlabels(const char *p, const char
    coldelim = '\t', const char rowdelim = '\n') {
  string instring(p);
  istringstream is(instring);
  readwithoutlabels(is, coldelim, rowdelim);
};
int get(const int row, const int col, string& val) {
  if ((row < 0) || (col < 0) || (row >= max_rows) ||
      (col >= max_cols)){
    val = "";
    return -1;
  }
  if (!cell[row][col].used()) {
    // Works for row, col greater than lowest_available too
    val = "";
    return 0;
  }
  val = cell[row][col].getval();
  return 1;
};
template <class T>
int get(const int row, const int col, T& val) {
  string thestring;
  int retval = get(row, col, thestring);
  istringstream is(thestring);
  is >> val;
  return retval;
}
int get(const string& row, const string& col, string& val) {
  int r;  // Integer value of row
  for (r = 0; r < lowest_available_row; r++) {
    if (rlabel[r].getval() == row)
      break;
  }
  if (r == lowest_available_row) {
    val = "";
    return 0;
  }
  for (int c = 0; c < lowest_available_col; c++) {
    if (clabel[c].getval() == col) {
      val = cell[r][c].getval();
      return 1;
    }
  }
  val = "";
  return 0;
};
```

```cpp
template <class T>
int get(const string& row, const string& col, T& val) {
    string thestring;
    int retval = get(row, col, thestring);
    istringstream is(thestring);
    is >> val;
    return retval;
}
int getrowlabel(const int row, string& val) {
    if ((row < 0) || (row >= max_rows)) {
        val = "";
        return -1;
    }
    if (!rlabel[row].used()) {
        val = "";
        return 0;
    }
    val = rlabel[row].getval();
    return 1;
};
int getcollabel(const int col, string& val) {
    if ((col < 0) || (col >= max_cols)) {
        val = "";
        return -1;
    }
    if (!clabel[col].used()) {
        val = "";
        return 0;
    }
    val = clabel[col].getval();
    return 1;
}
int find(const string& val, int& row, int& col) {
    int r, c;
    for (r = 0; r < lowest_available_row; r++)
    for (c = 0; c < lowest_available_col; c++) {
        if (cell[r][c].used()) {
            if (cell[r][c].getval() == val) {
                row = r;
                col = c;
                return 1;
            }
        }
    }
    return 0;
};
int findrowlabel(const string& val, int& row) {
    for (int r = 0; r < lowest_available_row; r++) {
        if (rlabel[r].used()) {
            if (rlabel[r].getval() == val) {
                row = r;
                return 1;
```

```cpp
        }
      }
    }
    return 0;
  };
  int findcollabel(const string& val, int& col) {
    for (int c = 0; c < lowest_available_col; c++) {
      if (clabel[c].used()) {
        if (clabel[c].getval() == val) {
          col = c;
          return 1;
        }
      }
    }
    return 0;
  };

  int writewithlabels(ostream& out, const char coldelim = '\t',
      const char rowdelim = '\n',
        const bool includelastrowdelim = 1) {
    out << coldelim;  // Top left corner: No label
    int i, j;
    for (i = 0; i < lowest_available_col-1; i++)
      out << clabel[i].getval() << coldelim;
    out << clabel[i].getval() << rowdelim;
    for (i = 0; i < lowest_available_row; i++) {
      out << rlabel[i].getval() << coldelim;
      // Don't output column delimiters after last entry on a row
      int last_used = lowest_available_col-1;
      while ((last_used >= 0) && !cell[i][last_used].used())
        last_used--;
      for (j = 0; j < last_used; j++)
        out << cell[i][j].getval() << coldelim;
      if (last_used >= 0)
        out << cell[i][j].getval();
      if ((i < lowest_available_row-1) || includelastrowdelim)
        out << rowdelim;
    }
  };
  int writewithlabels(string& ostring, const char
      coldelim = '\t', const char rowdelim = '\n',
        const bool includelastrowdelim = 1) {
    ostringstream os;
    int retval = writewithlabels(os, coldelim, rowdelim,
      includelastrowdelim);
    ostring = os.str();
    return retval;
  }
  int writewithoutlabels(ostream& out, const char
      coldelim = '\t', const char rowdelim = '\n',
        const bool includelastrowdelim = 1) {
    int i, j;
```

```cpp
      for (i = 0; i < lowest_available_row; i++) {
        // Don't output column delimiters after last entry on a row
        int last_used = lowest_available_col-1;
        while ((last_used >= 0) && !cell[i][last_used].used())
          last_used--;
        for (j = 0; j < last_used; j++)
          out << cell[i][j].getval() << coldelim;
        if (last_used >= 0)
          out << cell[i][j].getval();
        if ((i < lowest_available_row-1) || includelastrowdelim)
          out << rowdelim;
      }
    };
  int writewithoutlabels(string& ostring, const char
        coldelim = '\t', const char rowdelim = '\n',
          const bool includelastrowdelim = 1) {
      ostringstream os;
      int retval = writewithoutlabels(os, coldelim,
          rowdelim,includelastrowdelim);
      ostring = os.str();
      return retval;
    }
  void print() {
      cout << "\t";  // Top left corner: No label
      int i;
      for (i = 0; i < lowest_available_col; i++)
        cout << clabel[i].getval().substr(0,6) << "\t";
      cout << endl;
      for (i = 0; i < lowest_available_row; i++) {
        cout << rlabel[i].getval().substr(0,6) << "\t";
        for (int j = 0; j < lowest_available_col; j++)
          cout << cell[i][j].getval().substr(0,6) << "\t";
        cout << endl;
      }
    };
  };


  // Test program
  int main(int argc, char *argv[]) {
    if (argc < 2) {
      cerr << "Usage: stable files_containing_Infinity\n";
      return -1;
    }
    stable intab;
    stable outtab;
    stable parsetab;
    for (int a = 1; a < argc; a++) {
      string prefix;
      parsetab.readwithoutlabels(argv[a], '.', '/');
      parsetab.get(parsetab.nrows()-1,0,prefix);
      // Get string between last '/' and first following '.'
      ifstream i(argv[a]);
```

```cpp
        if (i == 0) {
          cerr << "can't open" << argv[a] << endl;
          return -1;
        }
      intab.readwithoutlabels(i, ' ');
      i.close();
      int row, col;
      if (intab.find("Infinity",row, col)) {
        string value;
        intab.get(row,col-1,value);
        string suffix;
        intab.get(row-1,0,suffix);
        outtab.set(prefix,suffix,value);
        }
      }
    int endcol = outtab.ncols();
    int btpccol;
    int bestcol;
    outtab.findcollabel("\"btpc5a",btpccol);
    double btpcval, bestval, currval;
    string bestname;
    for (int r = 0; r < outtab.nrows(); r++) {
      bestval = 64;  // Ridiculously high
      for (int c = 0; c < outtab.ncols(); c++) {
        if (c == btpccol)
          outtab.get(r,c,btpcval);
        else {
          outtab.get(r,c,currval);
          if (currval < bestval) {
            bestval = currval;
            bestcol = c;
          }
        }
      }
      outtab.getcollabel(bestcol, bestname);
      ostringstream os;
      if (bestval < btpcval) {
        bestval = (btpcval-bestval)*100.0/bestval;
        os << bestname << "beats BTPC by" << bestval <<
            "percent\n";
      }
      else {
        bestval = (bestval-btpcval)*100.0/btpcval;
        os << "BTPC beats" << bestname << "by" <<
            bestval << "percent\n";
      }
      outtab.set(r,endcol,os.str());
    }
  outtab.setcollabel(endcol,"Result");
  outtab.print();
  ofstream o("losslesstable");
  outtab.writewithlabels(o);
```

```
o.close();
cout << "Reparse of last file:\n";
string lastfile;
parsetab.writewithoutlabels(lastfile,'.','/',0);
cout << lastfile << endl;
}
```

17.12 Refactoring

Rule 2

**Don't expect to get the representation
right first time. Look for opportunities to
improve code by *re*-representation.**

We can refactor the code in two stages. First, keeping everything in one
file, we look for redundancies, repetitions and unused code.

Clearly we can use **findrowlabel()** and **findcollabel()** in **set()**. As
well, it is possible to consolidate code for **read** and **write**, despite their
differences. This in turn causes consolidated use of **istream** for the **in**
argument.

To improve the power of the **stable** type we really need functions
that will allow us to step through occurrences of a particular **string** in
the **stable**. We add functions for **findbefore** and **findafter**, and then
immediately add template versions so we can search for other types represented as strings. Here are the appropriate definitions:

```cpp
int findafter(const string& val, int& row, int& col) {
   if ((row < 0)||(col < 0)||(row >= lowest_available_row)||
       (col >= lowest_available_col))
     return 0;
   for (int r = row; r < lowest_available_row; r++) {
     // First loop for start row starting at col
     for (int c = col; c < lowest_available_col; c++) {
       if (cell[r][c].used()) {
         if (cell[r][c].getval() == val) {
           row = r;
           col = c;
           return 1;
         }
       }
     }
     // Remaining loops start at column 0
     col = 0;
   }
   return 0;
};
template <class T>
int findafter(const T& val, int& row, int& col) {
   ostringstream os;
   os << val;
   return findafter(os.str(), row, col);
```

```cpp
        }
    int findbefore(const string& val, int& row, int& col) {
        if ((row < 0)||(col < 0)||(row >= lowest_available_row)||
            (col >= lowest_available_col))
            return 0;
        for (int r = row; r >= 0; r--) {
            // First loop for start row starting at col
            for (int c = col; c >= 0; c--) {
                if (cell[r][c].used()) {
                    if (cell[r][c].getval() == val) {
                        row = r;
                        col = c;
                        return 1;
                    }
                }
            }
            // Remaining loops start at last column
            col = lowest_available_col-1;
        }
        return 0;
    };
template <class T>
int findbefore(const T& val, int& row, int& col) {
    ostringstream os;
    os << val;
    return findbefore(os.str(), row, col);
}
```

find() is similarly provided with a template version.

We might consider making **tcell** a nested class of **stable**, or making everything in **tcell** private with **stable** as a friend. But why prohibit use of this little gem? We leave it as a lightweight publicly available class.

The second step in refactoring is to divide the program up into pieces (at last). We convert our single large file into **stable.h**, **stable.cpp** and **teststable.cpp**, reorganizing to group functions together. In doing this we have to remember to leave all template code in the header (at least until the compiler has a working **export**). **stable.h** is commented up to document the use of public member functions, superfluous semicolons and default parameters are removed from **stable.cpp**, and we add name, date and basic version information to all three files.

With the program divided into parts it is possible to perform serious testing on **stable**. **teststable.cpp** now uses a header file **test.h** to simplify testing. Although this uses C/C++ preprocessor macros, it provides a handy test framework.

Testing reveals a conceptual problem in the design of **findafter** (and **findbefore**). It is not clear whether the search should start at the specified position or one after (before). A problem with requiring the caller to increment the position is that there may be no further columns on the current row and the call to **findafter** would then return -1. The problem with incrementing in the class is that the initial **col,row** location is not searched. But this is probably the better functionality, so we fix it in the implementation. We also have to fix find to the case of (0, 0).

Here are the final files from the project as discussed here (with a short version of **teststable.cpp** rather than the long version used for full testing). In fact, three other test programs were developed exercising different functions of the stable class in different ways. These were closely tied to the uses defined at the beginning of the chapter and so not only served to test operation but also yielded practical results.

Final **stable.h**

```cpp
// Header for stable class
// John Robinson 2003
// Version 0.1    25 August 2003
#include <string>
#include <iostream>
#include <sstream>
using namespace std;
// tcell provides the structure of each cell in the stable class.
// It may be used separately for other collections if desired.
// Lightweight: all member functions defined inline here.
class tcell {
    bool cellused;
    int rowindex;
    int colindex;
    string content;
public:
    tcell() {
        cellused = 0;
        content = "";
        rowindex = colindex = -1;
    };
    ~tcell() {
    };
    void setindices(const int r, const int c) {
        rowindex = r;
        colindex = c;
    };
    void setval(const string& cont) {
        cellused = 1;
        content = cont;
    };
    void delval() {
        cellused = 0;
        content = "";
    };
    const string getval() {
        return content;
    };
    const int row() {
        return rowindex;
    };
    const int col() {
        return colindex;
    };
```

```
            const bool used() {
                return cellused;
                };
            };
        // stable provides a 2D table of tcells plus row and column
        // labels, with facilities as outlined in member functions
        // below.
        class stable {
            // Comments on private members are for reference only. Users
            // should not infer implications about how stable is
            // implemented.
            int max_rows;  // Set on instantiation
            int max_cols;  // Set on instantiation
            int lowest_available_row; // Row numbers equal to or
                    // higher than this are not yet used.
            int lowest_available_col; // No rows have entries is this or any
                    // higher numbered column
            tcell *rlabel;  // Row labels are outsize the table itself
                    // in a 1D array
            tcell *clabel;  // Column labels are outsize the table itself
                    // in a 1D array
            tcell **cell;  // The array of pointers to arrays (rows) of cells
            // read() and write() functions are called by the public
            // interfaces and simply encapsulate the two alternatives of
            // reading/writing with and without labels
            void read(istream& in, const bool withlabels, const char
                    coldelim, const char rowdelim);
            int write(ostream& out, const bool withlabels, const char
                    coldelim, const char rowdelim, const bool
                    includelastrowdelim);
        public:
            stable(const int r = 2048, const int c = 256);
            // Users may instantiate with default parameters and not
            // incur an excessive memory hit. Only if the number of
            // columns is expected to be greater than 256 or the number
            // of rows greater than 2048 need the parameters be used
            ~stable();
            void reset();  // Clean up the table for reuse
            int nrows() {
                return lowest_available_row;
                }
            int ncols() {
                return lowest_available_col;
                }
            int maxrows() {
                return max_rows;
                }
            int maxcols() {
                return max_cols;
                }
            // set and get functions work on individual cells of the stable,
            // or on the row or column labels. They allow setting and
```

```cpp
// getting of any data types that have istream ostream
// operators defined to allow the types to be represented
// as strings.
// set returns -1 if outside stable size limits; 0 otherwise
// get returns -1 if outside size limits; 0 if row-col indices
// not used or label not found; 1 otherwise.
int set(const int row, const int col, const string& val);
int get(const int row, const int col, string& val);
// set/get the value at coords (col, row) in the table to val
// For get, if the value is unused, returns 0 and val = "".
template <class T> int set(const int row, const int col,
     const T& val) {
  ostringstream os;
  os << val;
  return set(row, col,os.str());
  }
template <class T> int get(const int row, const int col, T& val) {
  string thestring;
  int retval = get(row, col, thestring);
  istringstream is(thestring);
  is >> val;
  return retval;
  }
// As above, for arbitrary types that can be represented
// as strings
int set(const string& row, const string& col, const string& val);
int get(const string& row, const string& col, string& val);
// set/get the value with row, column labels as specified to val
template <class T> int set(const string& row,
      const string& col, const T& val) {
  ostringstream os;
  os << val;
  return set(row, col,os.str());
  }
template <class T> int get(const string& row,
      const string& col, T& val) {
  string thestring;
  int retval = get(row, col, thestring);
  istringstream is(thestring);
  is >> val;
  return retval;
  }
// As above, for arbitrary types that can be represented
// as strings
int setrowlabel(const int row, const string& val);
int setcollabel(const int col, const string& val);
int getrowlabel(const int row, string& val);
int getcollabel(const int col, string& val);
// Explicitly set/get the row/column label with given index

// Find functions look for a particular entry after or before
// the specified (col,row) coordinates, where "after"/"before"
```

```cpp
// defined according to raster ordering of stable. If found,
// return 1 and row, col set to coordinates. If not found,
// return 0 and row, col are garbage.
int findafter(const string& val, int& row, int& col);
// Find (col, row) coords of next occurence in table of val
// after the (col, row) values on input.
template <class T> int findafter(const T& val, int& row,
      int& col) {
   ostringstream os;
   os << val;
   return findafter(os.str(), row, col);
   }
// As above, for arbitrary types that can be represented
// as strings
int findbefore(const string& val, int& row, int& col);
// Find (col, row) coords of previous occurence in table of val
// before the (col, row) values on input.
template <class T> int findbefore(const T& val, int& row,
      int& col) {
   ostringstream os;
   os << val;
   return findbefore(os.str(), row, col);
   }
// As above, for arbitrary types that can be represented
// as strings
int find(const string& val, int& row, int& col);
// Find coords of first occurence of val in table
template <class T> int find(const T& val, int& row, int& col) {
   ostringstream os;
   os << val;
   return find(os.str(), row, col);
   }
// As above, for arbitrary types that can be represented
// as strings
int findrowlabel(const string& val, int& row);
int findcollabel(const string& val, int& col);
// Find index of row/col label with value val

// Reading functions allow reading from a stream or a string. The
// input is parsed according to the specified row and column
// delimiters. The newly read information overwrites
// anything that was previously in the stable.
// Reading can either be withlabels -- meaning the source
// stream/string has labels at the top and left hand side or
// withoutlabels. In the latter case, the labels will be set to
// empty strings.
// Writing functions are symmetric to reading functions
// with an additional parameter that says whether to output
// the final rows rowdelim at the very end.
void readwithoutlabels(istream& in,
   const char coldelim = '\t', const char rowdelim = '\n');
void readwithoutlabels(const string& instring,
```

```
            const char coldelim = '\t', const char rowdelim = '\n');
        void readwithoutlabels(const char *p,
            const char coldelim = '\t', const char rowdelim = '\n');
        void readwithlabels(istream& in,
            const char coldelim = '\t', const char rowdelim = '\n');
        void readwithlabels(const string& instring,
            const char coldelim = '\t', const char rowdelim = '\n');
        void readwithlabels(const char *p,
            const char coldelim = '\t', const char rowdelim = '\n');

        int writewithlabels(ostream& out, const char coldelim = '\t',
            const char rowdelim = '\n',
            const bool includelastrowdelim = 1);
        int writewithlabels(string& ostring, const char coldelim = '\t',
            const char rowdelim = '\n',
            const bool includelastrowdelim = 1);
        int writewithoutlabels(ostream& out, const char
            coldelim = '\t',
            const char rowdelim = '\n',
            const bool includelastrowdelim = 1);
        int writewithoutlabels(string& ostring, const char
            coldelim = '\t',
            const char rowdelim = '\n',
            const bool includelastrowdelim = 1);

        // print is a debugging function that prints out a truncated
        // version of each item in the table.
        void print();
    };
```

Final **stable.cpp**

```cpp
// Implementation of stable class
// John Robinson 2003
// Version 0.1    25 August 2003
// Only public functions not defined here are nrows() and
// ncols() and the template versions of set/get/find. All these
// are in stable.h header file.
#include "stable.h"
#include <string>
#include <iostream>
#include <fstream>
#include <sstream>
using namespace std;

// Read and write are private member functions that support
// the public withlabels and withoutlabels versions.
void stable::read(istream& in, const bool withlabels, const
        char coldelim, const char rowdelim) {
    // Read from file with or without labels.
    reset();
    string line;
```

```cpp
int startindex;  // -1 for with labels, because first row and
                 // column are before zeroth row and column;
                 // 0 for without labels.
if (withlabels)
  startindex = -1;
else
  startindex = 0;
int r = startindex;
while(!in.eof()) {
  getline(in,line, rowdelim);
  if ((line == "")&&in.eof())  // Nothing after final rowdelim
    break;
  int c = startindex;
  string temp;
  string::size_type currpos, delimpos;
  currpos = 0;
  while ((delimpos = line.find(coldelim, currpos))
            != string::npos){
    temp = line.substr(currpos,delimpos-currpos);
    if ((r ==-1)&&(c==-1)) {
      if (currpos != delimpos) {
        cout <<
"Warning: readwithlabels from file that doesn't begin with \n\
  a column delimiter. The first entry will be lost.\n";
      }
      c++;
    }
    else if (r == -1)
      setcollabel(c++,temp);
    else if (c == -1) {
      setrowlabel(r,temp);
      c++;
    }
    else
      set(r,c++,temp);
    currpos = delimpos+1;
  }
  temp = line.substr(currpos);
  if (r == -1)
    setcollabel(c,temp);
  else if (c == -1)
    setrowlabel(r,temp);
  else
    set(r,c,temp);
  r++;
}
if (!withlabels) { // Reset all labels
  for (r = 0; r < lowest_available_row; r++)
    rlabel[r].setval("");
  for (int c = 0; c < lowest_available_col; c++)
    clabel[c].setval("");
}
```

```cpp
        }
int stable::write(ostream& out, const bool withlabels, const
      char coldelim, const char rowdelim, const bool
      includelastrowdelim) {
  int i, j;
  if (withlabels) {
    out << coldelim;  // Top left corner: No label
    for (i = 0; i < lowest_available_col-1; i++)
      out << clabel[i].getval() << coldelim;
    out << clabel[i].getval() << rowdelim;
  }
  for (i = 0; i < lowest_available_row; i++) {
    if (withlabels)
      out << rlabel[i].getval() << coldelim;
    // Don't output column delimiters after last entry on a row
    int last_used = lowest_available_col-1;
    while ((last_used >= 0) && !cell[i][last_used].used())
      last_used--;
    for (j = 0; j < last_used; j++)
      out << cell[i][j].getval() << coldelim;
    if (last_used >= 0)
      out << cell[i][j].getval();
    if ((i < lowest_available_row-1) || includelastrowdelim)
      out << rowdelim;
  }
}
// Remaining functions are public. See stable.h for usage
stable::stable(const int r, const int c) {
  max_rows = r;
  max_cols = c;
  lowest_available_row = 0;
  lowest_available_col = 0;
  rlabel = new tcell[max_rows]; // Row labels
  clabel = new tcell[max_cols]; // Column labels
  cell = new tcell *[max_rows]; // All the cells
  for (int i = 0; i < max_rows; i++)
    cell[i] = 0;
}
stable::~stable() {
  for (int i = 0; i < max_rows; i++) {
    if (cell[i])
      delete [] cell[i];
  }
  delete [] cell;
  delete [] clabel;
  delete [] rlabel;
}
void stable::reset() {  // Clean up the table for reuse
  for (int i = 0; i < max_rows; i++) {
    if (cell[i]) {
      for (int j = 0; j < max_cols; j++)
        cell[i][j].delval();
```

```cpp
        }
      }
    lowest_available_row = 0;
    lowest_available_col = 0;
  }

  // In header files, set/get come in pairs. Here are the sets
  // are defined first, then all the gets

  // set functions
  int stable::set(const int row, const int col, const string& val) {
    if ((row < 0)||(col < 0)||(row >= max_rows)||
        (col >= max_cols))
      return -1;
    if (row >= lowest_available_row) {
      while (lowest_available_row <= row) {
        if (!cell[lowest_available_row])
          cell[lowest_available_row] = new tcell[max_cols];
        lowest_available_row++;
      }
    }
    if (col >= lowest_available_col)
      lowest_available_col = col+1;
    if (!cell[row])
      cell[row] = new tcell[max_cols];
    cell[row][col].setval(val);
    return 0;
  }
  int stable::set(const string& row, const string& col, const
      string& val) {
    int r;  // Integer value of row
    if (!findrowlabel(row, r)) {
    // Not already in array.
      if (lowest_available_row == max_rows)
        return -1;
      // Otherwise, use r = lowest_available_row and
      // move lowest_available_row up
      r = lowest_available_row++;
      rlabel[r].setval(row);
      if (!cell[r])
        cell[r] = new tcell[max_cols];
    }
    int c;  // Integer value of column
    if (findcollabel(col, c)) {
      cell[r][c].setval(val);
      return 0;
    }
    // Not already in array.
    if (lowest_available_col == max_cols)
      return -1;
    c = lowest_available_col++;
    clabel[c].setval(col);
```

```cpp
    cell[r][c].setval(val);
    return 0;
}
int stable::setrowlabel(const int row, const string& val) {
  if ((row < 0) || (row >= max_rows))
    return -1;
  if (row >= lowest_available_row) {
    while (lowest_available_row <= row) {
      if (!cell[lowest_available_row])
        cell[lowest_available_row] = new tcell[max_cols];
      lowest_available_row++;
    }
  }
  rlabel[row].setval(val);
  return 0;
}
int stable::setcollabel(const int col, const string& val) {
  if ((col < 0) || (col >= max_cols))
    return -1;
  if (col >= lowest_available_col)
    lowest_available_col = col+1;
  clabel[col].setval(val);
  return 0;
}

// get functions
int stable::get(const int row, const int col, string& val) {
  if ((row < 0) || (col < 0) || (row >= max_rows) ||
      (col >= max_cols)){
    val = "";
    return -1;
  }
  if (!cell[row][col].used()) {
    // Works for row, col greater than lowest_available too
    val = "";
    return 0;
  }
  val = cell[row][col].getval();
  return 1;
}
int stable::get(const string& row, const string& col, string& val) {
  int r;  // Integer value of row
  for (r = 0; r < lowest_available_row; r++) {
    if (rlabel[r].getval() == row)
      break;
  }
  if (r == lowest_available_row) {
    val = "";
    return 0;
  }
  for (int c = 0; c < lowest_available_col; c++) {
```

```cpp
        if (clabel[c].getval() == col) {
          val = cell[r][c].getval();
          return 1;
          }
        }
      val = "";
      return 0;
      }
  int stable::getrowlabel(const int row, string& val) {
      if ((row < 0) || (row >= max_rows)) {
        val = "";
        return -1;
        }
      if (!rlabel[row].used()) {
        val = "";
        return 0;
        }
      val = rlabel[row].getval();
      return 1;
      }
  int stable::getcollabel(const int col, string& val) {
      if ((col < 0) || (col >= max_cols)) {
        val = "";
        return -1;
        }
      if (!clabel[col].used()) {
        val = "";
        return 0;
        }
      val = clabel[col].getval();
      return 1;
      }

  // find functions
  // Note that the basic find uses findafter
  int stable::findafter(const string& val, int& row, int& col) {
      if ((row < 0) || (col < 0) || (row >= lowest_available_row) ||
          (col >= lowest_available_col))
        return 0;
      for (int r = row; r < lowest_available_row; r++) {
        // First loop for start row starting at col+1
        for (int c = col+1; c < lowest_available_col; c++) {
          if (cell[r][c].used()) {
            if (cell[r][c].getval() == val) {
                row = r;
                col = c;
                return 1;
                }
            }
          }
        }
        // Remaining loops start at column 0
```

```cpp
        col = 0;
      }
    return 0;
    }
  int stable::findbefore(const string& val, int& row, int& col) {
    if ((row < 0) || (col < 0) || (row >= lowest_available_row) ||
        (col >= lowest_available_col))
      return 0;
    for (int r = row; r >= 0; r--) {
      // First loop for start row starting at col-1
      for (int c = col-1; c >= 0; c--) {
        if (cell[r][c].used()) {
          if (cell[r][c].getval() == val) {
            row = r;
            col = c;
            return 1;
            }
          }
        }
      // Remaining loops start at last column
      col = lowest_available_col-1;
      }
    return 0;
    }
  int stable::find(const string& val, int& row, int& col) {
    row = 0;
    col = 0;
    // Have to test (0,0) before using findafter().
    if (cell[0][0].used() && (cell[0][0].getval() == val))
      return 1;
    return findafter(val, row, col);
    }
  int stable::findrowlabel(const string& val, int& row) {
    for (int r = 0; r < lowest_available_row; r++) {
      if (rlabel[r].used()) {
        if (rlabel[r].getval() == val) {
          row = r;
          return 1;
          }
        }
      }
    return 0;
    }
  int stable::findcollabel(const string& val, int& col) {
    for (int c = 0; c < lowest_available_col; c++) {
      if (clabel[c].used()) {
        if (clabel[c].getval() == val) {
          col = c;
          return 1;
          }
        }
      }
    }
```

```
    return 0;
  }

// Read and write functions
void stable::readwithoutlabels(istream& in, const char
    coldelim, const char rowdelim) {
  read(in, false, coldelim, rowdelim);
  }
void stable::readwithoutlabels(const string& instring, const
    char coldelim, const char rowdelim) {
  istringstream is(instring);
  readwithoutlabels(is, coldelim, rowdelim);
  }
void stable::readwithoutlabels(const char *p, const char
    coldelim, const char rowdelim) {
  string instring(p);
  istringstream is(instring);
  readwithoutlabels(is, coldelim, rowdelim);
  }
void stable::readwithlabels(istream& in, const char coldelim,
    const char rowdelim) {
  read(in, true, coldelim, rowdelim);
  }
void stable::readwithlabels(const string& instring, const
    char coldelim, const char rowdelim) {
  istringstream is(instring);
  readwithoutlabels(is, coldelim, rowdelim);
  }
void stable::readwithlabels(const char *p, const char coldelim,
    const char rowdelim) {
  string instring(p);
  istringstream is(instring);
  readwithoutlabels(is, coldelim, rowdelim);
  }

int stable::writewithlabels(ostream& out, const char coldelim,
      const char rowdelim, const bool includelastrowdelim) {
  return write(out, true, coldelim, rowdelim,
    includelastrowdelim);
  }
int stable::writewithlabels(string& ostring, const char coldelim,
      const char rowdelim, const bool includelastrowdelim) {
  ostringstream os;
  int retval = writewithlabels(os, coldelim, rowdelim,
      includelastrowdelim);
  ostring = os.str();
  return retval;
  }
int stable::writewithoutlabels(ostream& out, const char coldelim,
      const char rowdelim, const bool includelastrowdelim) {
  return write(out, false, coldelim, rowdelim,
    includelastrowdelim);
```

```cpp
    }
int stable::writewithoutlabels(string& ostring, const char
    coldelim, const char rowdelim, const bool
    includelastrowdelim) {
  ostringstream os;
  int retval = writewithoutlabels(os, coldelim, rowdelim,
      includelastrowdelim);
  ostring = os.str();
  return retval;
  }


// print is a debug function really
void stable::print() {
  cout << "\t";  // Top left corner: No label
  int i;
  for (i = 0; i < lowest_available_col; i++)
    cout << clabel[i].getval().substr(0,6) << "\t";
  cout << endl;
  for (i = 0; i < lowest_available_row; i++) {
    cout << rlabel[i].getval().substr(0,6) << "\t";
    for (int j = 0; j < lowest_available_col; j++)
      cout << cell[i][j].getval().substr(0,6) << "\t";
    cout << endl;
  }
}
```

Final **test.h**

```cpp
// Testing functions for stable class (v0.1)
// John Robinson, 25 August 2003
#include <string>
#include <iostream>
using namespace std;
template <class T> int testval(T actual, T expected, char *name) {
  if (actual == expected)
    cout << " OK ";
  else
    cout << "**ERROR** ";
  cout << name << " = " << actual;
  if (actual != expected)
    cout << "(expected " << expected << ")";
  cout << endl;
  }
int test(int actual, int expected, char *name) {
  if (actual == expected)
    cout << " OK ";
  else
    cout << "**ERROR** ";
  cout << name << "() returned " << actual;
  if (actual != expected)
    cout << "(expected " << expected << ")";
  cout << endl;
```

```cpp
    }
template <class T> int test(int actual, int expected, char *name,
    T param1) {
  if (actual == expected)
    cout << " OK ";
  else
    cout << "**ERROR** ";
  cout << name << "(" << param1 << ") returned " << actual;
  if (actual != expected)
    cout << "(expected " << expected << ")";
  cout << endl;
  }
template <class S, class T> int test(int actual, int expected,
    char *name, S param1, T param2) {
  if (actual == expected)
    cout << " OK ";
  else
    cout << "**ERROR** ";
  cout << name << "(" << param1;
  cout << "," << param2 << ") returned " << actual;
  if (actual != expected)
    cout << "(expected " << expected << ")";
  cout << endl;
  }
template <class R, class S, class T> int test(int actual, int
    expected, char *name, R param1, S param2, T param3) {
  if (actual == expected)
    cout << " OK ";
  else
    cout << "**ERROR** ";
  cout << name << "(" << param1 << "," << param2;
  cout << "," << param3 << ") returned " << actual;
  if (actual != expected)
    cout << "(expected " << expected << ")";
  cout << endl;
  }
template <class Q, class R, class S, class T> int test(int actual,
    int expected, char *name, Q param1, R param2, S param3,
    T param4) {
  if (actual == expected)
    cout << " OK ";
  else
    cout << "**ERROR** ";
  cout << name << "(" << param1 << "," << param2;
  cout << "," << param3 << "," << param4;
  cout << ") returned " << actual;
  if (actual != expected)
    cout << "(expected " << expected << ")";
  cout << endl;
  }
template <class P, class Q, class R, class S, class T> int test
    (int actual, int expected, char *name,
```

```cpp
                     P param1, Q param2, R param3, S param4, T param5) {
    if (actual == expected)
      cout << " OK ";
    else
      cout << "**ERROR** ";
    cout << name << "(" << param1 << "," << param2;
    cout << "," << param3 << "," << param4;
    cout << "," << param5 << ") returned " << actual;
    if (actual !=expected)
      cout << "(expected " << expected << ")";
    cout << endl;
  }
// See info cpp for explanation of syntax of next line
#define TEST(a,b, ...) test( b ( __VA_ARGS__ ), a , #b ,
    __VA_ARGS__ )
#define TESTVAL(a,b) testval( b , a , #b )

Final teststable.cpp (short version)

// Test program for stable class
// John Robinson, 25 August 2003
// All commentary in this file is done by couts!
#include <string>
#include <iostream>
#include <fstream>
#include <sstream>
#include "test.h"
#include "stable.h"
using namespace std;
const string firsttab[3][4] = { "a1", "a2", "a3", "a4", "b1", "b2",
     "b3", "b4", "c1", "c2", "c3", "c4" };
const string firsttabcollabels[4] = { "One", "Two", "Three",
     "Four" };
const string firsttabrowlabels[3] = { "Alpha", "Beta", "Gamma" };

int main(int argc, char *argv[]) {
  if (argc < 2) {
    cout << "Usage: stable\
    scratchfile_for_intermediate_results\n";
    return -1;
    }
  cout << "VALUE SETTING TESTS\n";
  cout << "First we'll set up a small table size 4 by 3 to hold:\n";
  stable first(3,4);
  for (int r = 0; r < first.maxrows(); r++) {
    for (int c = 0; c < first.maxcols(); c++) {
      TEST(0, first.set, r, c, firsttab[r][c]);
      }
    }
  cout << "In doing that we've exercised maxrows() and\
     maxcols()\n";
  cout << "As well as set(int, int, string)\n";
```

```
cout << "Now we'll set up the row labels:\n";
for (int r = 0; r < first.nrows(); r++)
   TEST(0, first.setrowlabel, r, firsttabrowlabels[r]);
cout << "And the column labels:\n";
for (int c = 0; c < first.ncols(); c++)
   TEST(0, first.setcollabel, c, firsttabcollabels[c]);
cout << "In doing that we've also exercised nrows()\
     and ncols()\n";
cout << "And now a selection of sets and set{row|col}label ";
cout << "outside the bounds of the table\n";
cout << "The sets should all return non-zero. The row/col\
     sets ";
cout << "should each return 0 once only\n";
int badcoords[4][2] = { -1,0, 2,6, -3,-3, 3,4};
// Next three arrays are expected returns
int expset[4] = {-1, -1, -1, -1};
int expcol[4] = {0,-1,-1,-1};
int exprow[4] = {-1,0,-1,-1};
for (int i = 0; i < 4; i++) {
   int r = badcoords[i][0];
   int c = badcoords[i][1];
   TEST(expset[i],first.set,r,c,"text");
   TEST(exprow[i],first.setrowlabel,r,"text");
   TEST(expcol[i],first.setcollabel,c,"text");
   }
cout << "The current state of the stable is:\n";
first.print();
cout << "Now change the changed labels back using\
     find{row|col}label\n";
int r, c;
if (first.findrowlabel("text",r) != 1)
   cout << "Couldn't find the row label!\n";
else
   first.setrowlabel(r, firsttabrowlabels[r]);
if (first.findcollabel("text",c) != 1)
   cout << "Couldn't find the col label!\n";
else
   first.setcollabel(c, firsttabcollabels[c]);
cout << "And now the table is back to:\n";
first.print();
cout << "GETTING BY LABEL TESTS\n";
string val;
for (r = 0; r < first.maxrows(); r++) {
   for (c = 0; c < first.maxcols(); c++) {
      TEST(1,first.get,firsttabrowlabels[r],
         firsttabcollabels[c], val);
      TESTVAL(firsttab[r][c],val);
      }
   }
string garbage("garbage");
TEST(0,first.get,firsttabrowlabels[0],garbage,val);
TEST(0,first.get,garbage, firsttabcollabels[0],val);
```

```cpp
        cout << "FINDING TESTS\n";

        TEST(1,first.find,"b3", r, c);
        TESTVAL(1,r); TESTVAL(2,c);
        TEST(1,first.find,"a1", r, c);
        TESTVAL(0,r); TESTVAL(0,c);
        TEST(0,first.find,"garbage", r, c);
        TESTVAL(0,r); TESTVAL(0,c);
        r = 2; c = 1;
        TEST(0,first.findafter,"b3", r, c);
        TESTVAL(2,r); TESTVAL(0,c);
        r = 0; c = 3;
        TEST(1,first.findafter,"b3", r, c);
        TESTVAL(1,r); TESTVAL(2,c);
        r = 2; c = 1;
        TEST(1,first.findbefore,"b3", r, c);
        TESTVAL(1,r); TESTVAL(2,c);
        r = 0; c = 3;
        TEST(0,first.findbefore,"b3", r, c);
        TESTVAL(0,r); TESTVAL(3,c);
        cout << "READING AND WRITING TESTS\n";
        cout << "Saving to scratchfile\n";
        ofstream out1(argv[1]);
        first.writewithlabels(out1);
        out1.close();
        stable second;
        cout << "And reading back in to large table\n";
        ifstream in1(argv[1]);
        second.readwithlabels(in1);
        in1.close();
        cout << "Written and read table:\n";
        second.print();
        cout << "Now re-reading second table without labels\n";
        ifstream in2(argv[1]);
        second.readwithoutlabels(in2);
        in2.close();
        second.print();
        cout << "So, the previous table including labels is now part\
            of the actual table\n";
        cout << "Saving this as a table without labels...\n";
        ofstream out2(argv[1]);
        second.writewithoutlabels(out2);
        out2.close();
        cout << "And re-reading as if it had labels\n";
        ifstream in3(argv[1]);
        second.readwithlabels(in3);
        in3.close();
        second.print();

        cout << "NUMERICAL TESTS\n";
        TEST(0,second.set,5,2,200);
        TEST(0,second.set,3,5,100);
```

```cpp
    TEST(0,second.set,2,2,100);
    cout << "After insertion of numeric values outside the "
    cout << "original table limits, the table is:\n";
    second.print();
    cout << "Finding value 100:\n";
    TEST(1,second.find,100, r, c);
    TESTVAL(2,r); TESTVAL(2,c);
    cout << "Finding next value 100:\n";
    TEST(1,second.findafter,100, r, c);
    TESTVAL(3,r); TESTVAL(5,c);
}
```

Appendix:
Comparison of algorithms for standard median filtering

Martti Juhola, Jyrki Katajainen, and Timo Raita

© 1991, IEEE. Reprinted, with permission, from *IEEE Transactions on Signal Processing*, Vol. 39, No. 1, January 1991, pp. 204–208.

Abstract—**In standard median filtering we search repeatedly for a median from a sample set which changes only slightly between the subsequent searches. We review several well-known methods for solving this running median problem, analyze the (asymptotical) time complexities of the methods, and propose simple variants which are especially suited for small sample sets, a frequent situation. Although we have restricted our discussion to the one-dimensional case, the ideas are easily extended to higher dimensions.**

I. Introduction

A signal is here a sequence $S = (x_1, x_2, \ldots, x_n)$ of discrete values from a finite range. In median filtering [1]–[8] we slide a *window* of $m = 2k + 1$ samples and output

$$y_i = \mathrm{median}(x_j | j = i - k, \ldots, i + k).$$

The value y_{i+1} is determined by rejecting the oldest sample x_{i-k}, inserting x_{i+k+1} and searching the median again.

Manuscript received January 4, 1989; revised March 8, 1990. The work of M. Juhola and J. Katajainen was supported by the Academy of Finland.

M. Juhola and T. Raita are with the Department of Computer Science, University of Turku, SF-20520 Turku, Finland.

J. Katajainen was with the Department of Computer and Information Science, Linköping University, Linköping, Sweden. He is now with the Algorithm Theory Group, Department of Computer Science, Lund University, S-22100 Lund, Sweden.

IEEE Log Number 9040407.

We study data structures and algorithms for this *running median problem*. The view we have chosen rules out hardware-oriented techniques [6], [9], and [10], which are based on the bit representation of samples. Most methods can be extended to higher dimensions by replacing each sample by a vector of samples. We devise a data structure W for the set S supporting the following operations:

create(S)   returns W containing $x_1, x_2, \ldots, x_{2k+1}$,

findmedian(W)   returns the median of the set of samples in W,

replace(x, y, W)   removes the item y from W and adds x into W. To know x and y we must have a structure Q for the time dependent operations: *new(t)* returns the next sample value x_{i+k+1}, *oldest(t)* returns the sample added to the set at time $t - k$.

Create is used only once in the initialization. The two others are applied once at each step. Therefore the total time complexity of the latter ones is important. *Replace* is time dependent and could, be accomplished by:

insert(x, W) adds a new sample item x into W and *delete(y, W)* removes the item y from W.

These operations imply that we should maintain the samples within a queue [11]–[14]. Often, however, the data structure which supports the time dependent operations is indistinguishable from the "main" structure W.

We can find efficient data structures for time independent *findmedian*, *insert*, and *delete* operations For heaps [15] or search trees [14], each of them requires only $O(\log m)$ time. Thus they are theoretically optimal as to the asymptotic time complexity. Applying a sequence of $3k + 2$ (*findmedian*, *replace*) pairs we can sort a set $(x_1, \ldots, x_{k+1})$ as follows. Let $-\infty$ $(+\infty)$ denote a dummy element smaller (respectively, larger) than any of x_i. Using these dummy elements we can sort the elements x_i $(i = 1, \ldots, k + 1)$ by applying *findmedian and replace* operations. First, we form a set of $2k + 1$ elements by performing k (*findmedian*(W), *replace* $(-\infty, W)$) operation pairs followed by $k + 1$ (*findmedian*(W), *replace*(x_i, W)) pairs $(i = 1, \ldots, k + 1)$ where the *findmedians* can be considered as empty. After this the $k + 1$ (*findmedian*(W), *replace*($+\infty$, W)) operations give the elements x_i in sorted order. By surrounding each block of $k + 1$ samples of the original signal with the dummy elements as described, we have a sequence the length of which is still $O(n)$. For sorting k items we have an information theoretic lower bound $\Omega(k \log k)$. Thus, we have a lower bound $\Omega(n \log m)$ for filtering a signal of length $O(n)$ with a window size m. However, this is no more valid when the input is restricted to integers, since the complexity of algorithms is then expressed in n, m, and U, the size of integer universe. However, the performance of many algorithms is better than the theoretical results would suggest, since we can take advantage of the characteristics of the problem (e.g., the "continuity" of the signal).

II. Alternative Methods for Implementing Median Filtering

We shall present five ways to organize the data for efficient median filtering. We start with a structure in which no dependencies between the samples are stored. Then we gradually bring in facilities for characterizing more of the dependencies. In consequence, *replace* becomes slower and *findmedian* faster as we advance. The first two methods do not use sorting techniques in order to accomplish the median search [5], the next two methods do. In the last section we take advantage of the fact that the samples are integers from a given range.

All algorithms need a structure which holds the samples inside the window in time order ("age"). We use an array of length m with a time pointer indicating the current oldest value. This is called a *time ring* (the oldest value is replaced at each step by a new value). The time independent operations are performed in structure W. Each slot of the ring contains a pointer to the corresponding item in W.

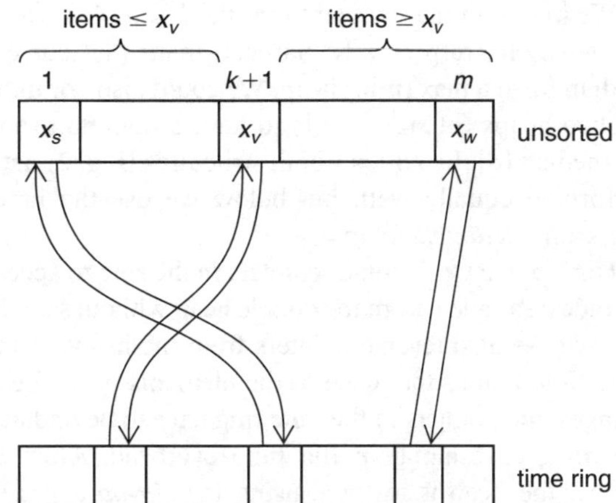

Fig. 1.　The invariant maintained in the unsorted permutation array.

A. Unsorted Set

Let W be a sample array. We add pointers to the elements of W so that each entry of W contains a link to the time ring and vice versa (Fig. 1). Now *create*, *findmedian*, and *replace* are implemented as follows:

create(S) initializes W by selecting the median of the first m samples.

findmedian(W) returns the median value from the $k + 1$st location of W.

replace(x, y, W) replaces the oldest item y by the new item x using the time ring. If x and y are on the same side of the median, only the corresponding slot of the ring is updated. Otherwise, the current median replaces y, x is stored to the middle slot, and the new median is the maximum or the minimum from the side of insertion.

Typically, *findmedian* is performed repeatedly by a linear-time median search algorithm, or some of its variants [11]–[14]. We have taken a shortcut by modified straightforward algorithms such that knowledge about the magnitude of a sample is recorded by comparing the new sample to the current median and placing it to the left or right half of W depending on the comparison. The median is always in the $k + 1$st position of W. The total time required by a (*findmedian*, *replace*) operation will be $O(m)$, with a small constant factor.

B. Heap-Based Methods

For larger windows, the above method may be improved by implementing the subsets on both sides of the median as priority queues. Heaps are practical because no extra storage is needed (see [8], [12], [13] for *min* and *max* heaps).

We maintain the invariant that the k samples which are smaller (respectively, larger) than the current median form a max (min) heap. We could also combine the two heaps (of size $k + 1$) to have a common root, the median [8]. In our tests both structures (Fig. 2) have performed equally well, but below we use the latter one, called a *double heap*.

For *replace* we maintain pointers in the ring to access the oldest sample y from the double heap without searching. We use also reverse pointers from the heaps to the ring. Note that as the place of the elements in the heap changes, the pointers in the time ring have to be updated accordingly, resulting in run time overhead. After an update, the heap is restored using the *sift-up* and *sift-down* operations [12], [13]. The operations are:

create(S) divides S into two subsets by a linear-time selection algorithm and constructs their max and min heaps using Floyd's method [11]–[14] (locations $-k, \ldots,$ 0 of the linear array W are reserved for max heap, locations $0, \ldots, k$ for the min heap).

findmedian(W) returns the median from the root of the double heap.

replace(x, y, W) replaces y by x in the heap. To maintain the heap property we *sift-down* or *sift-up* x. If x is moved into the root, we continue to *sift* x in the other heap (if necessary). The corresponding pointers in the time ring are updated accordingly.

The double heap is created in $O(m)$ time in the worst case [11]–[14]. *Findmedian* can be done in constant

and *sift* in $O(\log m)$ time. Replace can also be made in $O(\log m)$ time assuming that we use pointers to both directions. The total time of (*findmedian, replace*) is $O(\log m)$.

C. Sorted Set

Sorting works well for the median selection [16]. Insertion sort performs well if the window size is small ($\leqslant 20$), but otherwise a hybrid of quicksort and insertion sort should be used [17]. We have two ways to maintain the sorted order:

a) Sorting is delayed as long as possible. If y and x fall on the same side of the median, only the replacement is made, since the previous median is still valid. If they are on different sides, sorting is made anew.

b) Item y is replaced by x and "bubbled" (comparing pairs of two consecutive samples) to the right slot.

In terms of the former alternative the window operations are as follows:

create(S) sorts the samples of S to form the sorted array.

findmedian(W) returns the median value from the sorted array.

replace(x, y, W) replaces y by x, and if x and y are both either greater or smaller than the median, nothing else need be done. Otherwise, W is sorted. *Create* takes $O(m \log m)$ time in the worst case. *Findmedian* is a constant time operation, whereas *replace* uses $O(m)$ time.

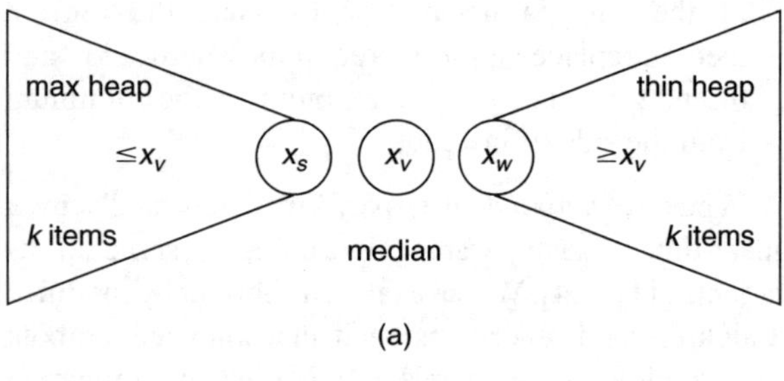

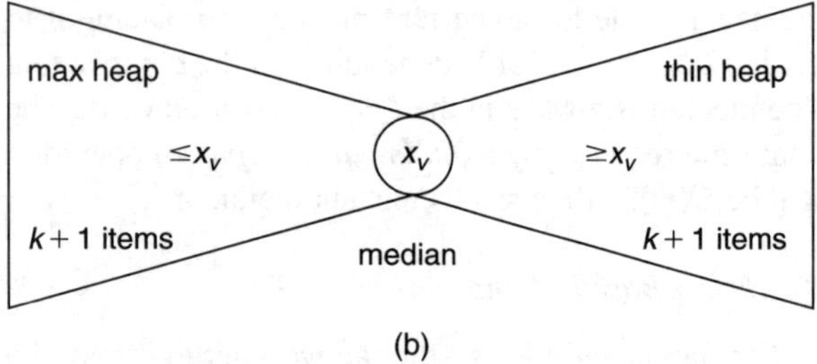

Fig. 2. Alternative ways for implementing the heap-based methods (within an array of length $m = 2k + 1$): (a) max heap of size k, the median, and min heap of size k; (b) max and min heaps of size $k + 1$ with a common root containing the median.

D. Search Tree Based Methods

The sorted order can also be maintained by using a *search tree* which facilitates efficient insertion and deletion. The data structure is easily extended to support also *findmedian*. A weight-balanced binary search tree performs *insert*, *delete*, and *findmedian* in $O(\log m)$ time [14] resulting in $O(n \log m)$ total time. Thus the search trees are theoretically as efficient as heaps.

For most signals, the difference between two consecutive samples is small. Let $d_i \leqslant m$ be the *distance* in elements over which the sample x_i is moved in the sorted sequence measured from the position of x_{i-1}. A search for the slot which is d_i positions away from the latest insertion takes time $O(1 + \log d_i)$ using *finger trees* [14]. Thus the filtering can be done in $O(n + \sum_1^n \log d_i)$ time assuming that we use two fingers (pointers to leaf nodes): *median* and *latest insertion finger*. Since d_i is usually small, it is enough to implement only the lowest level of the tree, i.e., to use a doubly linked list. This weakens the theoretical complexity to $O(n + \Sigma_1^n d_i)$ since *findmedian* and *replace* take $O(1)$ and $O(d_i)$ time. The small difference between x_i and x_{i+1} suggests, however, that *replace* takes almost constant time.

We maintain two orders—the time order and the sort order—in a single structure by using an array of length m as follows: each entry is of type (v, s, g), where v is the sample, s an index to the next smaller sample, and g an index to the next greater sample. Samples are stored in the structure, which is treated as a ring in time order (Fig. 3). The operations are:

create(S) sorts the first m samples in S to initialize the s and g pointers. The sorting is carried out with m inserts (cf. *replace*).

findmedian(W) returns the median value obtained by the median finger.

replace(x, y, W) replaces y by x using the latest insertion finger after which the s and g chains are updated accordingly. An appropriate node is found by comparing x to the previously inserted sample and advancing either s or g chains depending on the comparison. Both links and the latest insertion finger are updated. If the median is between x and y in the s (and g) chain, it is changed by moving the median finger one node towards the inserted sample along the s or g chain. The same scheme is followed if the sample to be deleted is the median itself.

E. Integer Data

Previous methods run for an infinite universe of samples. If samples are taken from a restricted universe, say from integer range $0 \ldots U - 1$, we can use more efficient methods. With the *radix method* [10] the operation pair (*findmedian, replace*) can be executed in $O(\log U)$ time. Even an asymptotically faster solution is possible with the *vanEmdeBoas trees* [18]–[21]. However, a practical implementation of this data structure is complicated and therefore it was rejected.

The *histogram* [22] is theoretically much weaker, but in practice it behaves very well. It employs an array *Samples* with index range $[0, U - 1]$. Each sample acts as an index to this array. Each location contains an integer expressing how many items have hit that slot. The median is a table index. If *Samples[median]* is greater than one, the items are thought to be ordered and a counter is used for the position of the median in this sequence. A description of the operations is as follows:

create(S) sets $Samples[i] \leftarrow 0$ for i $= 0, 1, \ldots, U - 1$. After this we set $Samples[x] \leftarrow Samples[x] + 1$ for each x in S. The minimum sample *min* is determined. The first median is found by starting from *Samples[min]* and summing consecutive components until the sum is at least equal to $k + 1$. The counter is set accordingly.

findmedian(W) returns the median index value.

replace(x, y, W) sets $Samples[y] \leftarrow Samples[y] - 1$ and $Samples[x] \leftarrow Samples[x] + 1$. If x and y are on the same side of the median, it does not change. On the other hand, if x is larger than the median, but y smaller, we check whether the counter is equal to *Samples[median]*. If so, we search the next nonzero element with a higher index. The counter is set to one and the median index is attached to this element. If the counter is less than *Samples[median]*, it is incremented by one. A similar procedure is applied when the median decreases.

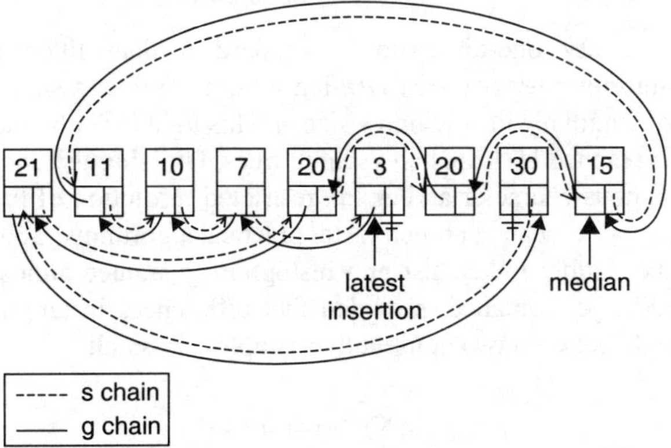

Fig. 3. The doubly linked list representation with two fingers. The number inside a node is the value of the corresponding sample. The links between the nodes indicate the order of magnitude between the samples. The samples are stored in the array in time order.

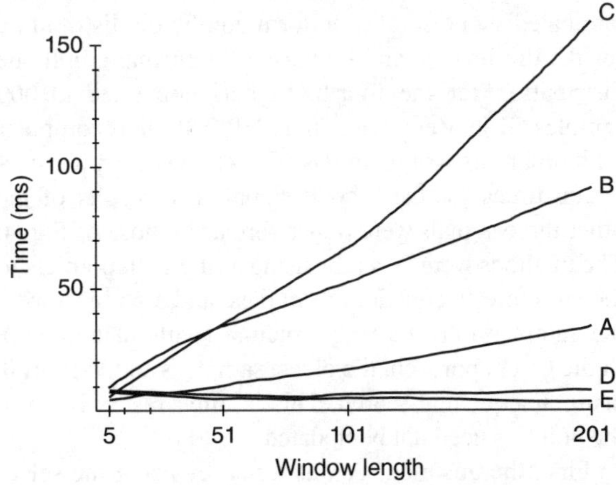

Fig. 4. Execution times of the saccadic eye movement signal with window lengths from 5 to 201: (A): unsorted set with the time ring containing the pointers; (B): two heaps with the time ring containing the pointers; (C): "bubbling" method; (D): doubly linked list; (E): histogram.

TABLE I

CHARACTERISTICS OF THE STANDARD MEDIAN FILTERING METHODS

| Method | Theoretical Time Complexity (Large Windows) | | Practical Time Complexity (Small Windows) | | Space Complexity | Practical Time Ranked in Tests |
	Create	*Median + Replace*	*Create*	*Median + Replace*		
A	$O(m)$	$O(m)$	$O(m)$	$O(m)$	$O(m)$	3
B	$O(m)$	$O(\log m)$	$O(m)$	$O(\log m)$	$O(m)$	4
C	$O(m \log m)$	$O(m)$	$O(m^2)$	$O(m)$	$O(m)$	5
D	$O(m \log m)$	$O(1 + \log d_i)$	$O(m^2)$	$O(d_i)$	$O(m)$	2
E	$O(m \log \log U)$	$O(1 + \log \log D_i)$	$O(U + m)$	$O(D_i)$	$O(U + m)$	1

Standard median filtering methods: (A) unsorted set with the time ring and pointers, (B) two heaps with the time ring and pointers, (C) "bubbling" method, (D) doubly linked list, and (E) histogram. Here m denotes the window length, d_i a distance in elements over which sample x_i is moved in the list from the position of x_{1-i}, D_i a difference of values of two consecutive medians, and U the universe of values of samples. Note that for (D) the time of *Median + Replace* is "amortized," i.e., some operation may take $O(m)$.

The complexity of *create* is $O(U + m)$. Despite the constant time *findmedian* operation, *replace* takes $O(D)$ where $D \leq U$ is the difference between the consecutive medians. Thus the filtering is done in $O(U + m + \sum_1^n D_i)$ time where D_i denotes the D for the ith sample.

III. COMPARISON OF THE METHODS

We studied the performance of the methods by using various signals. Two eye movement signals [23] were recorded electro-oculographically at 400 Hz with 12-b resolution. *Nystagmic eye movements* are of a bias "sawtooth" form. *Saccadic eye movements* form steps. The third and fourth signals were made with a test function of ramps, steps, peaks, and sinusoidal waves. It was contaminated by noise of a uniform amplitude distribution for the third signal and by noise of a normal amplitude distribution for the fourth. Signals contained 10 000 samples. Tests were made in an HP 200 microcomputer with programs written in Pascal [24]. Fig. 4 depicts execution times for the second signal. The results of the other three signals were rather similar to those of Fig. 4. The methods were tested by using a ring with pointers to the structure W containing samples, and also by storing the samples into the ring structure (without pointers). In the latter approach, the oldest sample is found from W using a sequential search. This is often faster, because the pointers need not be updated.

First, the unsorted set was tested by using the selection procedure [13], but it was slower than any of those in Fig. 4. Next we implemented the operations of the unsorted set from Section II-A (Fig. 4, curve A). Several heap variants were also tested [24]. The two heaps with the time ring of the pointers was the best (curve B). It did not perform well for small windows since the maintenance of pointers takes considerable time, which was the crucial disadvantage of this theoretically fast method. The 'bubbling" version (Section II-C, b)) without pointers is the fastest sorting-based method (curve C). Even this is slow for large windows. The doubly linked list (curve D) is inferior to the histogram (curve E) except for small windows.

We also tested the algorithms by using much larger noise amplitudes. The results were then close to those presented. Therefore the histogram method is recommended if the range of the samples is known and "sufficiently' restricted, and the window is moderately large. Otherwise, it is preferable to use a doubly linked list and thus obtain a fast general-purpose standard median filter. The comparison of the methods is presented in Table I.

IV. CONCLUSIONS

Every one-dimensional standard median filtering algorithm uses at least $\Omega(n \log m)$ time to filter a signal of length n with a window size m. This tight lower bound is reached by heaps and search trees. Utilizing the continuous character and/or the restricted resolution of the signal, we can get much faster practical algorithms. Both the doubly linked list and histogram guarantee almost $O(n)$ performance, provided that differences in amplitude between two consecutive samples are small.

ACKNOWLEDGMENT

The authors are grateful to O. Nevalainen for the present topic of investigation and to I. Pyykkö for supplying eye movement data. They also wish to thank J. Astola, R. Karlsson, and J. Teuhola for helpful discussions.

REFERENCES

[1] N. C. Gallagher and G. L. Wise, "A theoretical analysis of the properties of median filters," *IEEE Trans. Acoust., Speech, Signal Processing*, vol. ASSP-29, pp. 1136–1121, Dec. 1981.

[2] T. A. Nodes and N. C. Gallagher, "Median filters: Some modifications and their properties," *IEEE Trans. Acoust., Speech, Signal Processing*, vol. ASSP-30, pp. 739–746, Oct. 1982.

[3] P. Heinonen and Y. Neuvo, "FIR-median hybrid filters," *IEEE Trans. Acoust., Speech, Signal Processing*, vol. ASSP-35, pp. 832–838, June 1987.

[4] J. Astola, P. Heinonen, and Y. Neuvo, "On root structures of median and median-type filters," *IEEE Trans. Acoust., Speech, Signal Processing*, vol. ASSP-35, pp. 1199–1201, Aug. 1987.

[5] J. B. Bednar and T. L. Watt, "Alpha-trimmed means and their relationship to median filters," *IEEE Trans. Acoust., Speech, Signal Processing*, vol. ASSP-32, pp. 145–153, Feb. 1984.

[6] J. P. Fitch, "Software and VLSI algorithms for generalized ranked order filtering," *IEEE Trans. Circuit Syst.*, vol. CAS-34, pp. 553–559, May 1987.

[7] M. O. Ahmad and D. Sundararajan, "A fast algorithm for two-dimensional median filtering," *IEEE Trans. Circuits Syst.*, vol. CAS-34, pp. 1364–1374, Nov. 1987.

[8] J. T. Astola and T. G. Campbell, "On computation of the running median," *IEEE Trans. Acoust., Speech, Signal Processing*, vol. ASSP-37, pp. 572–574, Apr. 1989.

[9] G. R. Arce and P. J. Warter, "A median filter architecture suitable for VLSI implementation," in *Proc. 22nd Annu. All. Conf. Commun., Contr., Computing* (University of Illinois, Urbana, IL), Oct. 1984, pp. 172–181.

[10] E. Ataman, V. K. Aatre, and K. M. Wong, "A fast method for real-time median filtering," *IEEE Trans. Acoust., Speech, Signal Processing*, vol. ASSP-28, pp. 415–420, Aug. 1980.

[11] N. Wirth, *Algorithms + Data Structures = Programs*. Englewood Cliffs, NJ: Prentice-Hall, 1976.

[12] E. Horowitz and S. Sahni, *Fundamentals of Computer Algorithms*. Rockville, MD: Computer Science Press, 1978.

[13] G. H. Gonnet, *Handbook of Algorithms and Data Structures*. Reading, MA: Addison-Wesley, 1984.

[14] K. Melhorn, *Sorting and Searching*, vol. 1., *Data Structures and Algorithms*. Berlin: Springer, 1984.

[15] M. D. Atkinson, J. R. Sack, N. Santoro, and T. Strothotte, "Min-max heaps and generalized priority queues," *Commun. Ass. Comput. Mach.*, vol. 29, pp. 996–1000, Oct. 1986.

[16] F. Pasian, "Sorting algorithms for filters based on ordered statistics: performance considerations," *Signal Processing*, vol. 14, pp. 287–293, Apr. 1988.

[17] R. Sedgewick, "Implementing quicksort programs," *Commun. Ass. Comput. Mach.*, vol. 21, pp. 847–857, Oct. 1978.

[18] P. van Emde Boas, "Preserving order in a forest in less than logarithmic time and linear space," *Inform. Process. Lett.*, vol. 6, pp. 80–82, June 1977.

[19] P. van Emde Boas, R. Kaas, and E. Zijlstra, "Design and implementation of an efficient priority queue," *Math. Syst. Theory*, vol. 10, pp. 99–127, 1977.

[20] M. H. Overmars, "Computational geometry on a grid—an overview," Dep. Comput. Sci., University Utrecht, Utrecht, The Netherlands, Tech. Rep. RUU-CS-87-4, Feb. 1987.

[21] D. B. Johnson, "A priority queue in which initialization and queue operations take $O(\log \log D)$ time," *Math. Syst. Theory*, vol. 15, pp. 295–309, 1982.

[22] T. S. Huang, G. J. Yang, and G. Y. Tang, "A fast two-dimensional median filtering algorithm," *IEEE Trans. Acoust., Speech, Signal Processing*, vol. ASSP-27, pp. 13–18, Feb. 1979.

[23] M. Juhola, "On computer analysis of digital signals applied to eye movements," Dep. Comput. Sci., University Turku, Turku, Finland, Rep. A47, Nov. 1987.

[24] M. Juhola, J. Katajainen, and T. Raita, "Comparison of algorithms for standard median filtering: Extended version," Dep. Comput. Sci., University Turku, Turku, Finland, Rep. B 44, July 1989.

Index

-, 60

--, 65

!, 63, 83

!= , 63, 83

", 56

cannot be overloaded, 138

#define, 102, 108

#include, 54–56, 68

%, 60

& for reference parameter, 72

& for bitwise AND, 82

& for address-of, 84

&&, 63

(, 60

), 60

* for multiplication, 60

* for contents-of, 85

*/, 62

*=, 65

*dest++ = *source++ idiom, 87

. for accessing structure member, 83, 97

. for accessing class member, 124

. cannot be overloaded, 138

,, 83

/, 58, 60

/*, 62

//, 62

/=, 65, 84

: for conditional, 64

: for type derivation, 146

::, 123–125

?, 64

[, 74, 83

], 74, 83

^, 82

__FILE__, 183

__LINE__, 183

{, 83

|, 82

| |, 63

}, 83

~ for complement, 82

~ for destructor, 124–125

+, 59–60

+, 65

+= , 65

< , 63, *see also* <>

<< for put-to, 57, 84, 95, 169

<< for shifting, 82–84

<=, 63

<> for #include statement, 56

<> for template, 163

=, 59

-=, 65

==, 62

>, 62, *see also* <>

->, 83, 100

>=, 63

>> for get-from, 58, 59, 84, 95, 169

>> for shifting, 82–84

>>=, 84

2038 (year), 69

abs, 112

abstract (base) class, 117, 122, 156

abstract data type (ADT), 120, 210, *see also* example code in C++

abstract reasoning, 252

abstractions, 12, 51

academic collaboration and copying, 44

acceptance tests, 299

access function, 122

ACM (Association for Computing Machinery), 3, 250

acos, 112

actions, 31–32

Ada, 51, 270

addition, 60

address, 74, 84

adjectives, 32

ADT (abstract data type), 120, 210

ALGOL, 12

algorithm, *see also* example code in C++

   efficiency, 222

   existing, 25

all-at-once methodology, 239

alphabetization, 279

ambiguity of software, 37

analogies, 37, 44

analysis as part of design, 255

anchors, 194

AND, 63, 82

angle brackets, 56, 163

aphorisms, 235

apprenticeship, 35, 43

argc, 90
arguments/parameters of function,
    55
arguments to main, 90
argv, 90
arithmetic operators, 60
array, 73f
    address, 74
    bounds, 78, 174
        of linked lists, 203
        of pointers, 89, 195
artisan, 43
artisan view, 9, 35
ASCII, 56, 76, 104
asin, 112
assembly language, 27
assembly lines, 31
assignment, 63, 140
Association for Computing
        Machinery (ACM), 3, 250
associative array, 355
associativity of operators, 84
asynchronous FSM, 260
atan, 112
atof, 112
atoi, 91, 112
automata, 11
automatic variable, 92
auxiliary array, 230
average case analysis, 225
avionics system, 27
axiomatic basis (Hoare), 12

backslash, 55–56
backspace, 56
bad_alloc, 161
banana function, 350
base type, 122
BASIC, 11, 51
Bauhaus, 235
beauty, 35
beep, 56
bell curve, 78
Bentley, Jon, 246
best case analysis, 225
bias to confirmation, 252
big-Oh, 225
binary mode, 96
binary search, 221, 223, 226, 271,
        303
binary tree, 206, 221
biological systems, 31
bitwise Operators, 82
black box testing, 302
blocking of code, 179
Boehm, Barry, 239
books, 48
bool, 77
Borland C++, 53–54
bounds, 338
    array, 78, 174

braces, 57
brackets, 60
brainstorming, 237
break, 64
Brooks, Fred, 3, 9, 19–20
bsearch, 112
bugs, 172f
bureaucracies, 31

C, 51, 52, 55
C with Classes, 66
C#, xiii, 1, 51
C++, xiii, 1, 51f, 115f
    a better C, 52
    choice of for this book, 51
    idiom, 52
C99, 55
*CACM (Communications of the
        ACM)*, 16, 49
calibration, 33
call-by-reference, 72
call-by-value, 72
calloc, 112
carriage return, 56
case, 64
case studies, 310
    median filter, 310
    multidimensional minimization,
        329
    string table class, 353
cast, 69–70
catch, 160–161
catchment basin, 331
cctype, 98, 111
ceil, 112
central limit theorem, 78
cfloat, 111
changeability of software, 19
changeability of requirements, 23
changes of ownership of data,
        181
char, 60
chief programmer team, 37, 240
Church, Alonzo, 10
cin, 58–59
circular buffer, 313, 315
circular shift, 279
Citeseer, 250, 316, 333
clarity, 35, 184
class, 116, 120, 123–124
class (generic term), 121
class diagram, 146
Classic C, 55
clear, 169
clock radio, 266
close, 167
closed extensions, 254
cmath, 111–112
COBOL, 51
CODE (in object file), 110
code block, 62

code inspection, 300–301
cognitive psychology, 251
collaboration, 43
collections, 52, 194
collision resolution, 204
combinatorial mathematics, 11
COMEFROM, 16, 51
command prompt, 34
command-line interface, 90
comment, 18, 62, 185
commenting out, 62
*Communications of the ACM
(CACM)*, 16, 49
compilation, 53
compiler, 29, 53, 233
    Unix, 30
complement, 82
complete binary tree, 208
completely full binary tree, 206
complexity, 19
components, 32
compound statement, 62
computational complexity, 222f
computer (person who computes),
    329
computer science, xi, 2, 12
conditional: if, 62
conditions, 31–32
    in state diagrams, 260
    in state sketches, 270
console application, 54, 90
console outputter, 57
const:
    data member, 66
    function, 128
    correctness, 177
const_cast, 70
constraints, 245
constructor, 125
context-sensitive help, 182
control, 27–32
control-C, 59
copy constructor, 139, 144
copying, 44
copy-on-write, 145
copyright, 37
correctness, 14
cos, 112
cosh, 112
cout, 57
craft, 43, 49
craftsperson, 8, 35, 43
criteria, 25, 245
criterion function, 333
cross-compile, 53
cstddef, 61
cstdlib, 111–112
cstring, 100, 111, 113
C-strings, 52, 75
    vs C++ standard library **string**
        class, 76

ctime, 67, 113
cybernetics, 32
cyclic model for design, 236

DATA (object file), 110
data analysis, 353
data compression, 233
data flow diagrams, 258
data modelling, 329
data structure, definition, 119
data structures and algorithms, 1,
    29, 194f, 291
database, 93, 104, 172
debug.h, 183
debugging, 22
dec, 169
declaration, 59
decrement, 65
dedication, book, *See* to
default, 64, 182
default parameters, 73
defensive programming, 172
delete, 105, 181, 194
DeMarco, Tom, 258
dependent alternatives, 278
derived type, 122, 145
descent algorithm, 331
design, 8, 235f
    documents, 241
    methodology, 235
destructor, 125
device drivers, 27
dialog design, 251
dice, 79
differential equations, 329
Dijkstra, Edsgar, 12, 16, 19
disaster, 7
disk file, 93
divide by zero, 162
division, 60
division into files, 187
division tester, 67
documentation, 26, 241
doppelgänger, 187
DOS, 34
dot operator, 97
double, 60
double quotes, 55
double slash, 62
doubly linked lists, 197, 317
do-while, 65, 67
Dr. Dobbs' Journal, 49
dropper, 214
dynamic testing, 302
dynamic_cast, 70

efficiency, 222f
eigenvalues, 22
Eischen, Kyle, 37
embedded controllers, 27
emotion, 33

encapsulation, 115–116
endl, 169
ends, 169
engineer, 23
engineering view, 9, 23
enter-by-history, 264
enum, 60, 108
env, 90
eof, 169
equivalence classes, 302
ergonomics, 33
error handling, 30
error reporting, 183
error returns, 95
escape sequence character constant, 56
Essential Software Design, 3
event loop, 259
event-driven systems, 259
events, 31–32
evolutionary algorithms, 332
example code in C++:
  abstract data types:
    medianfilter, 323–327
    queue, 211
    simplex, 340–350
    stack, 123, 164
    stable – string table, 380–392
    String, 138
    table (using hash table), 293
    table (using unordered array), 217
    Vector, 129
  algorithms:
    binary search, 223
    insertion sort, 228
    mergesort, 229
    quicksort, 231
    selection sort, 227
    sequential search – unordered array, 223
    sequential search – linked list, 222
  programs:
    dice rolling simulation, 79
    division by zero exception tester, 162
    division tester, 67
    Fahrenheit–Celsius conversion, 66
    file copy, 95
    Hello World, 54
    linenumbers, 170
    location database, 126, 145–151
    memory allocation exception tester, 161
    mkframework (program framework), 190
    Roman number convertor, 98
    simplex optimiser, 340–351
    size of basic types, 61
    state sequencer, 289
    stable – string table class and tests, 380–397
    student marks classes and database, 106, 151, 157–160
    table ATD tester, 217
exceptions, 52, 160
exclusive OR, 82
executable, 53
exercises, 7
existence proof, 25
exit, 112
exp, 112
experimentation, 22, 256
experiment-driven development, 337
exponentiation, 61
export, 166, 323
extensibility in object-oriented programming, 117
Extensions in specification document, 254
extern, 93
Extreme Programming, xii, 1, 3, 27, 41, 299, 309, 355

fabs, 112
Fahrenheit–Celsius table, 66
false, 77
false = 0, 62
false minimum, 339
FAQs, 250
faults, 298
feedback systems, 31
file, 93
  copy program, 95
  structure, 185
finishing, 45, 309, 323
finite state machine (FSM), 259–272, 282, 288–290
float, 60
floating point exception, 162
floating point numbers, 58
floor, 112
flush, 169
for, 65–66
formal technique, 14
formfeed, 56
FORTRAN, 11, 51
foundations, 8
framework, 21
free, 112
free software, 44
freedoms, 254
friend, 132
FSM (finite state machine), 259–272, 282, 288–290
fstream, 95
fstreambase, 167
FTP, 250
function, 17, 70
  calls, 71

definition, 70
length, 186
prototype, 55

Gaussian, 78
gcount, 167, 169
get, 95, 167, 169
get from, 58, 95, 142, 169
getline, 169
Gill, xxx, *see also* dedication
glass box, *see* white box
global minimum, 331
global variables, 92–93, 116
GNU, 44
GNU/Linux, 53, 161
Gödel, Kurt, 10–11
golden section search, 331
Google, 250, 316, 332
GOTO, 12, 16, 65
goto in C++, 65
graph, 205
graphics, 233
Gries, David, 12
Gropius, Walter, 235
GUI, 34, 90, 257

habits, 7, 36, 43
hacker, 38
Handel-C, 52
hardware, 27, 29, 48, 52
hardware view, 9
Harel, David, 266
hash:
    function, 204
    key, 204
    table, 204, 220, 292
    table for table ADT, 293
header, 45, 188
header files, 146
heap, 209
heapsort, 232
Hello World, 54, 59
Helvetica, 7
hex, 169
hexadecimal, 56
high level languages, 26
histogram, 79
Hoare, Antony, 12
Hollywood, 38
Horowitz, Ellis, 15
human computer interaction (HCI),
    33–34, 250–253
Hungarian convention, 186

IBM, 19
IDE (integrated development
    environment), 53–54
ideation, 236, 238
idiom, 52
IEEE Computer, 49
*IEEE Transactions on Signal
    Processing*, 316

if, 62
ifstream, 95, 166–168
imitation, 43
implementation, 288f
inadequate checking of input data,
    181
incorrect code blocking, 179
incorrect initialization, 176
increment, 65
incremental development, 20, 37
incrementing, 65
indentation, 180, 185
informal review, 300
information hiding, 23, 278
inheritance, 117, 122, 145
inline, 125
inorder traversal, 207
input/output, 57
insertion sort, 227
inspector, 301
instantiation, 116
int, 54, 56, 60
integrated development
    environment (IDE), 53–54
Intercal, 51
internationalization, 76
Internet, 250
Interpolation, 329
interpreting shared items
    differently, 183
invalid memory access error, 174
invariants, 14
Inventor's Paradox, 256
invisibility, 19
ios, 166, 168
ios::binary, 96
iostream, 54, 57
isa, 117, 119, 146
isalpha, 111
isdigit, 99, 111
islower, 111
isprint, 111
isspace, 111
istream, 167–169
isupper, 111

Jackson, 262
Jackson System of Design, 258
Java, xiii, 1, 49, 51–52, 59–60, 78
Jones, Capers, 26
journal, 47
Juhola, Martti, 316

Kernighan, Brian, 15, 34
keystrokes, 69
Knuth, Donald, 16, 40, 49, 233

labels, 65
language lawyer, 240
language reference manual, 48
laws of arithmetic, 178
legacy code, 76

legal, 46
Leonardo, 47
level-order traversal, 208, 213
librarian, 240
library:
   containing books, 248, 333
   personal, 48
   software, 29, 53, 54, 117
lie, 29
line numbers, 57, 170
linear model for design, 236
linefeed, 56
linked list, 102, 196, 220
   compared to arrays, 196, 246
   of arrays, 203, 221
linking, 54, 110
Linux, 53, 161
LISP, 11, 51
literate programming, 16
literature view, 9, 14, 184
local minimum, 331
localize information, 21, 79,
   115–116, 171, 184, 187,
   278, 308
location class and database
   example, 126, 145–151
log, 112
log10, 112
logbooks, 46
logic errors, 182
Logical Construction of Programs,
   258
login, 30
long, 60
Loop Errors, 178
loop invariants, 14
loops, 65

machine code, 54
macro, 102
magic number, 32, 102
main, 54–55, 90
maintainability, 15
maintenance, 20, 38
Makefile, 187
malloc, 112
manipulator, 168
Mascitti, Rick, 66
math.h, 111
mathematical view, 9–10
Matlab, 22, 34
McCarthy, John, 11–12
Mead, Roger, 333
Media Handling Module, 285
median filtering, 310–327
member function, 122
memcmp, 113
memcpy, 113
memmove, 113
memory allocation, 105, 161
memset, 113
mentors, 43

mergesort, 229
message, 122
Metafont, 16
metaphor, 252
method, 122
methodology, 25, 235
Microsoft Windows, 46, 53, 96,
   161
millenium, 69
miniature language, 290
minimization, 329
mkframework, 189, 313
moderator, 301
modify not clone, 187
Modula 2, 14
modularity, 22
modularization, 116, 276
   documentation, 241, 285
modulus, 60
morale, 37
multidimensional arrays, 74
multidimensional optimization,
   329f
multifile structure, 188
multiple analogy view, 9, 37
multiple inheritance, 52, 166, 284
multiplication, 60
multiway tree, 206
my_strcpy, 76
my_strncpy, 77
Myers, Glenford, 298, 307, 309
Mythical Man Month, 19, 26, 49

'\n', 169
NAG (Numerical Algorithms
   Group), 233
namespace, 52, 54, 57, 277
NCSS (Non-Commentary Source
   Statements), 18
Neilson, Jakob, 33
Nelder, John, 333
Nelder-Mead, 334
new, 105, 181, 194
new (nothrow), 106
newsgroups, 49, 250
Newton-Raphson, 330
No Silver Bullet, 3, 19, 41
node in tree, 206
noisy function, 333, 338
non-analytic function, 331
Non-Commentary Source
   Statements (NCSS), 18
nonlinear equations, 330
normal curve, 79
Norman, Donald, 33
NOT, 63
Notations, 257
nouns, 32
NULL, 68
null-terminated strings, 76
Numerical Algorithms Group
   (NAG), 233

numerical analysis, *see* numerical methods
numerical integration, 329
numerical methods, 11, 233, 329, 333

object, 31–32, 121
object code, 54
object-oriented programming (OOP), 4, 115f, 171
object-oriented modularization, 282
oct, 169
off-by-1 error, 175, 338
ofstream, 96, 166–168
omega, 225
open addressing, 204
open extensions, 254
open source, 44, 52, 250
operating system, 27, 29, 53, 233
operator <<, 136
operator+, 137
operator overloading, 134
operator precedence and associativity, 83
operators that cannot be overloaded, 138
optimization, 331f
OR, 63, 82
order of magnitude, 224
ordered array, 292
ordered list ADT, 210
organic growth view, 9, 18
organizational behaviour view, 9, 36
OS X, 53
Osborn, Alex, 237
ostream, 167–169
out of scope, 126
overloaded operator, 118
overloading, 57, 132
constructors, 133
ownership, 37

pair (data structure), 216
pair programming, 37
paradigm, 115
parameterized types, 123
parameters, 32
parentheses, 60, 83
Parnas, David, 241, 243, 279, 286
Pascal, 14, 51
patents, 37
peek, 169
performance assessment, 307
periodicals, 248
Perl, 51
Philosophers, 9
pipelines, 46
piping, 46
pivot in quicksort, 230
pivot reference, 249

platform, 48
Plauger, P.J., 15, 34
play computer, 182
PLEASE, 51
pointer(s), 84
initialization, 176
or reference to a local variable, 180
to 0, 199
and addresses, 195
and arrays, 176
to a structure, 100
to a pointer, 90
polishing, 45
polite refusal, 29
Polya, George, 256
polymorphism, 122, 145, 151
pool, 116
pop, 124–125
Post, Emil, 10
post-decrement, 65
post-increment, 65
Postorder traversal, 208
pow, 112
precedence, 60, 84
precedence and associativity of operators, 83
preconditions, 182
pre-decrement, 65
pre-increment, 65
preorder traversal, 207
presentation conventions, 6
principles and rules, 8, 38
print, 57
cout, 54
printf, 55
priority queue, 213
probability theory, 78
problem, 23, 244
statement, 244f
problems with new and delete, 181
procedural language, 115
procedural programming, 52, 115
procedural/algorithmic problems, 248
producer, 301
program:
arguments, 89
definitions of, 9
examples, *see* example code in C++
hangs, 29
size, 18
programming:
definition of, 2
environment, 53
errors, 173
in the large, 48–49
in the small, 48
language, 14, 51
methodology, 25
Programming Pearls, 16, 49, 246

project in an IDE, 150, 187
project managers, 36
protected, 123
prototype, function, 55, 70
prototyping, 21, 257, 313, *see also*
    rapid prototyping
psychology, 33, 37, 251
public, 121, 123, 125, 146
public interfaces, 184
pure virtual, 117, 156
push, 124–125
put, 95, 167, 169
put to, 57, 95, 169
putback, 169

qsort, 113, 177, 315
QT, 52
query, 56
queue, 211
quicksort, 12, 230

rand, 67–68, 113
RAND_MAX, 67
random number generators, 69
random number tests, 323
random numbers, 67, 321
random variables, 79
rapid prototyping, 20, 313, 355,
    See also prototyping
read, 167, 169
readability, 17, 186
reader, 301
recipes, 1
    and surprises, 49, 333
reconfigurable hardware, 52
records, 93
recursion, 71–72
recursive_ipower, 72
redundant code, 17
refactoring, 14, 378
reference, 72
reference counting, 52, 145
regression testing, 305
reinterpret_cast, 70
relational operators, 62
relocatable address, 110
representation, 13, 51
requirements specification, 241, 253
requirements, change in, 23, 276
researching possible solutions, 255
restructuring, 22
return, 54, 64, 72
reuse, 75
rigidity, 252
riser, 214
robot, 27
Rock, Scissors, Paper, 82
Rockwell, 6
Roman number convertor, 98
Romance languages, 76
root node in a tree, 206
roots of an equation, 330

Rosenbrock, 350
rules for software design, 8, 38
Russell, Bertrand, 10–11

Saint-Exupéry, Antoine de, 235
scepticism, 239
Science Citation Index, 249
scientific and engineering
    programming, 1, 34, 235
scope, 92
scope resolution operator, 125
scribe, 301
scripting, 34
SDL, 261
searching, 11, 222, *see also* binary
    search
seed, 68
seekg, 169
seeking, 170
seekp, 169
selection sort, 227
self-documenting, 45
semicolons, 18, 58
sentinel, 103, 220
separate chaining, 204
sequential search, 222, 225
setfill, 169
setprecision, 169
setw, 169
shell, 34
shift, 82
Shneiderman, 251
short, 60
shortest-path graph algorithm, 12
SIAM, 250
sifting, 353
sign language, 257
signal conditioning, 310
silence, 29
silver bullet, *See* No Silver Bullet
simplex, 333f
simplex class, 340–350
Simula, 51
simulated annealing, 332
simulation, 78, 257
sin, 112
single quote, 55
sinh, 112
size_t, 61, 113
sizeof, 61
sliding window function, 311
slippery problems, 245
Slots and Signals, 52
Smalltalk, 115
soft logbook, 47
software:
    as construction, 38
    as Hollywood, 38
    as university, 38
    crisis, 26
    definition of, 3, 8
    design, 1f

engineering, xii, 2, 5, 19–20,
26–27, 41, 43, 48–49, 239,
286
libraries, *see* library, software
tools, 34, 46
solitary testing, 300
solving systems of equations, 329
sort key, 226
sorting, 11, 226f
Source Code Control Systems, 53
specification, 253
Specification Description
Language, 261
Sperling, George, 257
spiral model of design, 239
sqrt, 112
square root, 330
srand, 67–68, 113
stable – string table, 353
stack, 120,123, 164, 210
system, 110
standard library, 52, 111, *see also*
function and class names
standardization, 38
state, 31–32
diagram, 260
sequencer, 281, 288–289
transition diagram, 260
statechart, 261–268, 272, 274, 280
statesketch, 268–272, 335
static, 92
static analysis, 300
static_cast, 70
statistical testing, 304, 313, 321
statistics, 79
std, 54, 57
std::cout, 57
stdio.h, 55
stdlib.h, 111
stepwise refinement, 20
strcat, 113
strchr, 113
strcmp, 113
strcpy, 77, 113, 174
stream manipulators, 169
streams, 166
streams library, 110, 118
string – C++ standard library
class, 188, 192f, 354f
vs C-strings, 52, 76
String – example class, 118,
138–145
strlen, 113
strncpy, 100, 113, 174
Stroustrup, Bjarne, 66, 177
struct, 97
structured programming, 12, 16, 26
structured text processing, 353
structures, 97
stubs, 20, 101
student marks example, 100
style, 8, 17, 35, 172, 184–187

subroutines, 17
subscripting operator, 140
subtraction, 60
superstate, 264
surgical team, 240
switch, 64, 182
symbol table, 54
synchronous FSM, 259
system, 113
system:
crash, 174
heap, 111
programming, 52
stack, 110
test, 299
theory, 31–32
view, 27

tab, 56
table (ADT), 216, 291, 293
table-driven, 259
tan, 112
tanh, 112
tautology, 13
TDD (Test Driven Development),
8, 299, 355
team organization, 37, 240
technical journals, 48
technology, 24
telephone switches, 26
telephone test, 186
tellg, 169
tellp, 169
templates, 163–164, 323, 367
temporal filter, 312
terminal nodes, 206
Test Driven Development (TDD),
8, 299, 355
test plan, 241, 299
testing, 22, 30, 298f
TₑX, 16
TEXT (Object file), 110
text editor, 197
text mode for files, 96
textual principle, 5, 17
this, 142
Thompson, Ken, 30
thresholds, 32
throw, 160
time, 67–68, 113
Times Roman, 6
timesharing, 26
To:, *see* Gill
tolower, 111
tool building, 45
top-down design, 20, 277
toupper, 112
traversal, 207, 213
tree(s), 205–210
applications, 208
balancing, 221
definition, 206

Trojan horse, 30
Trolltech, 52
true, 77
true = non-zero, 62
truncation errors, 178
try, 160
Turing, Alan, 10–11
Turing Award, 19, 30
tweaking, 33
type, 31–32, 60, 120, 210
    conversion, 70
    extension, 117
typedef, 108, 163, 323

UML (Unified Modelling
    Language), 1, 241–242, 258,
    268, 274, 283–284
unallocated memory, 77
uncertainty (scaling), 255
uninitialized pointer, 174
Unix, 30, 34, 46, 53
unordered array, 292
unsigned, 60
URLs:
    ACM collected algorithms, 233
    Brooks, Fred, 19
    C/C++ sibling rivalry, 114
    C++ FAQs, 114
    Deitel, 53
    Dijkstra, Edsger, 12
    extreme programming, 3
    FAQs (general), 250
    floating point arithmetic, 193
    free software, 44
    GNU, 44
    GOTO considered harmful, 16
    Hello World, 59
    Hoare, Antory, 12
    Knuth, Donald, 16
    Leonardo da Vinci, 47
    McCarthy, John, 12
    median filtering on vision list,
        316
    Norman, Don, 33
    open source software, 44
    Principa Cybernetica, 42
    Russell, Bertrand, 10

Skiena, Steven on algorithms,
    233
Trolltech, 52
Turing, Alan, 10
Von Neumann, John, 11
use cases, 245, 354
user interface, 251
user interface builders, 53
user-centered view, 9, 33
using, 54, 57
using namespace std;, 57
    and modularization, 277

Van Dyke, Carolyn, 15
variable type errors, 178
variables, 59
Vector example, 129
verbs, 32
VHDL, 52
video analysis, 312, 328
virtual, 147, 156
virtual function, 117, 156
virus, 174
Visual C++, 53–54
visualization, 182, 337
void, 56, 73
Von Neumann, John, 11

wasting memory, 194
waterfall model, 239
web (system for literate program-
    ming), 16
werewolf, 4
while, 65
white box testing, 305–306
Whitehead, Alfred North, 10–11
whitespace, 58
Wiener, Norbert, 31
wimp, 29
Windows, *see* Microsoft Windows
Wirth, Nicklaus, 14
worst case, 225
write, 167, 169
ws, 169

Y2K, 69
Yahoo, 250, 316